WOVEN THREADS

Frontispiece. Bearer (No. 21) in Kilt, Procession Fresco, Corridor of the Procession, Knossos (photo A. Chapin. Archaeological Museum of Heraklion, Hellenic Ministry of Culture and Sports –Archaeological Receipts Fund)

WOVEN THREADS

Patterned Textiles of the Aegean Bronze Age

edited by
MARIA C. SHAW AND ANNE P. CHAPIN

with contributions by
*Elizabeth J. W. Barber, Giuliana Bianco, Brendan Burke, Emily C. Egan
and Suzanne Peterson Murray*

Oxford & Philadelphia

First published as a hardback in 2016. Reprinted in paperback in the United Kingdom in 2022 by
OXBOW BOOKS
The Old Music Hall, 106–108 Cowley Road, Oxford, OX4 1JE

and in the United States by
OXBOW BOOKS
1950 Lawrence Road, Havertown, PA 19083

© Oxbow Books and the individual contributors 2016

Paperback Edition: ISBN 978-1-78925-734-2
Digital Edition: ISBN 978-1-78570-059-0

A CIP record for this book is available from the British Library

Library of Congress Cataloging-in-Publication Data

Woven threads: patterned textiles of the Aegean bronze age / edited by Maria C. Shaw and Anne P. Chapin.
    pages cm
  Includes bibliographical references.
  ISBN 978-1-78570-058-3 (hardback)
 1. Textile fabrics, Ancient--Aegean Sea Region. 2. Decoration and ornament, Ancient--Aegean Sea Region. 3. Bronze age--Aegean Sea Region. I. Shaw, Maria C., editor. II. Chapin, Anne Proctor, editor.
  NK8907.W68 2015
  746'.041--dc23
                        2015035986

All rights reserved. No part of this book may be reproduced or transmitted in any form or by any means, electronic or mechanical including photocopying, recording or by any information storage and retrieval system, without permission from the publisher in writing.

Printed in the United Kingdom by Short Run Press Ltd

For a complete list of Oxbow titles, please contact:

UNITED KINGDOM
Oxbow Books
Telephone (01865) 241249
Email: oxbow@oxbowbooks.com
www.oxbowbooks.com

UNITED STATES OF AMERICA
Oxbow Books
Telephone (610) 853-9131, Fax (610) 853-9146
Email: queries@casemateacademic.com
www.casemateacademic.com/oxbow

Oxbow Books is part of the Casemate Group

Front cover: Bearer (No. 21) in Kilt, Procession Fresco, Corridor of the Procession, Knossos (photo A. Chapin. Archaeological Museum of Heraklion, Hellenic Ministry of Culture and Sports - Archaeological Receipts Fund)

Contents

Preface, by *Maria C. Shaw* vii

1. Spinning Ariadne's Thread: Sources and Methodologies 1
   Anne P. Chapin

2. Bronze Age Aegean Cloth Production: A Cottage Industry No More 17
   Brendan Burke and Anne P. Chapin

3. Patterned Textiles as Costume in Aegean Art 43
   Suzanne Peterson Murray

4. Palace and Household Textiles in Aegean Bronze Age Art..105
   Maria C. Shaw and Anne P. Chapin

5. Textile and Stone Patterns in the Painted Floors of the Mycenaean Palaces..........................131
   Emily C. Egan

6. Sailing the Shining Sea: Maritime Textiles of the Bronze Age Aegean 149
   Maria C. Shaw and Anne P. Chapin

7. String Lines, the Artist's Grid, and the Representation of Textiles in Fresco........................183
   Maria C. Shaw

8. Minoans, Mycenaeans, and Keftiu 205
   Elizabeth J. W. Barber

9. Observations, Summaries, and Conclusions 239
   Anne P. Chapin

Abbreviations... 257

Bibliography... 259

Preface

Should one – only half a century ago – have inquired into what was known about the appearance and dress of people in the Bronze Age Aegean, the answer would have been based on what we might "hear" about the subject in early literature, as in the Homeric poems, aided by depictions of frescoes found in Crete, the Greek mainland, and some of the Mediterranean islands. Then – as if in the sudden opening of a hidden treasure – there was the revelation of the excellently preserved Bronze Age frescoes brought to light at Akrotiri, on Thera, by Spyridon Marinatos's excavations on that island.

What is truly special – particularly for our purposes, here – is that, all of a sudden, and thanks to those discoveries, textiles began to acquire a narrative context which, along with our analyses of the human activities and gestures of the individuals portrayed, have added to our ability to better understand that past.

This book was originally conceived as a series of articles focusing on a number of depictions – sometimes fragmentary, and usually of people shown in association with patterned textiles. As the book progressed, Marie-Louise Nosch, of the Danish National Research Foundation's Centre for Textile Research, became interested and inquired concerning it. At about the same time further writing became difficult, so I called upon Professor Anne Chapin to help me complete it. With her usual alacrity and generosity, she agreed and, with time, invited others with specialties touching on various fabrics, as well as their manufacture and depiction, to broaden the spectrum of consideration.

Thus, to Anne Chapin – and to the aid of Marie-Louise Nosch – I remain indebted for making the completion of this volume possible, not to mention my husband, Joseph Shaw, who supported me throughout the progress of this project!

Maria C. Shaw

1

Spinning Ariadne's Thread: Sources and Methodologies

Anne P. Chapin

"Society is founded upon cloth."
Thomas Carlyle (1795–1881)

Woven textiles are produced by nearly all human societies. Generally defined as cloth or fabric produced by weaving, knitting, knotting, or felting, the term also refers to the fibers used to make textiles.[1] Loom-woven textiles first appear in the Levant in the Early Neolithic period during the 8th millennium BC, around the time when early farmers began to raise domesticated plants and animals, establish permanent settlements, and craft pottery. Weaving, together with its associated package of technological and social innovations, spread into Mesopotamia, Egypt, and Anatolia, and then into Greece, Europe, and central Asia.[2] These early textiles were put to a wide variety of uses: new forms of clothing could be made, since fabrics can be woven in all manner of shapes, sizes, and thicknesses and are thus more easily fashioned into clothing than tanned leather hides from animals. Textiles also served ceremonial, funerary, and household purposes, and were put to agricultural, industrial, and commercial use. Advances in seafaring and military technologies have also been linked to the development of textiles.[3]

This volume investigates evidence for patterned textiles (that is, textiles woven with elaborate designs) that were produced by two early Mediterranean civilizations: the Minoans of Crete and the Mycenaeans of mainland Greece (Fig. 1.1).[4] These two cultures prospered during the Aegean Bronze Age, *c.* 3000–1200 BC, contemporary with pharaonic Egypt (Fig. 1.2), and both could boast of specialists in textile production. Their cultures also developed lavish palaces, extensive cities, and prosperous towns run by extensive governmental bureaucracies whose economic transactions were recorded on clay tablets. These texts are written in the early scripts identified today as Linear A and B and demonstrate that the Mycenaeans spoke an early form of Greek (while the language of the Minoans remains unknown). The archaeological and documentary evidence thus indicates that even though formal written histories do not survive, these cultures were sophisticated in their literacy, their technologies, and their long-distance trade networks, which brought them into contact with Egypt and other civilizations of the Mediterranean.[5]

Archaeological evidence further suggests that Minoan and Mycenaean textiles were much desired as trade goods. Artistic images of their fabrics preserved both in the Aegean and in other parts of the Mediterranean seem to justify this desire, for they show elaborate patterns woven

Fig. 1.1. Map of the Aegean (A. Chapin and J. Silva)

with rich decorative detail and color. Unfortunately, only a few small scraps of textiles survive today (see Chapter 2), but evidence for their production is abundant, from flax seeds and woolly sheep bones to spindle whorls and loom weights, from crushed murex shells and dying vats to references to textiles on Linear A and B tablets. And, most important for this study, frescoes painted by Minoan and Mycenaean artists supply detailed information about a wide variety of now-lost textile goods. From the luxurious costumes of women and men to beautifully patterned wall hangings and carpets, to the more utilitarian pieces of decorated fabric used for the ikria (stern screens), deck furnishings, and sails of ships, images of textiles abound in Aegean art. These elaborate textiles (and the artwork that depicts them) are the subject of this volume. A review of surviving artistic and archaeological evidence indicates that textiles played essential practical and social roles in both Minoan and Mycenaean societies. The goal of this study is to show that Thomas Carlyle's well-known observation that "society is founded upon cloth" is as true of the Minoan and Mycenaean cultures of the Aegean as it is for modern society today.

	Traditional Chronology Dates BC	High Chronology Dates BC	Crete (Minoan)	Greek Mainland (Helladic)	Cycladic Islands (Cycladic)	
		AEGEAN CHRONOLOGY				
Early Bronze Age	Before 3000 to about 2000		EM I	EH I	EC I Grotta-Pelos Group Kampos Group	Early Prepalatial Period
			EH IIA		EC II Keros-Syros Group Kastri Group	
			EH IIB	EC II		
			EH III	EC III	EC III Phylakopi I Group	Late Prepalatial Period
Middle Bronze Age	2000–1625	2000–1725	MM IA	MH I	MC I	
			MM IB	MH II	MC II	Protopalatial Period of First ("Old") Palaces on Crete
			MM IIA			
			MM IIB			
			MM IIIA	MH III	MC III	Neopalatial Period of Second ("New") Palaces on Crete
			MM IIIB			
Late Bronze Age	1625–1525	1725–1625	LM IA	LH I	LC I	LC I: Theran Eruption
	1525–1450	1625–1500	LM IB	LH IIA	LC II	Late LM IB: widespread destruction across Crete
	1450–1425	1500–1425	LM II	LH IIB		Final Palatial Period on Crete: Mycenaeans at Knossos?
	1425–1300	1425–1300	LM IIIA	LH IIIA	LC III	Mycenaean palaces built; Knossos destroyed
	1300–1200	1300–1200	LM IIIB	LH IIIB		
	1200–1125	1200–1125	LM IIIC	LH IIIC		Postpalatial Period: Fall of Mycenaean civilization

Fig. 1.2. Chronological chart of the Aegean Bronze Age

Textiles in Greek mythology

Scattered among the rich stories that comprise the mythology of Classical antiquity are tales of spinners and weavers.[6] The Moirai, the three goddesses of fate, spin the thread of life, measure the thread, and finally cut it; and thus Classical Greek ideas about human mortality were expressed in textile terms. From Crete comes Ariadne, the daughter of King Minos of Knossos, who fatefully gave Theseus a ball of thread so the young Greek hero would not get lost in the labyrinth on his quest to slay the monstrous Minotaur.[7] From Greece is Penelope, faithful and resourceful wife of Odysseus and queen of Ithaca, who kept her many suitors at bay by promising to remarry only when she had finished a cloth which she wove by day and unraveled each night. Circe and Calypso, enchantresses encountered by Odysseus on his long journey home, were weavers, as are all women in the Homeric poems.[8] The most famous weaver of mythic tradition, perhaps, is Arachne, a girl from Lydia who learned weaving from Athena herself and foolishly challenged the goddess to a weaving contest, only to be turned into a spider so that she would weave forever.[9] Stories such as these hint at the importance of spinning and weaving in prehistory, but as Classical myths, their meanings are metaphorical rather than factual or historical. Direct evidence for the importance of woven textiles to early societies must be sought in the archaeological material.

Fine textiles and Aegean society

A starting point for investigating prehistoric Aegean interest in fine cloth may be found, perhaps unexpectedly, in two battle frescoes. The first, dating to the Neopalatial era, is the Shipwreck scene of Room 5 in the West House at Akrotiri, Thera.[10] This small but interesting vignette depicts a naval officer extending a staff in a gesture of power over naked enemies as they drift, lifeless and drowned in the sea (Fig. 1.3). The staff bearer's body language would seem to suggest that nakedness in battle was associated with cultural and military defeat.[11] The second fresco, Mycenaean in date, decorated a ceremonial hall in the palace at Pylos during the 13th century BC (Fig. 1.4).[12] This scene depicts Mycenaean Greeks, identifiable from their kilts, greaves, and boars' tusk helmets (which are characteristic of Aegean armor) fighting an enemy that is helmetless and wears only rough animal skins. Since they are clearly not Mycenaean, the identity of the skin-clad fighters remains something of a puzzle. Details of their costuming offer the most significant clues, particularly the animal skins whose white color and hairy markings identify them as sheep hides worn by tying two legs around a shoulder. In contrast to the neat cloth kilts and stitched black leather(?) lappets of the Mycenaeans, these costumes are rudimentary and could be made without fabric or sewing. The absence of textiles, then, categorizes the skin-clad warriors as un-Mycenaean – and what is more, as rough and uncivilized.[13] The Mycenaean Greeks themselves, then, seem to have associated cloth with civilization, and used the lack of cloth to represent a simple and backwards culture.

Though textiles distinguished Mycenaeans from their culturally "barbaric" opponents, textile production was actually widespread throughout Europe, the Near East, Egypt, and central Asia during the Bronze Age, and a great variety of textiles were being produced. This is evident in Egyptian art, where cloth and dress distinguish various ethnic groups (e.g., Asiatics and Libyans) from Egyptians.[14] Given the language and cultural differences between Minoans and Mycenaeans, one might hypothesize that Aegean artists would likewise differentiate among the various ethnic

Fig. 1.3. Shipwreck scene, from the north wall of Room 5, West House, Akrotiri (drawing A. Chapin and J. Silvia after Doumas 1992, pl. 29)

Fig. 1.4. Battle Fresco (22 H 64) from Pylos (restoration P. de Jong; courtesy of the Department of Classics, University of Cincinnati)

groups populating Crete, mainland Greece, and the islands, but recent investigation indicates that this was not the case. Instead, the peoples of the Aegean embraced pan-Aegean costuming traditions regardless of language or ethnicity.[15]

This broad cultural uniformity did not, however, extend to social status. Variations in cloth quality and pattern in the costumes depicted in the miniature frescoes of the Neopalatial Minoan period, which depict dozens, or even hundreds, of people in a single composition, preserve images of complex social interactions and hierarchies. The best-preserved miniature frescoes come from Room 5 in the West House of Akrotiri on Thera; the Shipwreck scene pictured in Figure 1.3 belongs to this pictorial program. The frescoes are dated to the Late Cycladic (LC) I period before the eruption of the Santorini volcano (c. 1630 BC in the high chronology), when the Minoan palaces on Crete were thriving. The paintings decorated the room's four walls just below ceiling level: the Flotilla Fresco from the south wall shows a naval procession leaving a Departure Town and heading to an Arrival Town.[16] Details of geology and setting suggest that the ships may be crossing the island's caldera from one promontory to another, and some scholars believe that the Arrival Town depicts Akrotiri itself, with the most decorated ship being captained by the owner of the West House.[17]

What is interesting for this inquiry is that of the hundreds of figures and the wide variety of costumes detailed in the West House frescoes, none wears clothing made of the elaborately patterned textiles investigated here (especially Chapter 3). Rather, the fabrics are unadorned and the costumes seem humble. The more prominent women in the Arrival Town wear simple striped clothing; distinguished men seated on ships wear plain cloth cloaks, some with black borders; working men wear loin cloths while still others are nude; and some townspeople of both sexes (peasants?) are dressed in hairy garments sewn from animal skins (Fig. 1.5). In contrast to these ordinary people, the adjacent "Priestess" Fresco, which is painted on a much larger scale, depicts a young woman wearing a fringed mantle over a delicately patterned short-sleeved garment typical of Neopalatial Minoan elite costume (Fig. 1.5). The interlocking circles create a

Fig. 1.5. Hierarchy of costume in the frescoes of the West House at Akrotiri: left, water bearers of the Shipwreck scene; center, a townswoman from the Arrival Town; right, the "Priestess" (drawings A. Chapin after Doumas 1992, pls 24, 27, 48)

complicated fabric design with a rich texture that stands in sharp contrast to the plain weaves of the townspeople and peasants. Hers is a fabric of wealth and social status. This evidence for hierarchy of fabric and costume suggests that the patterned textiles investigated in this volume were luxury items worn by a social elite to identify their special status. Interestingly, the same attitude toward cloth and clothing does not seem to translate to the large-scale depictions of two male youths who are also painted in Room 5. These boys, perhaps 12–14 years in age, are shown fully nude, which is in accord with Neopalatial social custom and artistic convention for subadult males.[18]

The artistic evidence

As indicated above, only traces of textiles survive from the Aegean Bronze Age. This study therefore draws evidence from other sources, primarily artistic but also including archaeological, and (to some degree) documentary. Among the artistic sources, glyptic art (that is, the decoration of seals and signet rings), figurines, ceramic decoration, and frescoes are all helpful. Seals and signet rings were frequently decorated with images of people wearing various costumes, but the scale is small and details can be difficult to discern.[19] Minoan and Mycenaean figurines are larger in size, but the human form is often rendered abstractly and details of costume (such as painted patterns) can be unclear.[20] Certain forms of ceramic decoration also reflect the weaver's art. The "Woven Style," for example, is a term used to describe a certain class of handmade Minoan vessels dating to MM IB (in the Protopalatial period) and painted with white, red, and orange designs on a dark ground. The motifs feature linear, circular, and interlocking designs, and seem inspired by woven patterns.[21] But other types of ceramic decoration can only be loosely linked to the textile arts,[22] and thus fresco painting remains as the primary visual source of information on Aegean textiles. Fortunately, the colors, patterns, and designs of elaborate textiles were depicted in great detail by Minoan and Mycenaean fresco artists.

Fresco painting itself developed as a significant art form about the time the first Minoan palaces were built on Crete, *c.* 1900 BC, and thrived until the collapse of Mycenaean palatial society, *c.* 1200 BC.[23] The first frescoes appeared in the early (Protopalatial) palaces of the Middle Minoan (MM) period, when plaster, the medium of fresco, was used primarily for functional reasons to cover walls and floors. Much of this plaster had a low lime content and was painted with simple monochrome decoration (red was especially popular), but nearly pure lime plasters were also developed in this period, and some of these were given the first fresco decoration in the Aegean.[24] At Knossos and Kommos, early frescoes imitated expensive building materials, particularly veined stone,[25] and a Protopalatial floor fresco from Phaistos suggests a carpet (Fig. 4.14). By the Neopalatial period of MM IIIB and Late Minoan (LM) I, frescoes had become representational as well as decorative, and painters were creating lively images of the world around them. Most famous are the undulating landscapes with flowers, birds, monkeys, and female figures (e.g., the frescoes of Ayia Triada on Crete and Xeste 3 in Akrotiri, Thera; see Fig. 3.8), but frescoes of charging bulls and athletes were also painted, as well as other subjects.[26] Even panoramic views of palace and town life were represented, as in the miniature frescoes of the West House, discussed above. These paintings are rich in detail, and numerous studies demonstrate the accuracy with which many objects were portrayed. Depictions of the costumes are especially informative – so much so that reproduction textiles have been woven and replica costumes have been fashioned on the basis of the detailed information provided by the frescoes.[27]

But by the end of the LM I period, *c*. 1500–1450 BC, things had changed: Minoan sites were destroyed all around Crete, with only Knossos surviving, and Mycenaean Greeks from the mainland took over the reins of power. New palaces, citadels, and other monumental structures were subsequently built at Mycenae, Tiryns, Pylos, and elsewhere, particularly in the LH III era, and they too were decorated in fresco. At first, Mycenaean patrons seem strongly influenced by their Minoan forebears in their choice of subject matter: processional women in fine costumes, bull leaping, and (now it is known) even naval scenes are found in LH II–IIIA1 painting, but then, as time passed, more typically "Mycenaean" subjects, favoring horses, chariots, war, and armaments, emerged in the artistic treatments of the LH III period, witnesses to the ongoing political concerns about warfare and defense. Throughout the Mycenaean era, the fresco painters recorded the world around them, although, it must be stated, with greater reliance on inherited pictorial formulas and less attention to naturalistic detail.[28]

On representation

A key assumption made throughout this volume is this: art is thought in material form. At the same time, as Ernst Gombrich so compellingly argued, representational art never holds a completely objective mirror to the physical world. Rather, it is shaped by the beliefs, customs, and practices of the culture and era in which the art is produced.[29] The cultural conditions contribute to the formation of what is called "period style," which is the mode or way of making things that is specific to a historical era. Egyptian art, for example, is easily distinguishable from Classical Greek art, no matter the subject depicted or the individual artist at work, because period style tempered and guided each individual artist's own personal production. Thus for thousands of years, Egyptian artists fashioned sculptures that followed their own cubic canon of proportion, whereas Classical artists made figures with a sense of naturalistic movement characteristic of ideal, organic form. Culturally influenced preferences in iconography (subject matter and symbols) also impact artistic production, and, together, style and iconography create artistic idioms that are distinctly identifiable for each culture and period.

Working within an artistic idiom, an individual artist makes choices, develops a personal style, and responds to the preferences of patrons. Thus, to represent a human figure, an artist confers with the patron, selects materials, decides on a composition, fashions a figure, and presents it in a social context. Each stage of art production involves negotiation, training, practice, knowledge, and skill. Decisions are made at every point in the process; the artist must solve technical problems (e.g., how to use the materials), decide what to represent (subject matter and iconography), and how to form the figure (artistic style). These choices in turn are shaped by a variety of influences that include the artist's own personal vision for the artwork as well as the artistic practices of the time, the function of the art, the desires of the patron, and its intended reception by an anticipated audience. Artist, patron, and audience alike, moreover, are influenced by a myriad of interconnected cultural, political, economic, material, and environmental factors that typically reach far beyond the actual time and place an artwork is produced and received. And thus, art is made when thought and belief intersect with materials and action. A nuanced study of art can reveal important information about the beliefs and cultural conditions that surround its making.

But, on the flip side, no artwork can be interpreted as a completely faithful, direct representation of the past. The limits of artistic evidence thus must be kept in mind when it is examined, as here, for evidence of lost textiles and past social customs.[30]

For this investigation, relatively simple means can be used to assess the value of any individual artwork as evidence. First is the completeness of the representation. In general, while frescoes preserve the color and patterns of depicted textiles in greater detail than do other artistic media, most of the frescoes investigated here are fragmentary. Nonetheless, modern restoration, when made after a careful review of the surviving evidence, can help overcome the limits of fragmentary preservation and provide a good sense of a composition's original appearance without being misleading.[31] Second is the level of naturalism artists bring to their representation, and again, today's scholars are fortunate that Aegean artists of the Late Bronze Age generally worked to achieve a fairly high standard of naturalism in their art. That is, even the least skilled artists painted recognizable human forms and included significant details of costume and attribute. Some Neopalatial artists working in the LM I period even seem to have pushed the limits of their artistic idiom in a singular quest for greater lifelikeness, as evidenced by images of children with the naturalistic proportions of youth (e.g., the three boys of different ages painted in Xeste 3, Akrotiri).[32] Where preservation is more complete (as at Akrotiri, Thera), the high level of figural naturalism and great artistic detail in the frescoes permits in-depth investigation of lost textiles and their cultural uses.

Thus, the ideal artistic evidence for this investigation derives from polychrome frescoes that are naturalistic and detailed, well preserved, and depict true-to-life textiles. Images of people in elaborate costumes, for example, offer excellent evidence for fabric patterns and costume design (Chapter 3), and experimental work confirms that what is depicted in art is both weavable and sewable.[33] Knowledge of these patterns can be applied to frescoes depicting textiles that are otherwise harder to recognize, such as those from household contexts (Chapter 4). But there is not always a direct translation from object to art, and fresco painting was influenced by cross-craft interaction, which occurs when craftsmen of different media engage one another and share ideas and information.[34] Interaction between fresco and textile technologies, for instance, is evidenced by the common use of indigo and murex pigments.[35] Likewise, transfers of ideas can result in new forms of iconography. Most of the female figures painted in Xeste 3 at Akrotiri, for example, are clothed in Minoan-style flounced skirts, but a few costumes are decorated with elaborate landscape motifs similar to those found in contemporary landscape art made in fresco and relief media (Figs 3.28, 3.30). A taste for landscape seems to have influenced the textile arts. And, as is typical of cross-craft interaction, influences ran both ways. The Crocus Panel from the House of the Frescoes at Knossos (Fig. 1.6) depicts a field of red crocuses bordered by an undulating band painted above an olive tree. Comparisons can be made to textile designs, such as those preserved on the faience dress models featuring crocus motifs found in the Temple Repositories at Knossos (Fig. 3.4), and also to landscape paintings with fields of flowers, as in the Crocus Gatherers Fresco of Xeste 3, Akrotiri (Fig. 3.25). The question is, does the Crocus Panel depict a textile (perhaps a wall hanging) or a landscape? Or a little of both? The evidence is not strong enough to argue persuasively either way; one can only say that the Crocus Panel as a landscape painting seems influenced by the textile arts.[36] Artistic currents flow in many directions.

Fig. 1.6. Crocus Panel from the House of the Frescoes at Knossos (restoration A. Chapin)

A methodology for investigating the artistic evidence

Some of the lines of inquiry followed in this volume might initially appear to be circular. After all, representations of textiles are identified from typically fragmentary artworks and then, to develop typologies, they are compared with other artworks, some of which are even more poorly preserved. How can anyone be sure that the frescoes actually depict textiles, particularly if human figures are not wearing them? Bluntly speaking, absolute certainty is not always possible, but the authors and contributors feel confident that this volume brings together solid evidence for many forms of elite Aegean textile production. Support is drawn from comparisons with historical materials (including the tools and techniques that were in use before the invention of modern, industrialized textile technologies) and with surviving traditions for hand-spinning, hand-weaving, and hand-sewing that are still practiced around the world today. These textile traditions – historical and living – offer conceptual models for the types of textiles that could have been produced during the Aegean Bronze Age, particularly in light of the growing scientific knowledge of the materials and technologies that were available to prehistoric textile workers. And, knowing what *could* have been made facilitates the recognition of what *was* made, as evidenced by the artworks themselves.

This process of investigating the bits of surviving artistic, archaeological, and documentary evidence in relation to the reconstructed whole of prehistoric Aegean textile production draws on hermeneutic philosophy and its associated circle of understanding. Hermeneutics is a theory of interpretation named after the Greek god Hermes, who delivered messages from the gods by first understanding the gods' divine intentions and then translating them into terms that could be understood by mortals. That is, Hermes interpreted the messages. Likewise, a key component of hermeneutics is its interpretive focus on a text and the relationship(s) between the parts and the whole of that text. When one reads part of a text (say, its beginning), the reader forms expectations about the rest of that text – expectations which are then continually informed and adjusted as the reader seeks to understand the author's intentions and meaning. Preconceived ideas about the meaning of the text ("prejudices") are an integral part of the process, and are continually adjusted as better understanding of the text is achieved. Once the whole of a text has been read, and understood for meaning, then the parts can be reinterpreted within the context of the text's entirety. This reinterpretation can be expanded to include, for instance, the social and historical conditions surrounding the creation of that text and biographical details of its authorship, among other influential factors. In this way, the understanding of the whole text changes as a result of continued investigation of its parts, and a better understanding of the parts comes from an enhanced understanding of the whole.[37]

This relationship between reader and text, and between the parts and the whole of a text, is identified as a hermeneutic circle, and reflects a continuing process of interpretation and reinterpretation. Hermeneutic thought appears in Classical antiquity in the writings of Aristotle and Plato; later, the idea of textual analysis was applied to scripture. Today, hermeneutics can be defined as a theory or philosophy of the interpretation of meaning, and the hermeneutic circle is a methodological approach that is applied to many fields of study in which objects, ideas, or individuals are investigated in relation to a whole.[38] In art history and archaeology, art or artifacts may be substituted for the text, and interpretation of a work of art or artifact then draws from investigation of its constituent parts in relation to its whole, particularly with respect to its social and historical contexts.[39] For the purposes of this study, elite Aegean textile production represents the whole of the "text," which must be reconstituted from analysis of its many parts, including the tools and technologies of early textile production, historical comparanda, and, especially, the individual artistic representations of the textiles produced in the Bronze Age and discussed in the coming chapters. Investigation of each of these parts contributes to a better understanding of the whole of Aegean textile production, just as a view of the whole facilitates an understanding of the parts.[40] And thus, what might seem at first glance to be a roundabout method of investigation is in fact necessary in order to make sense of these otherwise disparate pieces of evidence.

Contextual analysis

While the hermeneutic circle offers a method for approaching the artistic and archaeological evidence for elite Aegean textile production, it is the ultimate aim of this study to situate the material within its Bronze Age context and to infer from it the meaning(s) that elite patterned textiles held for the societies and people who produced them. This is to be done by investigating how elite textiles, as indicated by the artistic and archaeological evidence, functioned in their human and physical environments, diachronically, in relation to what is known of existing economic and social structures.[41] Such an analysis includes an assessment of the iconography of textile art with a view to interpreting the ideas conveyed by the signs and symbols of textile design, together with the roles this imagery may have played in Aegean prehistoric society. Contextual analysis also includes a spatial analysis of the types of buildings (and rooms) in which textile art was found, and the relationship of these structures with others in the same settlement, and even further afield. The meanings that are gained from such a contextual analysis are entangled, relational, flexible, fluid, and generally depend on the point(s) of view from which they are read. But overall, it is hoped that reading the evidence within its contemporary archaeological contexts will throw new light on the intellectual, emotional, mystical, and cultural significances that elite textiles held for those who made and used them in the Bronze Age. Thus the threads of evidence discussed below are themselves woven into a tapestry of meaning, and the patterns that emerge will (hopefully) be recognizable as individual and cultural expressions of the belief systems and ideologies that characterized the eras of their creation.

Organization

The artistic evidence that comprises the major focus of this study is presented as a typology. The physical evidence of Aegean textile production is surveyed in Chapter 2. In addition to spindle whorls and loom weights used to fashion different weights of fabric, Mycenaean Linear B texts refer to the people involved in various aspects of textile production – the flax or linen workers, the makers of heavy *te-pa* cloth (carpet?), and even the sewing women, "decorators" (of cloth), the "headband makers," and the male "finishers" who may have been responsible for adding the special fringe. Costumes made from intricately patterned textiles are explored in Chapter 3. These investigations provide strong support for the hypothesis that patterned textiles were intimately associated with elite social ranking and were used actively in cult and ritual performance throughout the Aegean Bronze Age. Chapters 4 and 5 consider frescoes imitating luxurious wall hangings, ceiling cloths, and floor coverings, and offer information for how textiles were displayed in elite households, villas, and palaces. Intricately woven textiles can also be identified in naval scenes, as shown by Mycenaean frescoes of ikria made of cloth, which imply a close connection between patterned textiles and maritime power (Chapter 6). How textile patterns were painted in fresco with the aid of string lines and grids is the subject of investigation in Chapter 7. Chapter 8 looks beyond the orbit of Greece to explore how Aegean textiles, as luxury trade goods, can be recognized in Egyptian contexts, both in the paintings that decorated the tombs of Egyptian nobles and in the art of the pharaohs. Chapter 9 offers a synthesis of the findings.

Terminology: the patterns of patterned textiles

The following glossary and accompanying diagram, Figure 1.7, is intended to clarify and define textile terms known to those practitioners of the textile arts but less familiar to the uninitiated.[42]

ALL-OVER – a continuous pattern in which the motifs are fairly close together and evenly distributed, as opposed to stripes, borders, and plaids.

BARRED BAND – narrow strip decoration featuring alternate bars, or blocks, usually of red and yellow, or blue and black.

Fig. 1.7. The patterns of patterned textiles (A. Chapin)

CHEVRON – a zigzag design in a strip arrangement, also called a herringbone.

DIAMOND PATTERN – a lozenge shape that has the form of a rhombus in which opposite sides are parallel and adjacent sides have different lengths; also known as a lozenge pattern.

DIAPER PATTERN – a geometric pattern creating a grid of squares; also known as a net pattern.

FLAME PATTERN – a sawtooth-like design arranged in an even row and with the individual "flames" having slightly curved edges.

FOUR-POINTED STAR – a star shape with four points arrayed in an imaginary circle.

INTERLOCK PATTERN – a repeated pattern in which the elements are fitted together so that one cannot be moved without affecting others.

LOZENGE – a diamond shape that has the form of a rhombus in which opposite sides are parallel and adjacent sides have different lengths; also known as a diamond pattern.

NET PATTERN – a geometric pattern creating a grid of squares; also known as a diaper pattern.

NETWORK OR NETWORK PATTERN – an all-over, repeated design; also called a rapport pattern.

OGEE – a double curve in the form of an elongated S.

QUATREFOIL – a geometric design with four lobes; resembles a stylized rendering of a flower or leaf.

RAPPORT OR RAPPORT PATTERN – an all-over, repeated pattern; also called a network pattern.

REPEAT PATTERN – a surface decoration composed of elements (motifs) arranged in a regular manner.

ROSETTE – a rose or flower-shaped decoration with five or more radiating petals. In a dot rosette, flower petals and center are replaced by dots.

SAWTOOTH – a jagged or zigzag pattern with slightly curved edges that resembles the angled teeth of serrated saws.

SCALE PATTERN – a repeated design of overlapping, rounded arches.

SCATTER FIGURE – an all-over design, such as polka dots.

STRIP FIGURE – a design made to fit into a narrow band or strip, such as a lily chain.

STRIPE – a line that can be solid or dashed, broad or narrow, and straight or wavy (rippled).

TRICURVED ARCH – a pattern made of two ogees, which each have a flattened S-shape (with convex and concave segments) and meet in a point to form the arch. Tricurved arches are often used in network patterns.

YO-YO – a geometric design in which two parallel wavy lines, mirror images of one another, create a strip figure with alternating narrow and wide spaces between the lines.

ZIGZAG – a line with a series of short, sharp, and alternating left and right angles.

Notes

1   Merriam Webster's *Encyclopædia Britannica Concise* defines textiles as, "Any filament, fibre, or yarn that can be made into fabric or cloth, and the resulting material itself. The word originally referred only to woven fabrics but now includes knitted, bonded, felted, and tufted fabrics as well. The basic raw materials used in textile production are fibres, either obtained from natural sources (e.g., wool) or produced from chemical substances (e.g., nylon and polyester). Textiles are used for wearing apparel, household linens and bedding,

upholstery, draperies and curtains, wall coverings, rugs and carpets, and bookbindings, in addition to being used widely in industry." See http://www.merriam-webster.com/concise/textile, accessed 16 September 2014.
2 *PT*, 126–144. On Neolithic developments, which emerge over many centuries, see McCarter 2007. For the Neolithic era in Greece, see Halstead 1999b; Perlès 2001. For the concept of the "Neolithic package," see Çilingiroğlu 2005. On the significance of cloth, see Weiner and Schneider 1989.
3 *PT*, 247–259; Good 2001; Gleba and Mannering 2012.
4 The foundation for the study of Aegean textiles within their broader Mediterranean, European, and Near Eastern contexts remains Elizabeth J. W. Barber's *Prehistoric Textiles* (*PT*). Iris Tzachili's (1997) investigation of Aegean weaving and weavers was unfortunately unavailable to the authors for this study. Important recent publications include the many volumes published by the Centre for Textile Research. This study, however, is the first to focus on Aegean *patterned* textiles.
5 For introductions to Minoan and Mycenaean cultures, see Preziosi and Hitchcock 1999; Betancourt 2007a; Schofield 2007. For more detailed reviews, see Shelmerdine 2008; Cline 2010.
6 Classic texts on Greek mythology include Bullfinch 1855; Hamilton 1942; and Graves 1955. More recently, see March 2009; Morford, Lenardon and Sham 2013.
7 See Morris 1992, 354–357, with further bibliography.
8 Pentelia 1993.
9 Tzachili 2012.
10 Doumas 1992, pls 26, 29. For a study of the iconography, see Morgan 1988.
11 An important exception in Aegean art is the depiction of many athletes and children as nude, or nearly nude. See, for instance, the Boxing Boys of Building Beta (Doumas 1992, pl. 79) and the runners of the Flotilla Fresco in the West House (Doumas 1992, pl. 38), both from Akrotiri, Thera. On youthful nudity in fresco painting, see below, note 18; Chapin 2012.
12 Lang 1969, 43–49, 71–72 (22 H 64), pls 16, 177, A, M.
13 Lang 1969, 44–45; Davis and Bennet 1999, esp. 111–112, 115; Chapin forthcoming.
14 See, for example, Baines 1996; Bard 1996.
15 Blakolmer 2012b.
16 The bibliography on this fresco is large. For excellent photos, see Doumas 1992, pls 35–48; for iconography, see Morgan 1988.
17 Strasser 2010; Strasser and Chapin 2014; Vlachpoulos 2015, with references to alternative interpretations.
18 Doumas 1992, pls 18–25. Paul Rehak (1998) commented on how male youths are typically nude in Neopalatial art whereas girls are depicted wearing clothes. For youths in Neopalatial art, see Davis 1986; Koehl 1986; Chapin 2009. For more on sex, gender, and costuming traditions in Aegean art, see Chapin 2012; for detailed study of the costumes in the miniature frieze, see Morgan 1988, 93–103; on the nude youths, see Chapin 2007.
19 For a recent study of seal iconography, see Crowley 2013.
20 On Minoan figurines, see Rethemiotakis 2001; for a recent review of Minoan figurine studies, see C. Morris 2009. On Mycenaean figurines, see Schallin and Pakkanen 2009.
21 MacGillivray 2007, 111–112, figs 4.5, 4.6.
22 Marcar 2007.
23 For overviews of Aegean fresco painting, see Immerwahr 1990; Chapin 2010; 2014.
24 On the rise of Minoan fresco painting, particularly in the Protopalatial period, see Blakolmer 1997.
25 *PM* I, 251–252, fig. 188a, b; Immerwahr 1990, 21–22; Shaw and Shaw 2006, 182, no. 96, 224–225, pl. 2.38a.
26 See Immerwahr 1990, 39–75.
27 See, for example, B. Jones 2001; 2003; 2007; 2009; 2012; Lillethun 2003; 2012.
28 The best and most detailed survey of Mycenaean painting remains Immerwahr 1990, 105–146. See also Chapin 2010, 2014.
29 Gombrich 1960.
30 For introductions to the basic terms and tenets of art history, see Kleiner 2012, 1–13; Stokstad and Cothren 2013, xxii–xli. For the methods of art history, see Hatt and Klonk 2006.
31 Though early efforts at fresco restoration may have been overly enthusiastic (e.g., the addition of women's heads to the "Ladies in Blue" Fresco from Knossos [Fig. 9.1]), recent restorations are scientific in quality and

are indispensable for recreating the original appearance of compositions. See, for example, the beautifully restored frescoes from Akrotiri, Thera (Doumas 1992). On the value of fresco restoration, see Cameron 1976.
32 Doumas 1992, pls 109–111; Chapin 2007; 2009.
33 For discussion of whether patterns depicted in art were weavable with available technology, see *PT*, 315–330; Spantidaki 2008. On recreated costumes, see above, note 20.
34 On cross-craft interaction, see McGovern 1989; on cross-craft interaction in plaster technologies, see Brysbaert 2007; 2008b, 179–181.
35 Brysbaert 2008b, 180, with further references.
36 Chapin and Shaw 2006.
37 Gadamer 1989.
38 Ramberg and Gjesdal 2013.
39 This approach addresses the "hermeneutic problem": how can one understand something (e.g., art) from the past which seems enigmatic and not easily intelligible, particularly when one does not share in the conditions that gave it meaning? See Hatt and Klonk 2006 for the methods of art history; Davey 2002 for a detailed discussion of hermeneutics as it relates to the study of art; and Hodder and Hutson 2003 for a review of interpretive models in archaeology.
40 Davey 2002.
41 A leading postprocessual theorist on contextual analysis is Ian Hodder. On the entangled relationship between humans and material culture, see Hodder 2012. For an overview of archaeological theory, see Trigger 2006.
42 Adapted from *PT*, Donahue 2006, and the *Artlandia Glossary of Pattern Design* (http://www.artlandia.com/wonderland/glossary/), accessed 15 May 2014.

2

Bronze Age Aegean Cloth Production: A Cottage Industry No More

Brendan Burke and Anne P. Chapin

Archaeologists and historians today are well aware of the importance of textiles and textile production in the Aegean during the Bronze Age. From the production of the raw materials necessary to manufacture the cloth, to the use of the finished textiles in social contexts (including households and palaces), textile production was a demanding industry that required hours of labor from a wide range of workers. This included the shepherds in the hills raising woolly sheep and farmers toiling away to grow the flax, and spinners, weavers, fullers, and "finishers" who decorated the fabrics. Unlike contemporary Egypt,[1] however, only a few small bits of cloth survive from the Aegean Bronze Age. Archaeological treasures in their own right, these scraps hint at the rich textile tradition that once characterized the Minoan and Mycenaean eras, now mostly lost. But by using multidisciplinary approaches that combine the study of surviving textile fragments, together with the tools used in cloth production, artistic renderings of cloth, and textual documents related to cloth and cloth production, it is possible to gain a better understanding of Aegean cloth manufacture, distribution, and consumption. We are fortunate that all these sources of information do survive from the Aegean Bronze Age, although in varying degrees. What follows is an introduction to this evidence, beginning with an overview of existing scholarship.

Scholarship on Aegean textile production

Monographs, dissertations, and excavation reports focused solely on cloth production were at one point unheard of in archaeological literature, but beginning with the works of Marta Hoffman, Jill Carington Smith, and Elizabeth Barber,[2] the field of ancient textile studies is now well-established: it currently boasts devoted newsletters, conferences and conference proceeding volumes, experimental workshops, and outreach research programs that connect very different fields with the study of textiles in the ancient world. The University of Copenhagen's SAXO Institute's Centre for Textile Research (CTR), publisher of this volume, has played a fundamental role in facilitating research by bringing many aspects of the study of cloth production to the attention of historians, archaeologists, and other scholars. The CTR's conferences, workshops, and research fellowships have fostered a highly collegial environment for multidisciplinary investigation into the topic. Archaeologists and historians engage with craft specialists and together they apply their knowledge to what is known about the production of cloth in the ancient world, often with surprising results. A large database of sites in the eastern Mediterranean with cloth-related tools has been published online, 'to record as many textile tools from as many types of sites as possible' and will soon be published as the *Tools and Textiles Database*.[3]

Excavators once dismissed loom weights and spindle whorls as uninformative artifacts, objects that were often discarded because they were too numerous or merely the domain of women. For example, in 1957, Rodney Young recorded in an excavation notebook from Gordion, "We finish digging the burned fill … [finding] more whorls, doughnuts [loom weights] and iron knives. To recount in detail would be tedious."[4] Over the last few decades, scholars trying to understand the Greek past have come to value the role of cloth in ancient economies. Today, attention is paid to the distribution of craft residues, and the corresponding analytic and comparative work brings this important industry to life. Each stage of cloth production can be mapped with specific reference to the two most frequently encountered forms of evidence for spinning and weaving: spindle whorls and loom weights. Findspots and proper documentation of primary contexts are key to understanding the technologies involved in the production of cloth. Most recently, evidence for early textiles has been collected from Bronze and Iron Age contexts across Europe in an effort to present the material in a systematic manner and make it readily available to the non-specialist.[5]

The role of cloth in the early Greek past, as a mass-produced staple commodity fueling Minoan and Mycenaean palace economies, and remembered as a noble craft and metaphor of female virtue, finds parallels in other regions of the world. Some archaeologists have approached the study of textiles as sociocultural anthropologists, building theoretical models from ethnographic parallels and comparing these observations to the archaeological record, particularly in the New World and in northern Europe.[6] Others have focused their research on specific classes of textile tools,[7] including specialized studies of the spindle whorl.[8] Some scholars working in the Near East, where records are abundant, have looked exclusively at textual documents related to textile production,[9] while others have relied on a combination of textual and archaeological evidence for studying cloth manufacture and labor within complex economies.[10] Barry Kemp and Gillian Vogelsang-Eastwood, for example, have thoroughly examined aspects of New Kingdom Egyptian textile technology and production methods, and they found that there was a complex interrelationship between public and private spheres when it came to cloth manufacture.[11] Scale is another issue: yields of raw materials and measurements of finished-product weights and sizes allow quantifications of productivity.[12] Once such assessments are achieved, it becomes possible to reconstruct the mobilization networks that manufactured the finished products. How much textile was needed by one household, or by one palace, and how much surplus could be produced for regional and extra-regional trade – these are the questions now being addressed in current scholarship.

The physical remains of Aegean textiles

Until recently, it was believed that no textiles survived from the Aegean Bronze Age. Now, quite happily, this is no longer true. Fragments of textiles have been identified at a variety of prehistoric sites, including Akrotiri on Thera; Chania and Mochlos on Crete; and Mycenae on the Greek mainland. The volcanic eruption on Thera, dated to the LM IA period, *c.* 1630 BC, carbonized and preserved pieces of textile material that were then found during the recent excavation of pillar pits for Akrotiri's new roof. From Pit 65N come about 50 fragments of wool thread;[13] and Pit 52, notable for the discovery a golden goat figurine and a pile of caprine (goat) horns, yielded fragments of a finely-woven linen cloth embellished with three forms of decoration: a hem sewn with a thicker thread; embroidery; and different types of fringe embellishments, sometimes brought together with a thread to form tassels.[14] These fragments confirm that the textiles depicted in contemporary frescoes at Akrotiri and elsewhere in the Aegean are depicted with realistic detail.[15] Also from Akrotiri, in Pit 1b, come bits of a tabby-woven barley sack made with spun thread and strips of plant fibers, found together with a net; and Pit 68a yielded a fragment of utilitarian cloth made with linen warp threads and weft threads of another, as yet unidentified, plant fiber.[16]

From Chania on Crete come fragments of a narrow textile band, carbonized by fire, about 6 mm in width; it was made with only three linen warp threads but it was densely woven with ten weft threads per centimeter, which covered the warp threads completely. Interestingly, the weft threads were made from dark-colored goat hair, together with supplementary threads of light-colored nettle fiber. The hairs are short, each only about 1.5 cm in length, and wrapped around the selvedge (the edge of the band) only once. The Chania textile, then, is not a true example of weaving, but rather, of plaiting.[17]

From Mochlos, fragments of tabby-woven textiles were preserved on the surfaces of some bronze objects. The threads are light in color, fine in quality, and even in diameter; they are likely made from plant material, probably flax, so the textiles may be linen.[18] Additionally, a piece of string associated with a pierced bronze plate was identified in material from the century-old excavations of LM III tombs at Kalyvia, on Crete. Analysis determined that the string was made from three two-ply threads and spun from bast fibers, possibly flax or hemp.[19] Finally, from the Greek mainland, Heinrich Schliemann reported finding "traces of well-woven linen" attached to sword blades in Shaft Grave IV at Mycenae, but no decoration is noted,[20] and fragments of tabby-woven linen cloth were found with bronze daggers and a spearhead in Tomb N of Grave Circle B at Mycenae.[21] Fragments of tabby-woven linen were also found in a Mycenaean tomb at Ayia Kyriaki on the island of Salamis, near Athens,[22] and calcified (probably linen) fragments were found with ceramic vessels in Tomb 1 of the Mycenaean cemetery at Pylona on the island of Rhodes.[23]

Most interesting is new evidence for a vertical (warp-weighted) loom made of wood that was discovered in the 2009 excavation of the *Casa delle Sfere Fittili* (the House of the Terracotta Spheres) at Ayia Triada in southern Crete. The house, named for the large number of loom weights found there in previous excavations, dates to the LM IB period and belongs to an area of the town that was closely associated with the famous Royal Villa. Holes in the pavement of a ground-floor room were found with traces of decomposed wood, which can be best understood as the remains of a fixed vertical loom measuring 110 cm wide. Additional calculations estimating the number of loom weights that would fit such a loom, and the type of thread that would have been used, suggest intriguingly that a medium–low density cloth would have been woven there – perhaps sail cloth. Specialized textile production, then, is indicated, and it is likely that it was conducted in relation to administrative activities at the nearby Villa.[24]

Fragments of other fibrous materials also survive. A possible blanket or doormat made of pressed pine needles and other organic materials (perhaps grasses or weeds) has been identified from MM IIB contexts on the floor of Room IL in the palace at Phaistos.[25] Similar matted material made of sedges (*cypracaeae*) was recently excavated from Akrotiri,[26] and bits of a red and blue blanket or mat were found under the body of the deceased in the Late Helladic tomb at Routsi.[27] Though these objects are not textiles *per se* (since the materials are neither spun nor woven), they do shed light on the variety of organic materials that were used in the Aegean Bronze Age to make a range of household furnishings. These finds also support the interpretation of some floor frescoes as painted imitations of matt-like floor coverings (discussed below, Chapter 4).

The earliest evidence for textile production

Aegean societies developed their textile industries from much older textile technologies. Indeed, the process of spinning fibers pre-dates by thousands of years nearly every other known craft or technology, including freestanding architecture, metallurgy, and ceramics. Karen Hardy has offered evidence that the earliest (and simplest) textile technology was the production of string or cord, and she notes that binding and attaching things is a cultural universal.[28] String production led to developments in looping and weaving, which enabled the production of bags and nets that could be used in hunting, fishing, and the collection of small items. Unfortunately, Palaeolithic evidence for string is rare, yet, ingeniously, Hardy looks at indirect evidence, such as perforated beads and net weights that were presumably suspended by some string or cord. A perforated bone point and a perforated wolf incisor, both from Repolusthöhle, Austria (*c.* 300,000 years ago), offer the earliest evidence for string.[29]

Direct evidence for early textile technology survives from the Upper Palaeolithic era, *c.* 45,000–10,000 years ago. Impressions in clay of twisted fibers forming a knotted net, approximately 25,000 years old, have been found at Pavlov in the Czech Republic.[30] The fibers were almost certainly from a plant, probably a bast fiber related to linen and hemp. Fibers of wild flax were discovered in 2009 in Upper Palaeolithic occupation levels (*c.* 30,000 years ago) of the Dzudzuana Cave in Georgia: some of the fibers were colored (turquoise, pink, and black to gray), others were twisted, and still others were spun.[31] These exciting remains suggest that textile technologies were well advanced much earlier than had previously been supposed. Towards the end of the Palaeolithic era, in what is known as the Epipalaeolithic (*c.* 9200–8500 BC), linseed remains from pre-farming contexts have been found in north Syria.[32] Flax is a highly useful plant resource, since linseeds are a good source of oil and nutrition, and its long fibers can be used for textiles.[33] Charred linseeds were found at Pre-Pottery Neolithic B (*c.* 8000 BC) sites in the Levant, including Ramad, Jericho, and Nahal Hemar.[34] These do not all ensure fiber use, but scraps of cloth from Nahal Hemar in the Judean desert conclusively demonstrate the production of linen textiles by the 8th millennium BC.[35] Wild flax was probably domesticated about this time and henceforward became dependent on human cultivation.

The earliest evidence for loom-woven cloth comes from clay impressions of textiles that were most certainly linen, found at the 7th millennium BC site of Jarmo in northern Iraq.[36] Neolithic linen is also known from Çatal Höyük and Çayönü in Anatolia,[37] and flax was reportedly cultivated widely in Anatolia before the 8th millennium BC.[38] Linen appears in the archaeological record of Egypt somewhat later, in the 5th millennium BC, and consists of pieces of coarse linen cloth found in the Faiyum oasis together with spindle whorls and flax seeds.[39] Cultivated flax seeds have also been recently identified in Early Neolithic levels at Knossos on Crete, and suggest that flax was part of a "package" of domesticated plants and animals brought to Crete by Early Neolithic settlers.[40]

Cloth production in the Aegean also begins in the Neolithic era: textile equipment is found at Knossos in the Middle Neolithic era, in the early 4th millennium BC.[41] Tools for cloth production are widely distributed throughout the site, which suggests that individual households were making cloth for their own needs. By the Late Neolithic, however, equipment for textile production is concentrated in certain areas of the site, which suggests a nascent shift towards craft specialization. Perhaps autonomous individuals or household units came together to produce cloth for group consumption and exchange. As was the case elsewhere, these early textiles were probably made from plant fibers, since fiber sources from animals had not yet been well developed. Only after c. 2000 BC were new breeds of fleecy sheep (which produce wool) introduced to Greece from Anatolia, at which time, the peoples of the Aegean began producing a variety of elaborate textiles in linen, wool, and perhaps other materials.[42]

The materials used to produce textiles

Thus, textiles in the ancient world were made from both plant and animal fibers, and the basic element for woven cloth was spun threads or yarns.[43] Felted textiles have also been found at Gordion and in Greece; they were made by applying heat and pressure to unspun wool (wool fibers have naturally occurring scales that make felting possible). Most wool and plant fibers, however, were manually spun – a time-consuming process that requires more human energy than does the actual weaving. Fibers are either spun clockwise or counterclockwise, which twists the thread to the left or the right. The general convention for describing the direction of spun thread is to call it Z-spun (clockwise) or S-spun (counterclockwise), referring to the direction of the slant in the middle of these letters. This descriptive information is useful: for example, most yarn in Egypt was S-spun, while contemporary European thread was Z-spun, so when even small fragments of cloth are found, arguments can be made for their possible origins. Much can also be learned from the analysis of early yarns, either by using the naked eye or by viewing threads under a microscope: fiber identification, processing methods, and traditions for textile production can all be evaluated. Thread counts and their diameters are other key components of fiber analysis. Distinguishing between plant and animal fibers, however, can be difficult, and sometimes, even when a fiber is no longer preserved, a cast (a pseudomorph) still allows for structural and manufacturing analysis of the threads.[44]

Flax

Domesticated flax (*Linum usitatissimum*) developed from wild varieties of flax that originally had wide distributions across western Europe, the Mediterranean, Caucasia, and southwest Asia (Fig. 2.1). Its value as a source of both oil and fiber contributed to its being one of the first plants to be domesticated, but its cultivation is one of the most labor-intensive chores in agriculture. The plant is fairly unaggressive and requires much weeding. With proper timing, though, the flax harvest yields ripe seeds for linseed oil, stalks with fibers suitable for the production of fine linen cloth, and processing debris that can be used for fuel, fodder, and fertilizer (Fig. 2.2).[45]

Fig. 2.1. Flax in a field (photo C. Dalziel, JoybileeFarm.com)

Well-watered, easily drained agricultural land is needed to grow flax, and harvesting requires strong manual labor, as flax stalks must be pulled out of the ground whole. Then seed pods must be removed in a process called rippling. The pods are crushed to separate the seeds from the protective cortex, and the seeds are either collected for future sowing or they are crushed with the pods to be used as chaff fodder. Cleaned seeds can also be pressed for linseed oil. The stalks, on the other hand, are retted (that is, left to rot). This is achieved by either placing them in standing water or spreading them out on a flat surface to collect dew and decompose (ground retting). In the retting process, which can last 10–14 days, the pectins that bind fibers to other parts of the plant are dissolved by enzyme action from bacteria or fungi. Next, the stalks are left to dry, and then, in a process called scutching, they are beaten and combed to break the pith away from the usable fibers in the center. The scutching produces short fibers (tow) and long fibers. The scutched tow yields a woody matter called shives that can then be used for fertilizer. The long fibers are hackled – that is, they are combed to remove the short fibers and to ensure that all the long fibers are parallel to one another. At this stage in the processing, the raw linen, or rove, is ready for spinning, which can be either dry or wet spun.[46] Treating the rove with boiling water and sodium carbonate or sodium hydroxide will further breakdown the fibers and remove any unwanted color, yielding softer, finer threads. Note that from 1000 kg of raw flax, the yield of useable fibers is less than 100 kg.

Flax production was once widespread across Europe and the Mediterranean Basin (including Egypt). An early depiction of the flax harvest can be found in the 19th Dynasty tomb of Sen-Nedjem

2. Bronze Age Aegean Cloth Production: A Cottage Industry No More

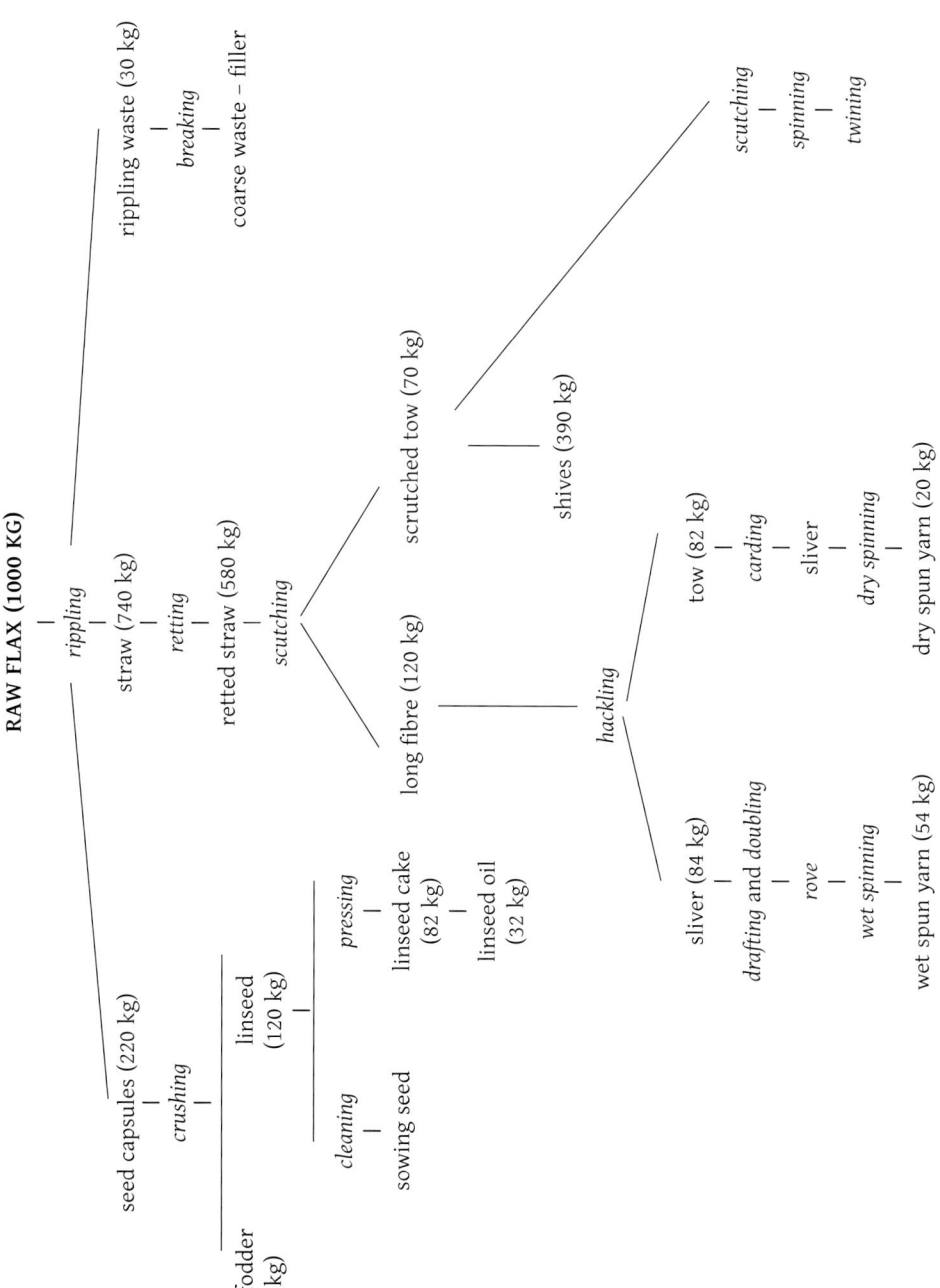

Fig. 2.2. Various phases of flax processing and uses (B. Burke after Salmon-Minotte and Franck 2005)

at Egyptian Thebes; another can be found in the Book of the Dead of Nakht, dating to the 14th century BC. In one of the panels of the Book of the Dead, tall green flax is shown growing closely together, which helps in the cultivation of the long straight fibers that are needed to produce the fine threads desired for linen manufacture.

Wool

Wool is a different story: sheep were first domesticated in the Near East at about 9000 BC, based on molecular dating, and were reared primarily for their meat, bones, and hides.[47] Prior to the 4th millennium BC, most sheep had fairly short hair of coarse fibers not well suited to spinning.[48] But thousands of years after flax and short-haired sheep were domesticated, some breeds of sheep were selectively bred for woolly fibers. Artistic evidence for this development appears in the Near East: a clay figurine from Sarab, Iran, dating to *c.* 5000 BC, shows what appear to be tufts of wool, and animal figurines decorated with woolly backs are found in Mesopotamia along with texts that show sheep were a concern to economic centers.[49] This change is also visible in the faunal record: sheep bones become more robust over time because a heavier musculature develops from carrying a weightier coat of wool.[50] Finally, early preservation of wool occurs in Lower Egypt in Predynastic contexts (before *c.* 3000 BC). Woolly sheep were introduced to Europe (including Greece) by *c.* 2000 BC, at which time wool production became important to both pastoral and agricultural economies. Fleecier sheep purposefully bred for use in textile production shows a shift in agricultural patterns that had significant social ramifications: the transition to exploiting wool, which is a fiber more efficiently produced than linen, initiated the development of large textile workshops and an attached labor class.[51] These changes can also be seen in the faunal record: the highest quality wool comes from young neutered males, called wethers, which are in greater evidence in the faunal record of the Aegean after *c.* 2000 BC. Also, in general females yield better wool than males, and palaeozoologists studying the age and sex ratios of bones have been able to distinguish between sheep that were raised primarily for their meat and those that were raised for the production of wool.

Silk

Silk is a natural protein fiber that can often be woven into fine textiles. It is produced by certain species of insect larvae when they build their cocoons; the most well-known species is the mulberry silkworm (*Bombyx mori*), which is now domesticated and raised in captivity. Many of the same processes and implements that were used in the Mediterranean for wool and linen manufacture are also found in silk production. All require an initial use of water (for boiling, soaking, and washing), followed by bleaching and dying, as well as carding and combing to remove impurities. These stages are of course followed by spinning and weaving. The technological assemblage used to manufacture silk is invisible to us in the archaeological record, so Bronze Age silk production is difficult to demonstrate, but significant evidence indicates that it was possible.[52]

The earliest evidence for silk in the Mediterranean region is an isolated strand from an Egyptian mummy dating *c.* 1000 BC.[53] The earliest silk fiber from a secure context comes from the Halstatt burials *c.* 700 BC,[54] and the earliest silk fiber from a Greek context is from the 5th century BC in the Kerameikos cemetery of Athens. Analysis has demonstrated that the silk fibers are from a native silk moth species.[55] Although physical evidence for silk textiles does not survive from the Late Bronze Age Mediterranean, new evidence suggests that a localized silk industry had developed in Harrapan India in the Bronze Age in the 2nd millennium BC.[56] These new finds are probably from a species of wild silk moth indigenous to the area, and silk was likely produced for a small elite class of consumer. The prevalent use of wool and cotton by the Harrapans is paralleled in the Mediterranean by the use of wool and linen, while silk may also have been produced for an elite niche market.

What evidence, then, is there for silk in the Bronze Age Aegean? From the excavations at Akrotiri, a white, crystalline cocoon, 44 mm long and 18 mm wide, was discovered in the House of the Ladies. Though calcified by time, analysis suggests that the cocoon was likely produced by the silk-producing moth, *Pachypasa otusi*.[57] Eva Panagiotakopulu suggests that the findspot within the house indicates that it had been deliberately brought indoors. And, as Trevor Van Damme observes, "this cocoon, therefore, represents the earliest and in some ways the best evidence for both the knowledge of wild silk and its potential applications."[58] Interestingly, its find context in the House of the Ladies, where frescoes probably depict the presentation of a sacral skirt (Fig. 3.33),[59] raise the possibility that the garment depicted in the fresco was woven from silk. A set of unusually light loom weights, with two holes for the attachment of threads rather than the more common single hole, also discovered at Akrotiri, could possibly be connected with the weaving of a light textile such as silk.[60]

Sea silk

Another possible supply of fine fibers comes from a seemingly unlikely resource, the sea. Unfortunately, the topic of sea fibers remains poorly investigated, and only one study investigates the possible use of sea silk in Aegean prehistory.[61] Sea silk comes from a fan mussel (*Pinna nobilis*),[62] which is the largest native bivalve in the Mediterranean and can reach a length of up to 120 cm (Fig. 2.3). The *Pinna* belongs to the family *Anisomyaria*, and lives in soft-sediment areas and beds of sea grass at depths from 0.5 m to 60 m. At least 20 different species of related mollusks produce sea-silk fibers, but *Pinna nobilis* is the most commonly cited. In recent years, fishing, trawling, and anchoring have greatly depleted beds of *Pinna*. Pollutants have also damaged their reproductive capabilities, and it is now listed as an endangered species currently under strict protection in the Mediterranean by a European Council Directive (92/43/EEC, Annex IV). In the past, people sought out this mollusk for food, for its large decorative shell, and for the fibers by which it attaches itself to the seabed. These fibers, which are long byssal threads, are secreted through the *Pinna*'s foot to fasten its "fist" to the underwater surface.[63]

Harvesting the byssal threads is attested in 19th century sources. Rakes were used to disconnect the bivalves from the seabed. The foot with the fibers attached was then cut off, washed in soap and water to desalinate it, and left to air-dry. Next, the fibers were cut from the foot, combed, carded, and spun. The thread produced from sea silk is extremely fine and reddish to golden brown in color. Some sample fibers have reached 1.75 inches (44.5 mm) in length, showing that spinning was necessary to produce a thread. It is reported that only 3 ounces (85 g) of useful fiber results from every pound of *Pinna* harvested.

Many Aegean sites have yielded finds of *Pinna nobilis*, from which sea-silk fibers could have been produced. Examples of the shell were collected by David Reese, who generously shared

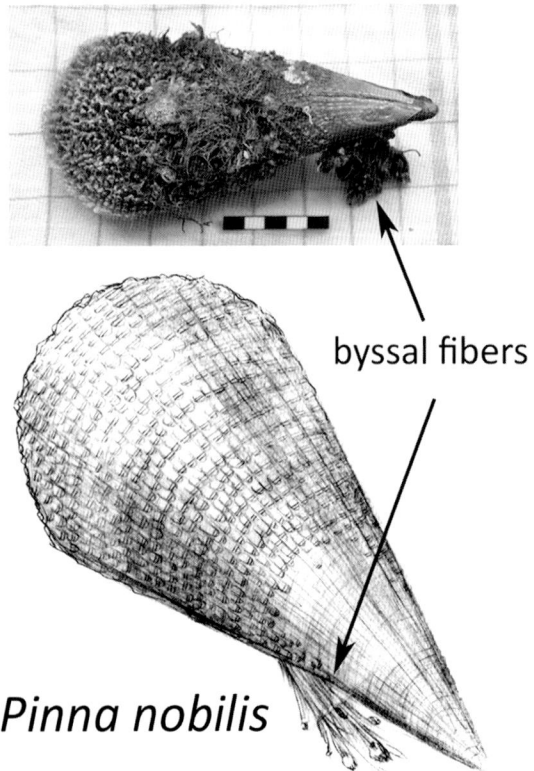

Fig. 2.3. Sea silk/Pinna nobilis (photo B. Burke; drawing G. Hill)

his information. On Crete, shells have been found in the recent Late Minoan excavations at Chania and in the Unexplored Mansion at Knossos. There are also reports of them in LH IIIC levels at Lefkandi, on the Greek mainland. Findings of the bivalve, however, are not proof of sea-silk production: the shell is very large and decorative, which suggests it could have been collected in antiquity for its value as a shell, as today. Some archaeological examples show wear marks on their edges, suggesting to Reese that they were also used as a tool of some kind.

It is possible that sea silk, or garments made from sea silk, are depicted in some Aegean frescoes featuring women engaged in ritual activity. The Young Priestess (Fig. 1.5) from Room 5 of the West House at Akrotiri is distinguished by her distinctive blue-shaven head, brightly painted red lips and ear, and a unique costume consisting of a one-shouldered robe worn over a short-sleeved garment. Some have suggested that the robe was dyed with saffron, and it has long been speculated that some of the young women at Akrotiri, particularly the Crocus Gatherers of Xeste 3 (Fig. 3.25), are actually wearing saffron-dyed costumes. The Young Priestess's garment, however, is rather shaggy and rust-colored, which suggests that it was made from a different material. Is it possible that her rather large billowy robe is made of *Pinna nobilis* sea-silk fibers? The West House was the findspot for a large number of *Pinna nobilis* shells, and so a connection is possible. In addition, the small container the priestess carries has been interpreted as an incense burner, yet she seems to be touching the material, which would not be possible if it were burning incense. The container itself looks very much like a shell, and its contents greatly resemble the small red strands of sea-silk fibers.[64]

A second possible identification of sea silk comes from a Mycenaean context, in the Room with the Fresco Complex (Room 31) at Mycenae (Fig. 3.43), where a female figure presents flame-like offerings to an altar painted with two female figures who probably represent deities. The offerant's gifts are usually identified as sheaths of wheat, but the color is not the golden yellow one might expect if the plants were wheat, but bright red, nor are kernels of wheat depicted. A better interpretation of the fresco would be that the offerant is presenting *Pinna nobilis* fan shells, which produce red fibers.[65]

Dyes

Information on dyes is unfortunately limited. Frescoes depict a variety of colors and richly patterned cloth – blues, reds, yellows, black, and white are most common – but how these colors were created, and which dyes were used, is uncertain. In contrast to wool, which accepts many dyes, flax takes color only with difficulty. However, to weave patterns, the fiber is dyed before it is woven, either as raw material or as hanks of yarn or thread.[66] What follows is information on the various dyes used to create vibrant colors:

- Reds: Madder (*Rubia tinctorum*), which creates a rich red color, was likely known to Aegean weavers; the plant-based dye was used to color the linen sail on a model ship from King Tuthankhamun's tomb in the Valley of the Kings, Thebes (discussed below in Chapter 6). It is native to Greece and the Levant, and grows wild on Crete today.[67] Henna and safflower are also possible sources of red dye, as demonstrated by mummy cloths dating to the 21st Dynasty.[68]
- Blues: In the late Egyptian era, blue cloth was probably dyed with indigo.[69] Interestingly, indigo has now been identified as an organic pigment in some Mycenaean frescoes.[70]
- Yellows: It is generally believed that saffron from native crocus flowers (*Crocus cartwrightianus*) was used to dye Aegean garments a rich, golden yellow color. Saffron gathering is depicted in the frescoes from Xeste 3 at Akrotiri (discussed below, Chapter 3).[71]
- Murex purple: The famous red-blue dye from sea-snails associated with the Levant was first exploited in the Middle Bronze Age on Crete. The dye could be made from three types of molluscs (snails) in the Mediterranean: *Bolinus brandaris, Hexaplex trunculus,* and *Stramonita haemastoma* (Fig. 2.4).[72] To dye a significant amount of wool purple required liquid dye from a great many snails. Some scholars exaggerate the miniscule quantity of dye yield from a snail. What must be remembered about the early dye studies is that the scientists are referring to the pure dye extract in its chemical form, rather than the practical solution which is readily usable for coloring spun fibers and woven textiles. Experimental archaeology has changed our understanding of murex purple dyeing and shown that the yield from just a few snails is quite significant.[73] Crushed murex shells found across Crete, at Kommos, Petras, and on the islands of Kouphonisi and Chryssi, off the south coast of Crete, show significant interest in murex dye in the Protopalatial Minoan era.[74] The dye was also used in fresco painting (as in Xeste 3 for the crocus flowers).

Other dyes are possible, and some are known from other ancient sources, but it is difficult to prove that they were used in the prehistoric Aegean. These include reds from insects (for example, kermes), iron-rich muds, tannins, and lichens; yellows from turmeric, pomegranate rind, and onion skins; and blue from woad.[75] Black and shades of brown and white found in finished textiles come from the undyed, natural fibers of sheep's wool; bleaching for a brighter white can be achieved with urine; and fine, undyed linen is of course pale in color. Except for murex purple, most of the dyes mentioned above are not colorfast, and the color needed to be fixed with a mordant (acidic mordants came from tannins, while basic mordants were from metals), but little information on mordants has been identified from Aegean Bronze Age contexts.[76]

Fig. 2.4. Hexaplex trunculus (murex) from a purple dye workshop (photo B. Burke)

Dye works

Dye works were used to clean the raw materials of textile production (e.g., wool) and to dye fibers. The installations required large quantities of water, water-tight pits, vats, or basins, mortars, pestles, and pounders (to process the dyes and mordants), and heat (fire pits or hearths and large pots) to facilitate the dyeing of the fibers.[77] Such facilities have been identified in the Early Minoan period at Myrtos: Fournou Koriphi, in southeastern Crete. Tubs and channels for liquids were situated in areas of the site near where large numbers of loom weights, together with spindle whorls, grind stones (querns), stone weights, and sheep bones were found; these finds indicate that the processing of fibers, probably wool, occurred there along with the spinning and weaving of textiles. A tub from Room 59 at Myrtos even bore traces of fatty residues, perhaps lanolin left over from wool washing.[78]

From Alatzomouri-Pefka, near the Minoan palace and town at Gournia in eastern Crete, comes exciting new evidence of a Middle Bronze Age (MM IIB) dye works consisting of nine vats cut into bedrock, found together with pounders, mortars, crushed murex shells, a well, and a rock-cut basin with a drain that was probably used to wash wool before it was dyed. Seven of the vats were probably used to dye the washed wool, as each was cut with a shallow trough onto which wool could be pulled to drain off excess dye. The number of vats, each about 1 m long, suggests that wool was dyed a variety of colors, each requiring its own basin to keep the colors pure, but other than murex purple (for which there is evidence in its shells), the dyes are unknown. Cook pots were probably used to heat the dye baths, and pithoi were found for storage, while cups and other fine wares demonstrate that dye workers lived at the dye works. Interestingly, very few loom weights were found, which suggests that the facility was dedicated to one stage of textile production – dyeing the wool – and not to spinning or weaving. The scale is palatial, and indeed, the Gournia palace is situated only a short distance away. A prism seal similar to others manufactured at Quartier Mu, Mallia, however, was found in the well and could have been owned by the Pefka site supervisor; its presence suggests a possible administrative connection with palatial authorities at Mallia.[79]

Recent excavations on the island of Chryssi, situated seven nautical miles off the south coast of Crete, offer insight into a Neopalatial dye works dedicated to the collection and preparation of murex purple. The site was abandoned suddenly at the end of the LM IB period, and one of the excavated buildings, House B1, was divided into domestic and production spaces. Household furnishings included cook wares, decorated vases, storage jars, fishhooks, knives, jewelry, and three sealstones. In the work spaces dedicated to murex purple production were found large quantities of murex shells, two hearths, ash from the fires, cooking pots, pounding stones, and a pounding platform, all of which suggest that murex mollusks were processed into dye in the workshop. Two shallow, rock-cut pools (fish tanks?) situated on the seashore, adjacent to the site, could have been places for gathering and keeping murex and even perhaps breeding them. The overall affluence of the Chryssi community attests to the prosperity enjoyed by those manufacturing the dye.[80]

Tools of textile production in the Bronze Age Aegean

The cleaned and dyed fibers needed to be processed into thread and woven into fabric. What follows is a brief description of the surviving tools used for these stages of textile production, namely, spindle whorls and loom weights.

Spindle whorls

Spindle whorls are usually small, fired clay objects of various shapes and weights that are used, together with a long slender rod called a spindle, to produce the tension-spun thread required by prehistoric weaving technologies (Fig. 2.5). In order to spin thread from fiber, the whorls

Fig. 2.5. White ground oinochoe by the Brygos Painter, from Locri, 490-470 BC; London, British Museum Vase D13, 1873,0820.304 (© Trustees of the British Museum, London)

are attached to the bottom of a spindle. Raw fiber (e.g., wool) is drawn out and hooked around a small notch at the top of the spindle, and then the whorl is spun manually. The resulting thread is thus subject to both tension and torsion, and winds itself around the rod. Elizabeth Barber estimates that it takes seven to eight hours to spin enough thread to weave on a loom for approximately one hour.[81] The bulk of the time required to produce textiles in the Bronze Age, then, was spent spinning.

Given this fact, surprisingly few spindle whorls survive from the Aegean Bronze Age. For example, excavations at the palace of Knossos (excluding the Prepalatial settlement) report finding no spindle whorls at all, in contrast to the hundreds of loomweights discovered there. The absence is especially striking when compared to the thousands of whorls reported from contemporary levels at Troy. The absence of spindle whorls from Crete can perhaps be explained by at least three possibilities: the first is that most spindle whorls were made of perishable materials such as wood; the second, that fibers were spun by a technique that did not employ the whorl-weighted drop-spindle (perhaps using an object not recognized in the archaeological record?), and the third, that tasks such as spinning became highly specialized in an organized Minoan textile industry and occurred away from excavated centers.[82] Unfortunately, not enough evidence survives to assess these hypotheses.

Looms

In contrast to spinning, plentiful evidence for weaving survives from the Aegean Bronze Age. A variety of looms were probably used, but there is firm evidence only for the warp-weighted loom. Nevertheless, there is indirect evidence for the back-strap loom, tablet (or card) weaving, and the band loom. Narrow woven bands are preserved in the early archaeological record from Mediterranean and west/central Asian contexts as early as *c.* 6000 BC. The narrow bands, which are also depicted in Aegean painting, were probably woven on small, portable looms that do not survive today.[83] What follows is a brief overview of these types of looms.

Back-strap loom

The back-strap loom is a simple, portable type of loom that is still used by traditional societies around the world. It consists of two sticks, or bars, between which warp threads are stretched. One bar is fastened to a fixed point (e.g., a tree or a post) with a rope, and the other is attached to the weaver by means of a strap wrapped around the weaver's back. Tension on the warp threads is created by the weaver, who leans back or forward as needed. A shuttle (a long, flat stick around which thread is wrapped) is used to pass the weft threads through the warp threads. The back-strap loom can be set up almost anywhere, and it can be moved from one place to another without difficulty, so these looms offer great flexibility and mobility for weavers. Complex patterns can be created with this simple device by using supplementary weft, but the width of the resulting cloth is limited by the weaver's arm length, and textiles woven in this manner tend to be rather narrow strips that can be used for belts, bags, carrying cloths, or even perhaps the flounces of Aegean-style skirts.[84]

Tablet weaving

Tablet weaving, also known as card weaving, produces narrow bands that can be used as borders, headbands, or belts. It is a simple form of portable loom that can be fixed upon a board or set up like a back-strap loom, with the primary difference being that each warp thread is run through a hole in a corner of a four-holed tablet, or card, made usually of perishable materials such as wood (Fig. 2.6). The tablets then become a set that looks a bit like a modern deck of cards; tablets are then turned (rotated front or back) to change the shed and create woven patterns. The bands produced are typically narrow and can be highly decorative.[85] It is possible that Aegean weavers practiced tablet weaving, since the complex patterns depicted on bands trimming Aegean costumes are similar to those produced by tablet weavers today. The four-holed cuboid loom weights (discussed below) look like thick versions of modern tablets, but it is difficult to recreate how such thick tablets (if that is what they are) could have been used in efficient tablet weaving.

Band loom

The evidence for band looms is slim, but this type of loom was needed to weave the heading band that secured the warp on warp-weighted looms, for which there is definitive evidence. The weft (horizontal threads) of the heading band would become the warp (vertical threads) on the warp-weighted loom. Elizabeth Barber surmises that there must have been separate band looms used in Bronze Age textile production,[86] and intriguingly, new investigations of the *Casa delle Sfere Fittili* (the House of the Terracotta Spheres) at Ayia Triada have revealed three holes in the floor of room 9 which can be explained as an installation for a band loom.[87]

Fig. 2.6. Tablet weaving (photo B. Burke)

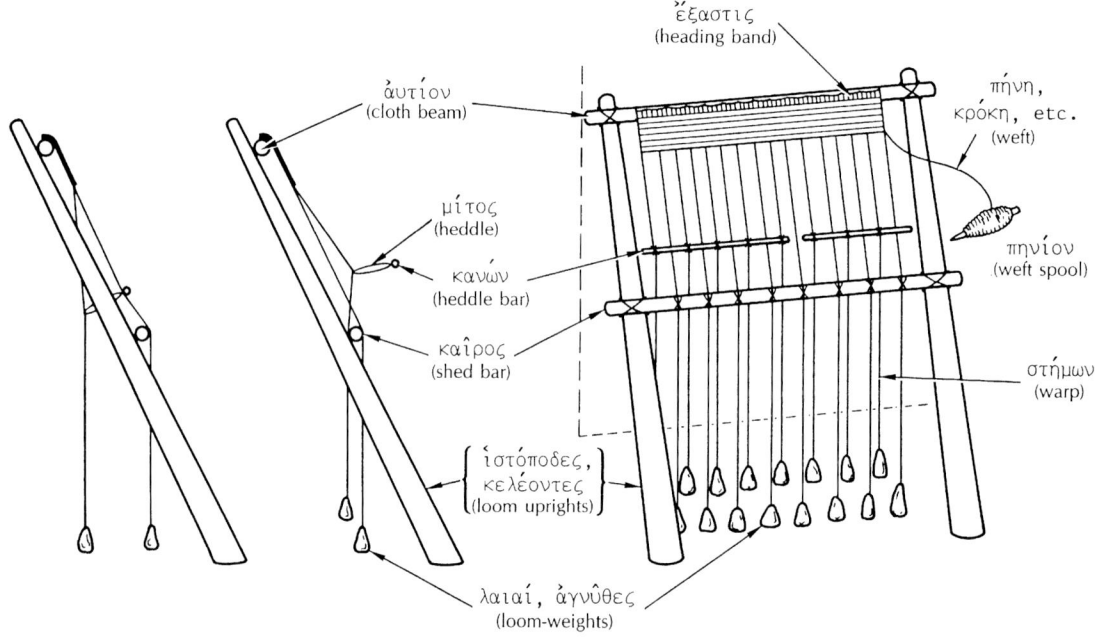

Fig. 2.7. A warp-weighted loom (drawing E. J. W. Barber)

Warp-weighted loom
Significant deposits of loom weights indicate that the warp-weighted loom was used for the production of the majority of textiles beginning as early as the Middle Neolithic period (Fig. 2.7). The basic technology for this type of loom is simple: a shed is created by leaning a simple frame against an upright wall. Rows of vertical warp threads hang downward and can be divided into any number of smaller sheds. Weft threads run horizontally through these sheds, and, by using a shuttle, a skilled weaver can create a variety of patterns with the interwoven threads. The woven design generally starts from the top, and the weave is packed upward. This technology is fairly simple, which may explain its great popularity, especially in comparison to the ground looms and two-beam vertical looms of Egypt and the Near East.[88] Recent investigations of the *Casa delle Sfere Fittili* at Ayia Triada, discussed above, may offer evidence for the a warp-weighted loom, situated beside the band loom reviewed above.[89]

Loom weights

Loom weights hang from the vertical warp threads at the bottom of a standing loom, thereby providing tension to the warp threads. The weights were attached in a variety of ways: they could be tied directly to warp threads, attached to a cord that was fastened to the warp, or tied to a thin bar that had warp threads connected to it. A series of recent scientific experiments with replica loom weights demonstrates that the weight of a loom weight defined the type of yarn being woven and the density of the warp threads. The thickness of the loom weights, which typically hang close together in a warp-weighted loom, together with the total number of loom

Fig. 2.8. Minoan loom weights: (a) spherical weight from Knossos; (b) discoid loom weight from Loom Weight Basement, Knossos; (c) cuboid weight from Petras (photos B. Burke)

weights being used, defined the width of a piece of fabric. The weight and thickness of a loom weight, then, determined the thread count and density of fabric woven on a warp-weighted loom.[90] There are three basic shapes for Minoan loom weights: spherical, discoid, and cubic (Fig. 2.8). The Mycenaean evidence for loom weights is unfortunately scarce and is not reviewed below.

Spherical loom weights
Spherical (melon-shaped) loom weights are, as the name suggests, approximately spherical in shape, about the size of an orange, and often fairly large and heavy (weights range from 86 g to 710 g).[91] Many have grooves running parallel to the suspension hole around the entire surface of the weight, which contributes to their melon-like appearance. The number varies but most often the grooves quarter the sphere.[92] These weights were produced on Crete in the Neopalatial era and just afterwards (MM–LM II), and are usually found in and around Minoan palaces in the northern and eastern parts of the island, but interestingly, they are absent from prominent Minoanized sites in the Aegean islands, such as Akrotiri on Thera, Ayia Irini on Keos, and sites on Kythera and Rhodes.[93] These oddly specific find contexts suggest that the weights were used for textile production serving the Minoan elite, perhaps using retainer workshops of full-time weavers employed by the ruling palace officials.[94]

Discoid loom weights
The second type of weight is the discoid loom weight, which is the most common Minoan loom weight and was used on Crete from EM II through LM III. It is so ubiquitous that Carington Smith suggests that it is just as culturally characteristic of the Minoans as is the Double Axe or the Horns of Consecration.[95] The discoid weights are pierced with suspension holes, and often the upper portion the weight is flattened and grooved. Deposits of discoid loom weights are known from many sites across Crete and are even found in Minoanized sites off-island such as Akrotiri,

Ayia Irini, Ialysos on Rhodes, and Iasos on the Anatolian coast.[96] But the most interesting is the deposit of over 400 discoid loom weights found by Sir Arthur Evans in the Loom Weight Basement of the East Wing of Protopalatial (Middle Minoan) Knossos.[97] Of these, 45 weights were weighed, drawn, and catalogued by Brendan Burke. The sample, representing perhaps 10% of the total found by Evans, was striking in its uniformity: all were flat, fired, and discoid in shape with a single suspension hole; they measured 9–10 cm in height and about 7.5–8.5 cm in width, and weighed 127–205 g. The similarity of these weights suggests that in the Protopalatial period, there was an attempt to regularize textile tools. And, if each loom used ten weights, and at least 400 loom weights were found, then the evidence suggests that there were perhaps 40 working looms at Knossos. The uniformity of the weights as well as their concentrated numbers suggest a regulated or standardized textile industry at Knossos, perhaps in the form of retainer workshops with full-time weavers working for the Minoan state, centered at the palace of Knossos.

Cuboid loom weights
The third major type of Minoan loom weight is the rectangular prism of cuboid form. These are usually pierced with four small suspension holes, one in each corner; thread wear suggests their use as loom weights, though it is difficult to reconstruct how they could have been used on Bronze Age looms.[98] It is possible that they were used in tablet weaving, which could explain the four-holed corners, yet how they would actually work, given their blocky nature and typically uneven size, is a problem that cannot be easily solved. The similar fabric and the appearance of cross marks, together with the fact that their measured weights fall within the range of other known loom weights, does, however, make it likely that they were used in weaving.

The earliest examples of cuboid weights were found in Middle Neolithic contexts at Knossos. Later, in the Middle Minoan period, cuboid weights[99] occur exclusively in eastern Crete and are not found west of Mallia. Carington Smith suggests that these weights offer evidence for population movements, and indicates the arrival of people who preferred discoid loom weights, so prevalent in Minoan palaces and settlements. Whether the distribution pattern of the loom weights reflects population movements is unclear but the geographical demarcation is distinctive.[100] For instance, over 100 cuboid weights were found at Palaikastro, in far eastern Crete, including several that were stamped;[101] 71 of the 100 belong to one deposit in House E 36, which is dated to the LM II (Postpalatial) period, and 15 of these had seal impressions or graffiti on them, suggesting organization, if not administration, of textile production.

Aegean textile industries: texts and contexts

Archaeological evidence can be used to trace the development of complex textile industries in the Aegean in both the Minoan and the Mycenaean eras. At Knossos, Prepalatial houses from the EM IIA period (*c.* 2600–2400 BC), built in the area of what would become the West Court, produced both spindle whorls and loom weights.[102] This means that both spinning and weaving activities were practiced at Knossos in tandem, and that there was no degree of specialization but rather independent, household production. Presumably individuals participated in all aspects of cloth production, from spinning raw thread to weaving finished textiles.

A similar picture emerges from the study of weaving technologies at Myrtos: Fournou Koriphi, a small, nonpalatial settlement of perhaps six households situated on the south coast of Crete and also dating primarily to the EM II period (*c.* 2600–2200 BC). The egalitarian nature of the architecture and the associated finds at Myrtos offer little evidence for social complexity,[103] yet the technologies for weaving are quite advanced: along with significant numbers of loom weights and spindle whorls are the earliest known spinning bowls – a technology that subsequently spreads elsewhere. Barber even suggests that spinners in Egypt and the Near East adopted the idea of fiber-wetting bowls from Minoan spinners.[104] She also suggests that 10th Dynasty wall paintings in Egypt that depict Minoan-style double-heart spiral patterns were probably inspired by Cretan textiles, and that some of the textiles produced at Myrtos could have been elaborately patterned, polychrome fabrics woven for export.[105] The large number of tools associated with cloth production, the location of Myrtos on the southeastern coast of Crete (which would have provided relatively easy communications with Egypt), and goods found on Early Minoan Crete originating from 10th Dynasty Egypt and the Near East all are suggestive of a trade relationship between Egypt and southern Crete.

After the Early Minoan period, at about 2000 BC and concurrent with the rise of the Minoan palaces, the archaeological distribution of textile equipment changes: no longer are spinning and weaving tools found together as they were at Prepalatial sites. This change, together with the rise of new textile technologies (such as purple dying from murex[106]), suggests multiple developments that may have contributed to the emergence of the Minoan palace system. And, with the appearance of Minoan palaces comes administrative interest in textile production, which can be detected even though administrative documents written in Linear A are both rare and undeciphered. As noted above, the 400+ discoid loom weights from the Loom Weight Deposit in the Knossos palace (and the 40 looms that were likely associated with the weights) offer good evidence for direct palatial involvement in textile production (weaving) during the Protopalatial era. Similarly, about 60 weights were discovered at Protopalatial Phaistos, where they were arranged in groups of about 12, for five looms.[107] Some 500 clay loom weights and 130 stone weights found at Quartier Mu, Mallia,[108] possibly indicate a palatial interest in weaving (though the exact nature of the relationship between the palace and the adjacent Quartier Mu remains unclear). Pietro Militello suggests that Protopalatial administrative participation in textile production was likely limited to satisfying the needs of the palatial elite for their own use, rather than for producing a surplus to be exported or traded.[109]

In addition to the tools of cloth production, we also can look at administrative documents for evidence of organized textile production.[110] On Crete, sealstones were carved with designs that signified the owner's identity and his participation in the Minoan administrative system. A certain type of three-sided sealstone, most examples of which come from Protopalatial contexts, is sometimes carved with weaving imagery: round objects, usually three to five in number, are shown suspended from poles and greatly resemble discoid loom weights; "tassels" or fringes running upward, perpendicular to the bar and the weights, can be best understood as warp threads from a loom (Fig. 2.9).[111] If this identification is correct, then these sealstones offer compelling evidence for weaving having been under the administration of Minoan authorities from the beginning of the palatial periods.[112]

In the Neopalatial era, Minoan palace officials continued their interest in textile manufacture, but much of the documentary evidence derives from archaeological remains of "villas" (this class of Minoan architecture shares many elements of elite palatial architecture, such as Minoan halls with *polythyra*, but they are built on a much smaller scale). Officials with close connections to the

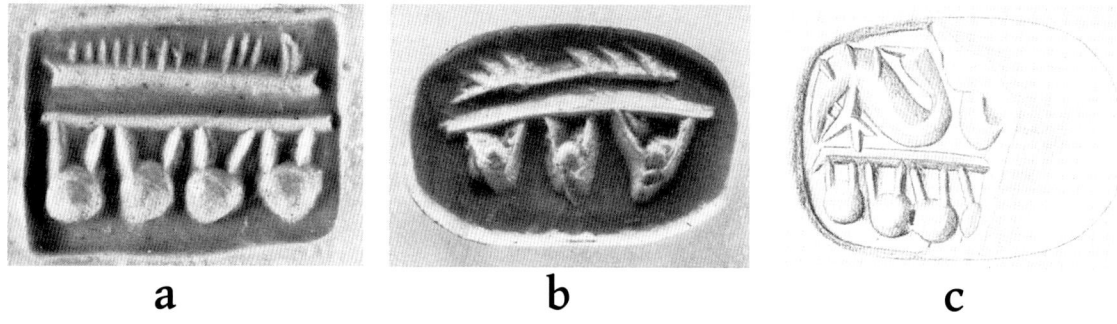

Fig. 2.9. Prism sealstones: (a) Nr. 151/Mallia 1823; (b) Nr. 125/Mallia 1797; (c) Nr. 124/Mallia 1795. CMS II.2, with permission, I. Pini

Minoan administrative system evidently lived and worked in these finely built structures.[113] From one, the Royal Villa at Ayia Triada, administrative involvement in the collection, distribution, and weaving of wool is indicated by a group of loom weights discovered in Room 27. Nearby, on a threshold between Corridor 9 and Vestibule 25 were found 45 noduli (a type of "sealing" used as a token or a receipt from an administrator), and a Linear A tablet (HT 24). The tablet records 36.5 units of wool (over 100 kg). We can imagine that the wool was either distributed or received; the administrators gave receipts for the transactions, and the information was recorded on HT 24.[114] Yet as Militello observes, there is no documented interest in the sheep or the herders that produced the wool, and thus administrative interest seems limited to the final stages of textile production. A decentralized system of textile production, then, seems likely.[115]

In sum, the analysis of the distribution of the tools and texts of textile production indicate that Neopalatial textile production was concentrated in certain houses, villas, and in the smaller palaces. Not every household evidently owned a loom, and intriguingly, the larger palaces (Knossos and Phaistos) offer no evidence at all for weaving in the Neopalatial era. For Militello, the archaeological evidence, together with surviving documentary evidence, suggests that the administrators of the Minoan palaces and villas were focused only on the final stages of textile production – that is, the weaving and manufacture of textile goods. The earlier stages of production, the cultivation of the flax and the raising of the sheep, were outside the control of the palatial administrations, and thus Minoan textile production overall can be classified as decentralized in its structure. Militello suggests further that the Minoan palaces might be better understood as places of cloth consumption rather than as centers of production, and that the Minoan elite were conspicuous consumers of textiles largely produced by those who were situated outside the palaces.[116] This concept will be touched on in subsequent chapters in this volume.

By the middle of the Late Bronze Age, c. 1400 BC, the Mycenaeans played a dominant role in the economic history of the Aegean both on the mainland and on Crete. Clay tablets inscribed by the Mycenaeans in their Linear B script have been deciphered as the earliest surviving form of written Greek. These tablets record a variety of transactions and offer good information on the Mycenaean economy. From the study of the thousands of Linear B texts that survive (mostly in fragmentary condition), it has been determined that a significantly large percentage of the tablets refer in some way to a well-organized textile industry. From such records we see that Mycenaean textile production, like its Minoan counterpart, was subject to the central authority of the palace.[117] But unlike the Minoan system, the Linear B tablets of the Mycenaean palaces show strong administrative interest in *all* phases of textile production, from sheep and flax field to finished cloth. Linear B texts document the numbers of sheep in various flocks and record amounts of the raw materials (wool and flax) collected from the towns and villages under the administration of the palaces. Once cloth was woven, palace scribes used a variety of ideograms and terms to make distinctions between different types of cloth, usually indicating the heaviness and amounts of wool used. The documents also provide information related to the textile personnel, including the allocation of food rations for dependent workers involved in this industry. All indications are that the craftsmen and craftswomen listed were highly specialized, as the tablets record specific occupations such as spinners, weavers, fullers, and finishers.[118]

Since very little evidence for textile activity has been found at Mycenaean palace centers, it seems likely that most cloth was made in outlying communities. While it was once thought that regional production was administered by palatial centers, this view of the heavy economic involvement of the palaces is changing in Mycenaean studies.[119] The recent discoveries of Linear B texts that deal with textiles at Iklaina, near Pylos, and Ayios Vassilios (just south of Sparta in the Peloponnese) suggest that administration of textile production was more dispersed than was previously believed. As Cynthia Shelmerdine states, "the redistributive model is inadequate ... a more productive model is a continuum, with individuals and groups involved in various ways and to various degrees with the central palatial administration, from full dependence to greater or less interaction to no contact at all."[120]

The actual extent of the palaces' involvement in cloth production, and other economic activities, is under constant review. But what we can say is that Linear B tablets from Knossos describe flock management (tablet series Da-Dg, Dn), sheep shearing (Dk/Dl), and wool allocations (Od) recorded by the palace scribes for outlying areas of the kingdom.[121] From tablets at Pylos and Knossos we learn that many of the female textile workers were categorized by their ethnic descriptions, which indicate their place of origin, such as Knidos *(ki-ni-di-ja)*, Miletus *(mi-ra-ti-ja)*, and Halikarnassos (Zephyra, *ze-pu$_2$-ra$_3$*) in Anatolia; this suggests a shared tradition of cloth working across these regions.[122] By correlating information on sheep-shearing, production targets, and workers' rations with various place names and ethnic descriptions, the Linear B texts provide insight into a highly structured and well-organized textile industry.

Mycenaean palaces, it should be stressed, were not all alike, and two major points distinguish textile production at the Mycenaean palaces of Pylos and Knossos. First, in comparison to Mycenaean Knossos, the palace economy at Pylos shows a greater degree of regionalized specialization among its various provinces and towns. The Linear B records found at Pylos reflect centralized control of craft activities at two major centers in Messenia, Pylos and Leuktron, each of which monitored activities dispersed throughout the kingdom.[123] As Killen summarizes, Pylos had 28 separate work groups, nearly half of which contained more than 20 women. Leuktron had six or seven work groups, one with 37 women.[124]

Second, linen production had a much larger role in the palace textile industry at Pylos than at Knossos. The various phases of linen cloth production are fairly well attested in the Pylian records, although they are far fewer in number than the wool records at Knossos. The linen ideogram SA occurs on the N-series of tablets from Pylos (Na, Ng, Nn) and is assessed from many different villages in both the Hither and Further Provinces of Pylos,[125] and in the tablets we find *ri-no,* which is almost certainly related to *linon* ("linen"). From the N-series tablets at Mycenaean Pylos, we see that textiles and related crops comprise 78% of the records, while food and livestock are found on only 16%. From the collection target tablet PY Ng 319, the Hither Province of Pylos (which is contrasted on the tablets with the Further Province) alone yielded nearly 50,000 kg of this fiber in one year. This emphasis on linen around Mycenaean Pylos is not surprising: in the 1950s this area of Messenia produced more than half of the linen in Greece.[126]

Conclusion

Bronze Age textile production was a highly specialized industry involving a wide range of laborers. For palace production, elites organized (to varying degrees) the manufacture of textiles both for their own use and for exchange. This kind of organization could occur only with centers that had close ties to rural hinterlands. The evidence presented here demonstrates that differences in social complexity and productive strategies translated into differential distributions of textile equipment.

Having looked at many examples of centralized textile production, both with textual and archaeological sources, what can we say about palace production? We have documentary records of cloth manufacture from Minoan and Mycenaean palaces. Many archaeologists have focused a great deal of attention on these major centers while not investigating smaller settlements, where activities such as pottery, metallurgy, and textile production are more likely to have been located. In the Minoan era, palace officials kept track of the later phases of textile production, particularly the weaving, and seem to have been less interested in the totality of production. Yet the "proprietors" of dye works at Pefka and Chryssi also owned sealstones, which suggests participation in the broader administrative system of Minoan Crete. In the Mycenaean era, administrative officials kept records of the entire process, from raw material to finished product. The production of cloth occurred in household settings, in smaller villages located in the hinterlands of the palaces, at settlements like Nichoria in the Peloponnese. The productive capacities of these places were assessed and the raw materials and finished cloth for the palaces were then collected as tribute (or more simply as taxes) by the palatial center. The palatial elite then used the textiles in a variety of ways: as clothing and furnishings for their palaces and households; as presents awarded to peers in elite gift exchange; as dedications to the divine within the religious sphere; and as commodities to be traded perhaps as an early form of currency.[127] Textile production at the palace, however, was primarily the work of women who were dependent to varying degrees upon the central authority, possibly against their will (perhaps as captives in war?). Palace-made cloth might be used to appease potentially hostile neighbors, impress valuable trading partners, and legitimize the elite power base. This circulation of textiles would have continued as long as the centralized economic system was working.

Notes

1. The dry environmental conditions are good for the survival of ancient cloth in Egypt. Famous examples include the linens found in the tomb of Tutankhamun, in the Valley of the Kings (Thebes). Likewise, a great deal is known about the textile industry at Amarna, childhood home of Tutankhamun. For more on the Egyptian textile industry, see Kemp and Vogelsang-Eastwood 2001.
2. Hoffman 1964; Carington Smith 1975; *PT*.
3. http://ctr.hum.ku.dk/tools/.
4. Young 1957, NB 67, 78, University of Pennsylvania Gordion Archives.
5. *Textiles*, v.
6. E.g., Parsons 1975; Smith and Hirth 1988; Parsons and Parsons 1990; Brumfiel 1991; Costin 1991; 1993; Andersson and Nosch 2003.
7. Carington Smith 1975, 1992; Dabney 1996.
8. E.g., Balfanz 1995; Crewe 1998.
9. Jacobsen 1970; Waetzoldt 1972; Dalley 1977; Ribichini and Xella 1985; Sollberger 1986; Szarzynska 1988; Völling 2008.
10. Murra and Morris 1976; Murra 1989; Stein and Blackman 1993; McCorriston 1997.
11. Kemp and Vogelsang-Eastwood 2001.
12. Anderson and Nosch 2003.
13. Moulherat and Spantidaki 2008; Spantidaki and Moulherat 2012, 189.
14. Doumas 2006; Moulherat and Spantidaki 2007; Spantidaki and Moulherat 2012.
15. Spantidaki and Moulherat 2012, 188.
16. Spantidaki and Moulherat 2012, 188, figs 7.1, 7.2.
17. Möller-Wiering 2006; Moulherat and Spantidaki 2009a; Spantidaki and Moulherat 2012, 189, fig. 7.3.
18. Möller-Wiering 2006.
19. Militello 2012b.
20. Schliemann 1880, 283.
21. Mylonas 1973; Spantidaki and Moulherat 2012, 189, 192, figs 7.4–7.6.
22. Moulherat and Spantidaki 2009b; Spantidaki and Moulherat 2012, 192–194.
23. De Wild 2001; Spantidaki and Moulherat 2012, 192.
24. Militello 2012b.
25. It was originally discovered in 1952 and only rediscovered in storage in 2005. Militello 2012b, 203–204.
26. Doumas 2006.
27. S. Marinatos 1967, A16; *PT*, 225.
28. Hardy 2008; see also Brown 1991.
29. Hardy 2008.
30. Adovasio, Soffer and Klíma 1996, 526–534.
31. Kvavadze *et al.* 2009.
32. Hillman 1975; Zohary and Hopf 2000, 127–131.
33. Zohary, Hopf and Weiss 2012, 103. Flax was probably first domesticated for its oil rather than its fibers.
34. Miller 2006, 48. For new genetic evidence for multiple domestications of flax, see Fu, Diederischsen, and Allaby 2012. See also Zohary, Hopf and Veiss 2012, 1–4, 101–106.
35. Van Zeist and Bakker-Heeres 1975; Schick 1988; *PT*, 11–15; McCorriston 1997, 519.
36. Adovasio 1983, 425–426.
37. McCorriston 1997, 519.
38. Nesbitt 1995, 75.
39. *PT*, 10; Wetterstrom 1993.
40. Sarpaki 2009, 226.
41. Evans 1968, 272; Carington Smith 1975, 182–184.

42 Other lab-based methods, such as DNA analysis, strontium- and isotope-tracing, can be employed in the analysis of fibers and faunal remains to determine the origins of woolly sheep and domesticated plants. See, for example, Frei *et al.* 2009; Brandt *et al.* 2011.
43 For a recent overview of textile manufacturing, see Gleba and Mannering 2012.
44 Unruh 2007, 167. Pseudomorphism is not fully understood but the basic process seems to be that molecules from corroding metal nearby migrate to the interstices of any adjacent organic material, leaving either a negative or positive cast of that material.
45 Salmon-Minotte and Franck 2005; Zohary, Hopf and Veiss 2012, 101–103.
46 Salmon-Minotte and Franck 2005.
47 Ryder 1969, 1983, 1993; Sherratt 1983; *PT*, 20–30; Hiendleder *et al.* 2002.
48 Ryder 1969, 1983, 1993; Sherratt 1983; *PT*, 20–30.
49 *PT*, 25; Waetzoldt 1972, 2007.
50 Payne 1973; McCorriston 1997, 521.
51 McCorriston 1997.
52 Panagiotakopulu *et al.* 1997; Van Damme 2012, 160–170.
53 Lubec *et al.* 1993; Van Damme 2012.
54 Hundt 1971.
55 Good 1995.
56 Good, Kenoyer, and Meadow 2009.
57 Panagiotakopulu *et al.* 1997, 420.
58 Van Damme 2012.
59 Murray 2004.
60 Tzachili 1990, 383.
61 Burke 2012. On sea silk in general, see McKinley 1998; Maeder 2002.
62 Linnaeus 1758.
63 The adhesive characteristics and water insolubility of these mussel excretions have been studied for glues or other bio-inspired technologies. The highly unusual mechanical property of byssal threads is that from one end they resemble soft rubber yet on the other end they are rigid like nylon, with a seamless transition. For an overview, see Burke 2012, 172.
64 Burke 2012, 174–175.
65 Burke 2012, 175–176.
66 *PT*, 225–235.
67 *PT*, 232.
68 *PT*, 227.
69 *PT*, 227.
70 Brysbaert 2008a.
71 Rehak 2004.
72 The term *murex*, first used by Arisitotle, refers to a broad family of snails that have been divided into more specific genera. For Minoan purple, see Burke 1999; 2010, 34-39; Brogan, Betancourt, and Apostolakou 2012.
73 Ruscillo 2006.
74 Burke 1999; Brogan, Betancourt and Apostolakou 2012.
75 *PT*, 229–235.
76 *PT*, 235–239.
77 *PT*, 239.
78 Warren 1972, 26–27, 53–54, 72, 75; *PT*, 240.
79 Betancourt, Apostolakou and Brogan 2012.
80 Apostolakou, Brogan and Betancourt 2012.
81 Barber 1997, 515.
82 Burke 2012, 50.
83 *PT*, 116.
84 Regensteiner 1970, 16–17.

85 *PT*, 118–122.
86 *PT*, 116–118.
87 Militello 2012b, 206–207, fig. 3.
88 Crowfoot 1936; Hoffmann 1964; Carington Smith 1975, 97–99; *PT,* 91–97; Kemp and Vogelsang-Eastwood 2001.
89 Militello 2012b, 204–206, fig. 4.
90 Mårtensson, Nosch and Strand 2009.
91 Analysis shows that some weights are not perfectly rounded, but have slightly flattened sides or "resting surfaces," as if they were placed on a flat surface to dry while the clay was still soft. The suspension hole is often at a 45 degree angle to this flat surface, perhaps indicating that the malleable clay ball was resting when the suspension hole was pierced through with a stick or finger at a 45° angle. Wear marks found at both ends of the suspension holes indicate that the weights were suspended with the hole on the horizontal plane, parallel to the ground.
92 These can be fairly broad grooves, 0.5 cm in width and depth, or they can be very faint and worn, perhaps indicating thread wear. Both the grooves and the interior of the central suspension hole were sometimes painted with dark red-brown iron-based paint used on fine wares, perhaps in order to prevent snagging when the weights were threaded.
93 Popham 1984, 249; MacDonald 2005, 67.
94 Burke 2010, 51–52.
95 Carington Smith 1975, 275.
96 See Burke 2010, 58, for further references.
97 *PM* I, 253.
98 It has been argued that these objects were not connected with weaving, since it is difficult to reconstruct the use of the four holes in each corner. Judith Weingarten suggests that these weights, along with stamped pyramidal "weights," might be tags attached to sacks or bags. While I wouldn't discount this idea, only some of the weights are stamped and when they are, it is in a manner similar to that found on definite loom weights. Cuboid weights, moreover, are fairly standardized in form and weight, which supports their use as textile tools. The few outliers in terms of weight noted by Weingarten (500+ g) are not unheard of for loom weights. Weingarten 2000b.
99 Burke 2010, 58–60.
100 Carington Smith 1975, 296.
101 Hutchinson 1939–1940; *CMS* V Suppl. 1A, nos 61; *CMS* II.6, 382–384; Weingarten 2000b, 491.
102 Wilson 1984, 214–219.
103 Warren 1972; Whitelaw 1983.
104 *PT*, 74–76.
105 Barber 1994, 109; 1997, 516, pl. cxciiia, b; 1998, 14.
106 Burke 1999.
107 Militello forthcoming.
108 Poursat, Prokopiou and Treuil 2000.
109 Militello 2007, 44.
110 Shoep 2002; Burke 1997.
111 John Younger (1980, 166–167) identifies this imagery simply as "Vertical Supports with Globular Attachments." Evans (1909, 131–132, figs 69a, 70b, 71a) had initially suggested that they depict vessels attached to poles for transport), but some prism seals are decorated with vessels and show lips, handles, and rims, all clearly delineated by the glyptic artist. The imagery on this group of sealstones is different.
112 Burke 2010, 43–48.
113 Militello 2007, 44.
114 Hallager 2002; Burke 2010, 49.
115 Militello 2007, 39.
116 Militello 2007.
117 Bennet 1990.

118 Killen 1962; 1964; 1979; 1984; 1988; 2007; Chadwick 1988; Bennet 1992; Palaima 1997; Nosch 2000; 2001a; 2001b.
119 See Halstead 1993; Pullen 2010; Shelmerdine 2013.
120 Shelmerdine 2013, 447.
121 Halstead 1981; 1991; 1992; 1999a; Bennet 1989, 27.
122 Chadwick 1988; Killen 1988; Nosch 2003.
123 Shelmerdine 1987.
124 Killen 1984.
125 Robkin 1979, 473.
126 McDonald and Rapp 1972, 240.
127 Burke 2010, 104–105.

3

Patterned Textiles as Costume in Aegean Art

Suzanne Peterson Murray

Introduction

Archaeological and epigraphical studies have borne out the fact that cloth production was an industry common to most, if not all, communities in the Aegean world, but while these sources attest to the existence of spinning and weaving equipment, the processing of wool and flax into fabric and garments (see Chapter 2, this volume), and even their subsequent distribution, they tell us little about the finished product. It is primarily through the depiction of cloth and costume in Aegean art that we can get an idea of the variety of these products of the loom and the degree of refinement that they could achieve. The extraordinary preservation of numerous wall paintings discovered at the site of Akrotiri on Thera has vastly enriched our stock of visual information about the textiles of the Aegean Bronze Age and provoked a great deal of discussion about social and cultural interrelations between Crete, the Cyclades, and Mycenaean Greece. Much of the ensuing research and debate has focused on costume, particularly the often lavish garments worn by female figures, from which characteristics of design, construction, color, decoration, and context provide clues to the pertinence of textiles in marking social status, gender roles, and religious traditions. The degree of detail shown in the representations of these garments has not only stressed the importance that artists placed on costume in relation to the subject depicted, but has given us a vivid picture of the skills of the weavers. With the discovery of these extensive mural cycles from Thera, the bits and pieces of poorly preserved fresco fragments with elaborate designs found at sites on Crete and the mainland could be more securely identified as textile depictions, and the evolution of costume iconography could be better examined. Studies by scholars such as Elizabeth Barber have verified that Aegean weavers were indeed technically capable of producing textiles as ornate and refined as those shown in the frescoes, so a close look at a selection of this evidence is warranted.[1]

In this chapter, I focus on representations of cloth that show distinctive and/or complex patterns, construction, and significance; and how these textiles are used in garments to express prestige and convey meaning in Minoan, Cycladic, and Mycenaean societies. A selection of examples of small-scale sculpture and engraved rings and sealstones are included to provide insights into how the textures of fabrics could be represented and the contexts in which they were used, but our primary focus is on the larger-scale representations in wall paintings of textiles used as costume. It is here that we are more fully informed and impacted by the vibrant colors and patterns of Aegean textiles and the striking ways in which they were intermingled and arrayed to produce distinctive and evocative garments.[2] The selection of examples is not intended to be a comprehensive catalogue of patterned cloth in Aegean fresco, but a representative sample of the more informative and distinctive depictions of elite clothing as it appears at multiple sites over the course of the Late Bronze Age.

In cases where a sufficient amount of a fresco composition is preserved to suggest a context, scenes in which elaborately patterned or decorated garments are worn often have a ritual context, as they depict richly dressed goddesses, priestesses, or other ministrants engaged in activities of epiphany, adoration, offering, or initiation. Comparanda from other artwork assist us in identifying the characteristic elements of such themes. Some types of garments can be categorized as festal attire, worn for ceremonial or cult activities as signifiers of a person's role and status within a scene; at times the decorative complexity of costume can be used to identify a hierarchy of prestige. In some cases, cloth is depicted as an offering in its own right, or it can be used as a sacred emblem.[3] Dresses and skirts made with intricately patterned fabrics are more characteristic of Minoan and Cycladic Island examples, while the garments depicted in frescoes from mainland (Mycenaean) sites show a tendency toward less elaboration in the garments' main cloth, with more attention being paid to the colors and patterns of their trims.

Before turning to specific examples of luxury clothing in Aegean art, it is useful to set forth some of the terminology used for types of garments and characteristics of design that are discussed in this chapter.[4]

Women's clothing
1) *Dress or chemise*: ankle- or calf-length; short-sleeved; open at the neck to waist level, and in some cases for the length of the garment. Decorative bands are added on the sleeves and along the edges of the deep V-neckline (see next entry). Some examples have a continuous bottom hem indicating that they are seamed at the front from the waist down, usually with a wide vertical band added.[5] There are a few examples that are loose and voluminous in the upper portion, but most display a body-hugging bodice.
2) *Bodice*: the short-sleeved garment covering the upper torso – this may refer to an independent item (blouse) or to the upper portion of the female dress/chemise. The distinctive Minoan bodice is tight-fitting with a V-shaped opening that curves around the breasts and joins at the waist. Bands in contrasting colors trim the exposed edges of the fabric (neckline and sleeve hems), run along the top of the shoulder, and continue down the length of sleeve, following the line of a likely seam. Some of these bands, which often contain their own decorative patterns, may have been header bands,[6] but most seem to function as separately woven strips used in a practical as well as an ornamental manner to mask seams and/or secure raw edges of the garment. In a few cases, vertical striping within the body and sleeves of a bodice may indicate the use of more complex seaming and piping.[7]
3) *Bell-shaped skirt*: full-length, sometimes with horizontal stripes or stippling; some have figurative panels.

4) *Long tunic*: similar to the chemise but closed at the neck, covering the chest; generally showing an A-line shape and usually retaining the vertical band down the center, at least in the skirt section.
5) *Flounced skirt or overskirt (kilt)*: a wraparound garment, knee- or calf-length, applied over a dress or below a bodice and secured with a cord or a wide belt/girdle. Multicolored, with tiers of flounces arrayed between larger sections of plain or patterned cloth. The flounces are marked by angled contours that project from the body of the skirt, and they are often represented in horizontal strips of red, blue, yellow, and white marked with vertical black bars. The flounced skirt is worn with the opening to the front, and is marked by a distinct dip of its lower corners and corresponding downward arc of the tiered flounces. In Mycenaean representations, the opening disappears, but the distinctive arc of the flounces is retained.

Men's clothing
1) *Kilt*: thigh-length, wraparound, usually with an arcing hem and a tight belt.
2) *Loincloth*: panels of cloth descend from waist at front and back in a sloping fit with high cut at hips.
3) *Tunic*: long (to ankles) or short (knee-length), with short sleeves and simple border bands.

Crete
We begin our look at representations of textiles in the costumes of the Aegean elite on Crete, where a vibrant Minoan culture produced complex palatial centers and villas, a rich array of arts and crafts, and, based on the visual evidence, elegant textiles. The sprawling palace at Knossos, the largest and most prominent on the island, served not just as a political hub and a focal point for cult activity, but also as a center for the collection of local produce and for the manufacture of luxury goods.[8] Production records on clay tablets inscribed in Linear B attest to a pronounced focus on textile production as well (see Chapter 2). It is not difficult to imagine that the elites, whose social status enabled them to devote or demand the extra hours of labor required to produce intricately woven textiles and construct complex garments, used such garments as indicators of their prestige.

While the products of the loom have not survived, illustrations in Minoan art of men in fine kilts and ladies in multicolored skirts and decorated bodices testify to the skill of Aegean weavers and the complexities of Minoan elite dress. Excavations have shown that the palace at Knossos, unlike others on Crete, was redecorated several times during the Neopalatial period with frescoes of women and men in richly colored and patterned garments.[9] Hints of what must have been a pageant of imagery in the rooms and corridors of the palace can now only be detected in assorted pieces of painted plaster and modern reconstructions. Although the fresco evidence is fragmentary, it preserves vivid and tantalizing details of fabrics dyed in bold colors and woven with intricate designs. One can well imagine that the palace set the standard for high society and the *haute couture* that signaled elevated status. Much of this finery was lavished on the elaborate costumes of women, but Knossos is also a source for rarer examples of men wearing showy textiles.

Knossos in the MM III–LM IB Periods

Among the early examples (MM III–LM IA) from the palace are fragments depicting a man in a yellow loincloth with a delicate dot-pattern in red (Fig. 3.1), and the elegant bodice of the "Lady in Red" (Fig. 3.2), its vivid red fabric incised with a net-pattern hatched in black, dotted at the interstices with white (beads?), and edged with decorative bands in a blue and black flame motif.[10] These fragments, although limited, give us glimpses of artists working in palatial contexts who are recording distinctive textile patterns. From the East Temple Repository of the palace, faience statuettes and miniature relief models of dresses reveal the striking fashions worn by the uppermost level of the female hierarchy, the priestesses, as well as those lavished on the goddesses they served.[11] Most depict a tight, breast-baring bodice above a wide belt or heavy girdle (perhaps of rolled cloth), and a long skirt. The famous "Snake Goddess" figurine from this deposit (Fig. 3.3), wears an elaborate bell-shaped skirt of multiple tiers or flounces. Over this at front and back is a curved, patterned apron that resembles the male loincloth in its construction and cut. The other examples show bell skirts marked with horizontal stripes or stitching, with a striking inset of crocus plants on the dress models (Fig. 3.4). These curious models, which depict the costume alone, presumably symbolize the

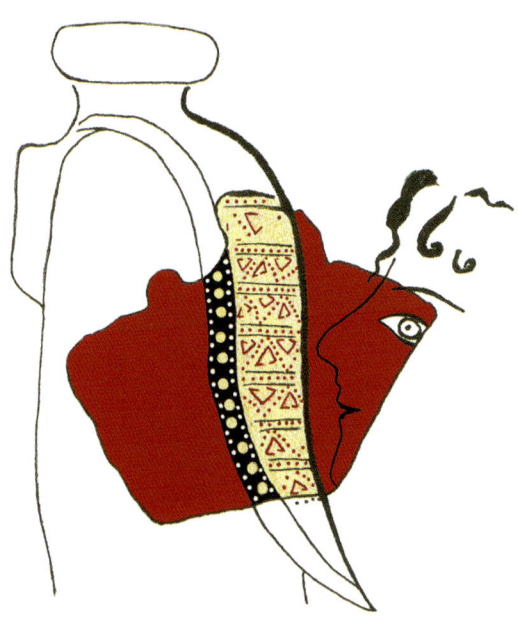

Fig. 3.1. Fresco fragment of a man in a loincloth, from the southwest area of the palace at Knossos (drawing S. P. Murray after PM II, fig. 485)

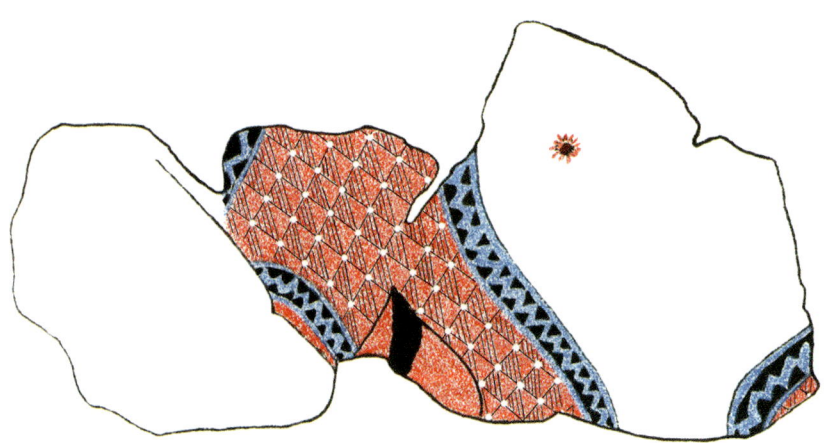

Fig. 3.2. Lady in Red, fresco fragment from Knossos (drawing S. P. Murray)

many gifts of lovely garments presented to the goddess of the palace shrine; gifts that signal the high value that must have been placed on fine textiles in Minoan society.[12]

We will return to these artifacts later in relation to the topic of figured textiles and to the role of cloth in Aegean cult; but let us turn to compelling evidence from this period at two other sites on Crete before returning to the later work at Knossos. Although badly damaged, images of large-scale female figures from the sites of Pseira and Ayia Triada provide substantial evidence for the extravagant textiles depicted in the MM III–LM I periods. These splendidly dressed ladies greatly enhance our knowledge about textile motifs, the nature of prestige attire, and the dissemination of religious iconography on Crete.

Pseira

Fragments of painted stucco relief, identified as parts of life-size female figures in rich clothing, were uncovered in 1907 in the debris of Building AC at Pseira, subsequently dubbed the "Shrine" due to the nature of this decoration.[13] Decades later, Maria Shaw undertook a thorough study of these fragments as part of the reinvestigation of this building. As the result of this study, she succeeded in

Fig. 3.3. Snake Goddess, faience figurine from the East Temple Repository, Knossos (photo A. Chapin. Archaeological Museum of Heraklion, Hellenic Ministry of Culture and Sports – Archaeological Receipts Fund)

Fig. 3.4. Faience figurine and dress models from the East Temple Repository, Knossos (drawing A. Chapin after PM I, figs 364, 382)

identifying two distinct female figures, Lady A and Lady B (Fig. 3.5), whom she dated to late LM IA or early LM IB, and she produced a detailed analysis of the textile designs represented on their costumes.¹⁴ I rely heavily here on her concise observations and excellent illustrations, which bring deserved attention to one of the most ornate depictions of textiles in Aegean art.

Lady B, the better preserved of the two figures, is convincingly restored as a seated female, facing left, with the upper torso turned to a frontal view (Fig. 3.6). She is dressed in a Minoan-style tight bodice whose white fabric is woven with a repeating design of pairs of narrow blue chevrons overlaid with black circles. Inserted in the space between each pair of chevrons is a running spiral, depicted with small red dots and a larger dot at each center. Each pair of chevrons opposes or mirrors the pair above it, creating diamond-shaped gaps that are filled with quadruple, interlocking spirals surrounding a blue four-pointed star or incurved lozenge (see details in Fig. 3.5).¹⁵ These spirals are also delineated by red dots, with a larger blue dot marking the center of each. The larger dots might be construed as beads (indeed the dot spirals as a whole could represent beadwork), but the painted decoration is flat and no relief is utilized here to highlight the differences in texture on the fabric.¹⁶ Enough of the sleeve is

Fig. 3.5. Stucco relief of two women, with pattern details, Pseira (restoration M. C. Shaw and G. Bianco)

preserved to identify the canonical trim of the Minoan bodice: narrow bands, attached along the front edges of the open bodice as well as across the outer contour of the shoulder and arm. In this case, the band is quite decorative as well – between two blue borders runs a line of white hooked spirals (a variant of the running spiral) outlined in black, again with a red dot terminating each spiral (Fig. 3.5). The use of border bands in the design and construction of garments is a distinctive feature in Aegean art and a vehicle for further decorative elaboration (see the typology section above).

The skirt, of which a large chunk of the upper section survives, depicts a bold pattern of even greater complexity (see inset detail in Fig. 3.5). Alternating fields of large blue and yellow lozenges with undulating contours, based on the tricurved arch motif, form a broad impression of a rapport design. Barber interprets this as an elaborate version of the "yo-yo" pattern, a textile motif commonly used in simpler forms.[17] The blue lozenges, with simple fillers of thin black chevrons and diamonds, tend to recede, while the yellow lozenges and their bold decorations stand out: white quadruple spirals with black outlines and large blue centers, framed by undulating ribbons of white discs on a black ground, which echo the contours of the yellow lozenge. A yellow rosette is overlaid on the interstices between the yellow lozenges in a way that

Fig. 3.6. Lady B, Pseira, as restored in the Heraklion Archaeological Museum (photo J. W. Shaw. Archaeological Museum of Heraklion, Hellenic Ministry of Culture and Sports – Archaeological Receipts Fund)

gives the impression of clasping and connecting the busy black and white ribbons in a loose net.[18] The skirt fragments also provide slight evidence for a girdle (only the raised relief of its lower edge survives) and for flounces, the latter preserved in a fragment showing a wide, multilayer band (Figs 3.5, 3.6) with a central row of ivy leaves (white outlined on blue) framed by a border of yellow bands with white disks and plain blue bands. The juxtaposition of all of these boldly patterned textiles for bodice and skirt is a dizzying extravaganza, but Shaw has astutely pointed out that the motifs echo each other in subtle and sophisticated ways through the repeated spiral, dot, and diamond/lozenge motifs. Nonetheless, as the reconstruction illustrates, the effect, particularly in large scale, would have been striking. The Minoans clearly liked an abundance of bold, intricate patterns.

In the case of Lady A (Fig. 3.5, left), little remains that is related to textile design other than a bodice sleeve whose fabric is decorated with a densely packed pattern of small white rosettes, outlined in black on a blue ground. A grid impressed in the plaster aided the artist in ensuring the regularity of the closely spaced flowers.[19] The sleeve is trimmed along its outer contour with a border of white rosettes on a yellow ground edged by blue and black bands. Another border at the front opening of the bodice is unfortunately indistinct, but may represent a yo-yo pattern using the same palette.[20] A few fragments designated as parts of a flounced skirt are too worn to determine any decoration and too sparse to determine whether this figure was seated or standing.[21] The lack of preservation gives us only a limited impression of Lady A's attire, but the design of her bodice, while quite decorative, does not convey the same level of opulence as the costume of Lady B. Our impression may be skewed by the comparative lack of evidence, or it may be a clue to a hierarchy between these two figures.[22]

The Pseira reliefs bring up a number of perplexing questions. Who are the characters in this scene, particularly the extravagantly dressed Lady B? Why are these richly attired ladies depicted in a house in a provincial town remote from the palatial orbit of Knossos? And, most pertinent to our study, do the painted details of the costumes represent actual textiles?

Maria Shaw, after examining the extant fragments of this scene and the drawings made in earlier studies, concluded that there is no evidence for more than two figures or for any indication of setting, but that the two large figures in relief, facing each other, would have dominated a single wall of a relatively small room. The restoration of Lady B's pose and the configuration of her costume are based in large part on an ivory plaque from Mycenae (Fig. 3.7), as is the unsubstantiated reconstruction of this figure in the Heraklion Museum, where she is shown as seated on a pile of rocks. Shaw's preferred reconstruction for the pair of figures, with Lady A standing and facing a seated Lady B, follows recurrent iconography in Aegean art that depicts a seated goddess (or her representative) being approached by one or more standing figures, who may fill the role of attendant, priestess, adorant, and/or offering-bearer.[23] This imagery is most often shown in glyptic representations and on rings, such as the gold ring from the Acropolis Treasure of Mycenae (Fig. 3.35), discussed below,[24] but it must also have been a feature of large fresco compositions whose influences are reflected and excerpted in these small items. Variations on this theme occur in the fresco of the seated goddess from Xeste 3 on Thera (Fig. 3.21) and in

Fig. 3.7. Ivory plaque, Mycenae; National Archaeological Museum 5897, Athens (National Archaeological Museum, Athens; photo W. M. Murray)

the cycle of processional female votaries depicted in Mycenaean wall-paintings (see below). Many of these scenes, even those on a miniature scale, show careful attention to depicting prestige garments, especially when they are worn by the seated female figures.

In spite of the abundance of female figures represented in Aegean art, it is notoriously difficult to distinguish which are goddesses, even among scenes that are clearly related to cult. Potential determinants are 1) a distinctive position in a composition (particularly a seated, preferably elevated pose), 2) being depicted as the recipient of offerings, and 3) elaborate dress and jewelry (especially if greater than other females in the scene). The goddess of the Xeste 3 fresco displays all of these characteristics,[25] but the identification of Pseira Lady B is more ambiguous. She is likely to have been seated (at what level is unknown), she wears lavishly decorative garments, and a companion figure faces her in a seemingly formal and isolated composition, all of which are suggestive of the VIP prestige of a goddess.[26] Lady A must remain a mystery, except as a well-dressed woman, since the details of pose and gesture that might have helped to define her role are lost.

And what of the elegant costumes on our Pseira ladies? If, in the eye of the Aegean artist, "fancy" equals "elite," then distinctive and elaborate costumes are a ready visual signal to denote illustrious figures or specific roles in a narrative. One must ask here whether these depictions accurately reproduce textiles, or whether the artist, in envisioning the extravagant attire of a goddess, is drawing upon decorative motifs from other mediums to illustrate and perhaps inflate his concept of "rich." The reliability of painted representations as a source of evidence for the production of luxury textiles is an issue that reverberates throughout this chapter, and the ornate examples from Pseira are a case in point. Clearly there are motifs here – rosettes, running spirals, chevrons, tricurved arches, lozenges, and net patterns – that are abundantly shared with designs on pottery, in ivory, metals, and jewelry, and even as painted décor framing other frescoes or trimming architecture. These decorative motifs (and more that we will see in the course of this chapter) were elements of a ubiquitous repertoire among Aegean artists and craftsmen, spread among many mediums, and there is every reason to expect these motifs to be reflected in, and in some cases generated by, textile production.[27] The technology was available, the techniques were known, and these motifs adapt readily to the rapport and strip patterns that are typical of textile design. The examples from Pseira illustrate the increased complexity and intricacy of design that could be attained by investing more time, more colors, and the float weave technique into the creation of luxury fabrics. I repeat here what Elizabeth Barber has stressed: all of these designs were quite achievable on the looms in use in the Aegean.[28] In the medium of painting, the methods used to render textiles may not produce precise replicas (nor do they need to), but I would suggest that they reflect familiar motifs, combinations of patterns, and impressions of rich and intricate work that would have been seen in the work of specialist weavers and displayed in the pageantry of ceremonies. For the Minoan viewer, who lived daily with the output of the spindle and the loom and the continuous process of producing simple cloth, these elaborate designs and exquisite fabrics communicated that the person or deity wearing them held a status that entitled them to garments demanding extensive hours of labor and expertise. In the more humble setting of Pseira, remote from the luxuries and spectacles of Knossos, the image of the splendidly dressed goddess would be all the more striking.

Ayia Triada

An interesting parallel to this theme, and one that provides more extensive information about costume and context, is found in a fresco composition from Ayia Triada.[29] Contemporary with the Pseira reliefs, this painting is dominated by two large-scale females represented in an outdoor setting filled with a variety of flowers, plants, and animals. The fresco in its original form appears to have completely covered three walls of a small interior room (Room 14) in a building generally identified as a villa.[30] The fire that destroyed the building in LM IB discolored the painting, and the upper portions of the women have largely disappeared (unassigned fragments nonetheless indicate that they wore much jewelry), but the overall composition and many of the details of the fabrics they wore are discernible. Pietro Militello's meticulous publication of the evidence and reconstruction of the scene (Fig. 3.8) indicate the following: the focal point of the composition, on the narrow east wall opposite the entrance, is a lively figure of a woman, depicted life-size and generally viewed as a goddess due to her elaborate costume and her association with a poorly preserved architectural structure (platform or shrine?) behind her.[31] Her pose, with flexed knees and tilted torso, make it uncertain whether she is seated on this construction (which fits the iconography for goddesses) or whether she crouches or dances in front of it (Fig. 3.9). She wears ornate festal attire in the form of a flounced skirt, the main fabric of which displays a field of interlocking quatrefoils alternating in white and blue.[32] Four-lobed filling ornaments within each quatrefoil (red on white, black on blue) add to the visual complexity, and red incurved lozenges fill the gaps between the interlocking motifs (Fig. 3.10).[33] This dizzying array is offset by a double tier of flounces at the knees and the hem, but these too enhance the impact of this garment. Each flounce is colored in blocks of blue, red, black, and white, conveying a colorful checkerboard effect. Above each pair of flounces is a band, decorated with a red hook motif, that seems to function as a narrower flounce in its own right, given its projecting tip. This is an unusual diversion from the usual *schema* of flounced skirts. At the waist is a narrow blue belt which repeats the red hook pattern (flanked by red borders), but little is visible of the bodice above it other than an indication of the usual V-shaped border bands (red and blue).[34] The skirt alone, however, with its vibrant interlocking quatrefoils, offset by the multicolor flounces, mark her as the attention-grabber in the room. Flanked on the south wall by a large, rambling scene of wild animals roaming through a rocky landscape, she is seen

Fig. 3.8. Fresco decoration restored to Room 14, Royal Villa, Ayia Triada (after Militello 1998, pls 2–4)

as the Great Goddess of nature in her realm, the mistress of animals (*potnia theron*), and the guardian of fertility.³⁵

On the north wall, in a gentler landscape filled with varieties of flowers, another woman kneels on the ground. Sufficient details of her garment are preserved to show that she does not wear a flounced skirt, but is attired more simply in a full-length dress or chemise of a light shade, nicely patterned with a dotted scale network drawn in red, each scale containing a small hatched lozenge (Fig. 3.11).³⁶ Visible between her thighs is a wide, plain red and blue band, which runs down the center of the dress. A more decorative version of this band, with a flame pattern (adder mark), a strip of red and blue squares, and perhaps a looped or rippled exterior edge, is preserved on a fragment of the outer sleeve, but other details of her pose and bodice are obscure.³⁷ The costume of this figure, with its simpler design, subtle pattern, and more subdued palette, presents a direct contrast to the colorful and complex weaves worn by the other and would have been a good visual indicator of the difference in status between the two women for a Minoan viewer familiar with both textiles and ritual. The woman in the dress would thus be subordinate to the goddess, as an adorant or ministrant in her realm (Militello titles her "*Adorante*"). Indeed, a brown area in front of her legs, against which she may be leaning, has been interpreted and restored as a baetyl, a large sacred stone (and here there may be two) that is illustrated on Aegean seals and rings with men or women clutching them as part of an ecstatic ritual to elicit the epiphany of the deity (Fig. 3.12).³⁸ The spectacular costume and active pose of the nearby goddess only adds to the dramatic effect.

The decorative patterns of these two garments find ready parallels in Aegean art

Fig. 3.9. Woman beside Shrine (goddess?), from the east wall of Room 14, Royal Villa, Ayia Triada (watercolor É. Gilliéron. Harvard Art Museums/Arthur M. Sackler Museum, Gift of Mrs Schuyler Van Rensselaer, 1926.32.53)

Fig. 3.10. Textile pattern on the skirt of the "goddess," Royal Villa, Ayia Triada (drawing S. P. Murray after M. C. Shaw 2000, fig. 3)

Fig. 3.11. Kneeling figure from the fresco of the north wall, Room 14, Royal Villa, Ayia Triada (watercolor É. Gilliéron. Ashmolean Museum, University of Oxford)

Fig. 3.12. Minoan gold ring with cult scene; Ashmolean Museum 1919.56, University of Oxford (drawing S. P. Murray after Barber 2012, pl. viiig and PM II, fig. 557)

in general as well as in other depictions of textiles.[39] The sleeves of the "Little Priestess" from Thera (Fig. 7.3) illustrate a simpler, but equally bold use of the blue and white quatrefoil, and a rectilinear variant of this motif, the interlocking cruciform pattern, appears on two of the men's kilts in the Procession Fresco at Knossos (Figs 3.18, 3.19, 7.10, 7.11), as well as on a fragment of a woman's skirt at Mycenae (Fig. 3.40).[40] This complex and distinctively Minoan motif even makes its way to paintings in Egyptian tombs, presumably as a result of textile exports.[41] In the Ayia Triada scene, this four-petal motif is an appropriate textile allusion to the abundant vegetation which the goddess ensures. The overlapping scale pattern of the adorant's dress is a basic network design that is found in many variations in Aegean art over the course of the 2nd millennium BC, including the more undulating version of the tricurved arch. As a textile motif, it occurs in paintings from Knossos (see below, Figs 3.18, 7.8), Chania, Mycenae, and Pylos (Figs 7.12, 7.13).[42] On the Harvester Vase, a carved stone vessel of ritual function found in the same villa at Ayia Triada, this design is prominently featured on the fringed cloak of a man heading up a line of others clad in loincloths – his cape clearly notes his prestige as the leader in what may have been a harvesting festival.[43]

A curious feature of the dress of the Ayia Triada adorant is the way in which the artist laid out the scale design like wallpaper, maintaining the grid of the pattern regardless of the change in direction warranted by woman's bent legs. It would appear that maintaining the integrity and clarity of the rapport design was a greater priority than visual naturalism. The fact that the dress has a fancy weave is more important to record than how it fits and moves as clothing. It is as if the artist conceived of the fabric as the original bolt straight from the loom. The resulting overall effect is somewhat rigid and monotonous, although this is alleviated to a degree by the delicacy of the curvilinear motif. In laying out the interlocking pattern of the goddess's skirt, the artist faced a more challenging design, but the overlay of flounces allowed him to separate the complex textile into two zones, each with its own incised grid.[44] The flounces, while functioning as an iconographic element of an upscale festal costume, also serve visually to enhance the sense of movement, distinguish the overlapping legs, and prevent the compelling textile pattern from overwhelming the figure.[45]

The visual effect of this room must have been stunning. The viewer in this small, dimly lit space was enveloped in an environment given over to nature suffused with the divine, surrounded by crocus, lilies, ivy, violets, myrtle, ragged rocks, leaping *agrimia*, cats stalking birds, and, commanding your attention at the center of it all, the flamboyant figure of the goddess with her swaying pose and exuberant skirt.[46] It is compelling to view this room as a small and personal shrine for elite use within the heart of a large building, rather than as a public display of prestige and ceremony.[47] Rooms decorated in a similar manner, engulfing the viewer in a vibrant world of nature, have been identified in nonpalatial buildings at Akrotiri (Spring Fresco, House of the Ladies), Phylakopi (see below, p. 62) and Knossos town (North Building, House of the Frescoes).[48] Not all of these include human/divine figures, but the lush, often hybrid vegetation bears an implicit message about the fertility and (super)natural largesse the goddess provides.[49] In the Ayia Triada scene, the extravagant decoration itself would serve as a tribute to the Great Goddess, and this tribute included "dressing" her in colorful, beautifully woven textiles. In their real form, such textiles and garments would have represented lavish and prestigious gifts.

Scholars such as Maria Shaw and Paul Rehak have seen in the elaborate scenes at Pseira and Ayia Triada a reflection of the spread of influence by Knossos, especially in religious iconography.[50] The decorative complexities of the garments of the Pseira and Ayia Triada ladies, and the emerging iconography of scenes related to female deities, can only make us wonder about the assorted and unconnected fresco fragments of female figures and extravagant costumes that were uncovered at the palace and that suggest pictorial programs prior to the LM IB destruction. Large-scale figures like the Lady in Red (Fig. 3.2) and the Ladies in Blue (Fig. 9.1), as well as miscellaneous fragments thought to show ornate textiles and jewelry (Figs 3.13, 3.14), indicate that the walls of

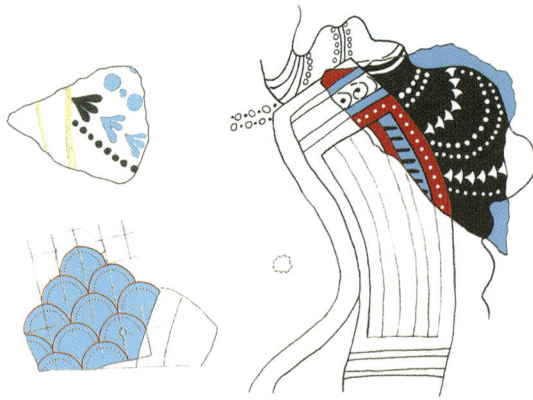

the MM IIIB–LM IB palace would have provided many illustrations of elite figures in prestige garments.[51] One such scene may have formed the early decoration of the west entrance passage into the palace, subsequently to be replaced by the hundreds of figures of the Procession Fresco, to which we now turn our attention.

Fig. 3.13. Fresco fragments of costume details, from beneath the Corridor of the Procession (left) and area north of the palace (right), Knossos (drawing S. P. Murray after PM II, figs 430, 431)

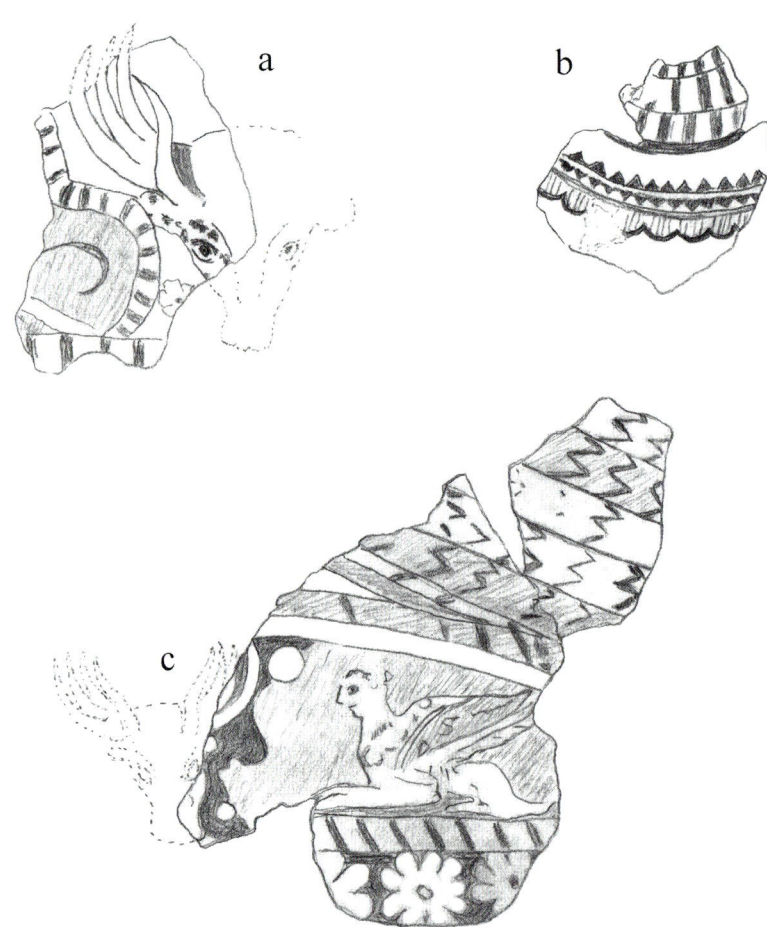

Fig. 3.14. Fresco fragments with textile and figural designs: (a, c) bucrania, (b) possible girdle, (c) sphinx, from the Northwest Fresco Heap, Knossos (drawing S. P. Murray after PM III, figs 25, 27)

Knossos in the LM II–LM III Periods
The Procession Fresco

When the palace was destroyed in the 14th century BC, an extraordinary painting of nearly life-size figures covered the walls of a long passage that led into the palace from the broad West Court and ultimately accessed the South Propylon and the formal staircase to the *piano nobile*.[52] A section of the painting was still clinging to the walls near the entrance when it was excavated, and additional fragmentary figures were found in the debris of the corridor and South Propylon, indicating the impressive extent of the composition. In most cases, only the lower portions of figures were preserved, but nonetheless we can get an impression of a cavalcade of people in ceremonial progression, some bearing rich vessels, and wearing a variety of formal and fine clothing. The portions of the fresco that were preserved at the west entrance into the palace (Fig. 3.15) indicate men and women moving down the corridor in overlapping groups, on both walls, with an interlude along the way focusing on a woman in ornate costume.[53] Only the feet (conventionally red for men, white for women) and lower edges of the longer garments are preserved. Two groups of men (Group A on the east wall and Group D directly opposite) wear long tunics of blue, yellow, or white, depicted with a taut A-line contour and decorated with a vertical band running down the front and broad bands in contrasting colors along the hem (some with small repeating patterns, e.g., flame, rosette, papyrus).[54] Each of the two groups is headed by one or more women who wear clothing indicative of the traditional festal skirts with multicolor flounces (there is too little preserved to tell much more).[55]

In the next sequence (Group B), the visible ankles of the men gathered around the ornate woman indicate shorter garments (probably kilts for most). The female figure here wears a long skirt (probably a dress or tunic) with a very elaborate border at the bottom (Fig. 3.16) – the only element of her attire which is preserved – depicting a row of colorful discs (paired in red and blue) above a band with an unusual repeating motif: a blue ellipse bisected by a squat red cross marked with a spherical yellow center (strips barred in red/yellow or blue/black also edge each row). Fugitive traces of the ellipse pattern may be discerned as well in a row above the line of discs, creating an intensely decorative impression that was probably referential as well. The oval design bears a distinct resemblance to the triglyph and half-rosette frieze (Fig. 3.17), a form of lavish architectural décor which survives at Knossos in stone reliefs (used as dadoes or revetments) and in fresco fragments depicting religious architecture, where it appears singularly as an emblematic device below the columns of a tripartite shrine (a row of colored discs, thought to indicate beam ends, appears above the columns).[56] The shrine depicted in the Temple Fresco from Knossos (Fig. 3.17), around which gaily dressed women and crowds of men are gathered, is probably the one that faced the Central Court of the palace, a focal point for ceremonial activities.[57] Thus the discs and the ellipse motif on this woman's dress may allude, in textile form, to architectural decoration associated with shrines (particularly that of the palace itself), and to her role in relation to it. These distinctive details of her garment, and her position in the composition, surrounded by clusters of men hovering around her, have led to the interpretation that she is a focal point in the progress of the ceremony associated with the procession – a goddess or priestess enacting the role of the goddess.[58] Once again, the elaborate nature of textile is used as a determinant of status in the scene. Christos Boulotis has proposed that the white strands that appear to the right of the distinguished woman represent the fringe of a garment or strip

Fig. 3.15. Procession Fresco (Groups A–B), Knossos; Ashmolean Museum, University of Oxford (PM II, fig. 450)

of cloth that the male attendants are offering as a gift to the goddess or are preparing to drape on her as part of a robing ceremony, an intriguing idea which refers to the importance of textiles as offerings (see further on this below).[59] He also identifies another type of male garment in the fresco: a calf-length skirt with a pointed back hem or "tail," similar to the garments generally identified as "hide skirts" in the offering scenes on the painted sarcophagus from Ayia Triada.[60]

Two large sections of the fresco from the corridor and the South Propylon, including the famous Cupbearer (Figs 3.18, 3.19), represent men in single file, wearing splendid, ornate kilts with wide belts. Here you get a sense of the spectacle of this once grand scene, with the march of men carrying luxury objects and wearing the finest textiles. I will discuss them here as a group, since they appear to have varied only in the details of their kilts and the objects they carried.[61] The short kilts feature fabrics with intricate patterns, hemmed with wide multicolor bands (plain or barred) that arc down to a point at the front, where a pendant network of strands ending in foliate ornaments is attached, perhaps as decorative weights. This pendant net could be an elaborate treatment of warp threads from the hem bands, with attached finials, as seen in simpler form in the decorative tassels on the sleeves of the female figures from Xeste 3 on Thera (Figs 3.22, 3.25).[62] The patterns of the textiles feature two varieties of the interlocking cruciform/quatrefoil motif with fillers of dotted rosettes or lozenges, a scale pattern (reminiscent of the Ayia Triada chemise), and an elegant design combining a net pattern with pendant ivy leaves.[63] On the latter, thick white dots atop each pendant leaf and at the interstices of the net grid may refer to beading. The leaves, outlined in white on a light blue ground, show added touches of yellow that may indicate gold highlights. This vegetal motif, with a dotted arc below the spiraling leaf, is akin to a group of gold ornaments found in a LH funerary context in Athens, which led Maria Shaw to suggest that the artist here may be representing metal ornaments attached to the fabric, although I wonder if it was the weavers themselves who were referencing jewelry motifs within their woven patterns for ornate fabrics.[64]

Fig. 3.16. Reconstruction of skirt border, Procession Fresco, Knossos (drawing S. P. Murray after PM II, fig. 456a)

Fig. 3.17. Triglyph and rosette friezes: (above) detail of shrine in the Grandstand Fresco, Knossos; (below) relief from northwest palace area, Knossos (shrine: drawing S. P. Murray after PM II, fig. 371; relief: drawing M. C. Nelson, after PM II, fig. 368)

Fig. 3.18. Bearers in kilts, with pattern details, Procession Fresco, from the Corridor of the Procession, Knossos (photo J. G. Younger; drawings M. C. Shaw and M. C. Nelson; adapted by S. P. Murray. Archaeological Museum of Heraklion, Hellenic Ministry of Culture and Sports – Archaeological Receipts Fund)

The kilted men wear tight belts that feature a convex curve at the top, as if padded or rolled over. Each is comprised of three horizontal bands in yellow and blue, with alternating patterns of spirals, rosettes, and bars. The coloring and rigid contour may indicate metals (gold and silver) with incised or inlaid decoration, or the belts may have been leather or textile with worked designs or attachments.[65] The men wear blue anklets (as do the women) and arm bands, presumably indicating silver jewelry, and the Cupbearer also wears a variegated stone lentoid seal on his wrist, all of which add to the image of wealth, high status, and ostentation on these figures.

Dating this fresco has been problematic, with opinions generally converging on a window ranging from LM IB–LM IIIA. Key to this issue, from an archaeological standpoint, is the damage that occurs to the palace in LM IB, followed by the presence of Mycenaeans running Knossos in LM II, ending with destruction of the palace in LM III, when the painting was still part of the décor. With regard to the clothing, the repertoire of ornate textile patterns in the kilts relates to our LM IB examples discussed above, while the other men's garments in the painting are akin to those shown on the LM IIIA Ayia Triada sarcophagus, particularly in the plainer fabrics and stiff A-line of the robes. A transitional date of LM II seems appropriate, as the image preserves some of the earlier traditions, which are often tenacious when it comes to sacral costume and royal pomp, while it also incorporates some new idioms, such as the seeming increase in male roles (and variety of dress) in association with ritual.[66]

Whether the Procession Fresco was intended to display, as one entered the palace, a theocratic message of power in a parade of palatial wealth, or the episodes of a major ceremony (perhaps originating in the West Court) that involved honoring and gifting the goddess, this long and grandiose scene expressed the pomp and pageantry of events at Knossos in which roles and prestige were signified through costume.[67] While the display no doubt included further ornamental flourishes in jewelry and hair, the main impact would have been in the array of luxury fabrics and ornate trims of the kilts, tunics, and skirts worn by the elite participants, as well as the parade of ritual vessels and offerings. In the four examples of patterned kilts that survived, we see the artist(s) striving for variety and contrast – the original scene of dozens, if not hundreds, of these figures must have illustrated a dazzling repertoire of the fanciest textiles of the day.[68]

Fig. 3.19. Cupbearer, from the South Propylon, Knossos, with pattern detail (photo A. Chapin; drawing: M. C. Shaw and M. C. Nelson, adapted by S. P. Murray. Archaeological Museum of Heraklion, Hellenic Ministry of Culture and Sports – Archaeological Receipts Fund)

The Cyclades

Whatever the extent of the palace's political and economic power may have been (a much-debated topic), the cultural influence of Knossos as a source of religious and artistic iconography, and as a key source of fashion and textile arts, is abundantly demonstrated not only at other sites on Crete, but throughout the Aegean and on mainland Greece. Among the Cycladic Islands, notably Melos and Thera, we see the strong influence of Neopalatial Crete in further themes of the goddess and her world of nature, teeming with plants and animals both real and imaginary, as well as in the associated use of distinctive textiles, some of such complexity that one wonders if they were imaginary too.

Phylakopi

Phylakopi is a good example in which we find all the elements of this theme. In the Pillar Crypt of this Bronze Age town on Melos were fragments of wall paintings that had fallen from an upper floor, depicting a lively seascape of flying fish in two friezes and, in a separate scene, a pair of facing women (Fig. 3.20).[69] The seated female, rendered about half life-size, wears a skirt which is itself a landscape, the much-damaged fragments indicating colorful rockwork and flying birds. Above the wide belt and its looped ties, she once had a bodice, but the surface is in very poor condition and only slight indications of red (borders?) are discernible. In her upraised hand is an item once thought to be a net (due to the marine theme in the room), but which is now interpreted as a swath of blue fabric.[70] The identification of this figure as a goddess seems indicated by her distinctively decorated skirt, her prominence in the scene, and the seated pose in which she appears to receive the blue cloth from a woman who bends toward her. The intricate pictorial representation on her skirt is more akin to fresco than textile, and might be considered purely artistic invention were it not for the discovery of a similar garment in a fresco from Thera (see below), as well as related comparanda from Knossos (Figs 3.4, 3.14). For the intricacies of such figurative designs, Barber has suggested various possible techniques: embroidery, appliqué, tapestry, or paint.[71]

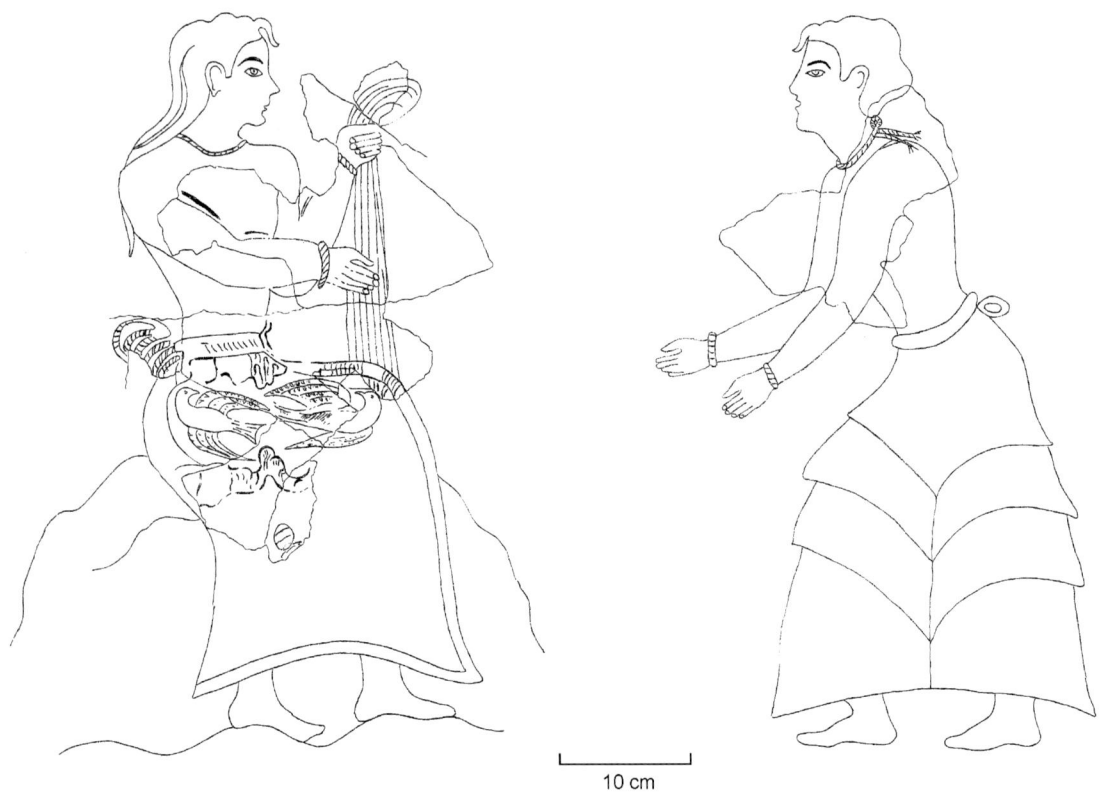

Fig. 3.20. Presentation Scene, Pillar Crypt, Phylakopi (drawing A. Chapin after Morgan 1990, fig. 8)

Only the upper torso of the other female in this scene survives, preserving a pose in which she bends forward slightly with her arms extended. The tied cord of a necklace appears at her neck, but a bodice, if it existed, is no longer visible. Her pose seems to indicate subservience to the seated female, and she may represent an adorant who has given her offering. Lyvia Morgan interprets this scene as an episode in a cycle related to rituals of robing the goddess or her surrogate, in which either cloth or a finished garment may be presented.[72] Once again, it seems that we see the goddess of nature, surrounded by, and in this case wearing, scenes representing her dominion over land and sea. As at Ayia Triada, the Pillar Crypt building at Phylakopi featured a room in which the wall decoration spotlighted the world of nature and paired it with an extravagantly dressed goddess and a female attendant. If any distinguishing elements of architecture were included with this goddess, they have not survived in recognizable form, but additional themes of flowers (lilies) and animals (monkeys), subjects which are connected elsewhere to goddess iconography, were featured in nearby rooms.[73] Morgan dates the painting to a period contemporary with the Ayia Triada fresco (LM IB or late in LM IA) and, based on the iconography of its decoration, ascribes a religious function to the Pillar Crypt complex. She also sees in this "Presentation Scene" and the nearby paintings a strong association with fresco cycles on Thera, particularly in the extensively decorated Xeste 3.[74]

Akrotiri, Thera

Excavations of the well-preserved LM IA town near Akrotiri have provided an abundance of wall paintings and produced an explosion in scholarship in the last four decades, much of it centered on the multiple paintings from Xeste 3.[75] Frescoes from this building and from the House of the Ladies have provided detailed representations of the clothing of women (and to some degree men) that have given us a much better picture of costume design, construction, and decoration, and many intriguing clues about the role of textiles in ritual activity and social custom.[76] Archaeological evidence has substantiated that the people of this town were very active in textile production, with heavy concentrations of artifacts associated with weaving and dye production found at the site.[77]

Xeste 3 was a large and well-built structure with dozens of rooms on two levels (and a third level in at least one section). Several rooms in the east half of the building yielded evidence of very extensive fresco decoration with multiple themes: ritual activities involving young women and girls (on two levels), mature women in procession, boys and a man bringing vessels and cloth, a marshy landscape, antic monkeys, men in hunting activities, and decorative friezes in floral, vegetal, and spiral motifs.[78] We shall concern ourselves primarily with the three scenes of women and girls, since this is where we gain the most insights about textiles and costume. The first of these decorated two walls (north and east) of an *adyton* (and *polythyron*) on the ground floor, the second covered the same walls of the storey above (Fig. 3.21). Since both of these spaces correspond to Room 3 on plans of the building, I will differentiate them by referring to one as the "Adyton Fresco" and the other as "Crocus Gatherers and Goddess." The third painting, also in the upper level, depicted pairs of mature women on each side of a corridor that connected Room 3 with a service staircase. The figures are slightly under life-size.

Fig. 3.21. The fresco decoration of Room 3, Xeste 3, Akrotiri (drawing A. Chapin after Doumas 1980, 295)

The Adyton Fresco
Three female figures are posed across the north wall (Fig. 3.22), directly above the descent into the *adyton*, and are linked to an image on the adjoining wall of a structure that appears to be a shrine; its door is decorated with spirals and lilies, and is topped by a distinctly Minoan religious symbol, the horns of consecration.[79] Two of the figures represent youthful adults, their long hair dressed with fillets, while the third shows the slightly smaller proportions and partially shaved head of an adolescent girl.[80] The elegant young woman at left, adorned with jewelry, garlands, and beautiful clothing, strides toward the shrine, a necklace swinging from her hand (Figs 3.22, 3.23). She wears a remarkable dress: the sheerness of the cloth is indicated by a pale wash of blue that leaves visible the line of her arm and torso beneath, and the loose and billowing fit conveys an impression of the lightness of the fine fabric. This fabric was delicately decorated with murex

3. *Patterned Textiles as Costume in Aegean Art* 65

Fig. 3.22. The Adyton Fresco, Room 3, Xeste 3, Akrotiri (courtesy the Thera Foundation – Petros M. Nomikos)

purple crocus blossoms, of which only the red stigmas are still readily visible.⁸¹ Heavier woven bands in darker blue (with simple striations or zigzags) trim the sleeve hems and neckline, places where reinforcement may have been desirable, but for the band that traditionally runs down the length of the sleeve a thinner, rope-like ribbon in yellow and red is used on the thin cloth. Long strands hang in an ornamental tassel from each sleeve – their blue color may indicate that they are derived from the sheer fabric or the hem band. Below her knees, the continuation of the sheer dress is visible beneath the kilt or overskirt that has been wrapped around her hips. In this latter piece of clothing we can recognize the tiers and patterns of the traditional Minoan festal garment: the flounced skirt, in this case an overskirt. The flounces here are not very pronounced, adding to the impression that this is a weightier garment, but the angled zones of color and pattern, meeting at the open front and topped by a wide belt, are reminiscent of examples from Crete. It features rather dark tones, in contrast to the light chemise, and is dominated by a black fabric with a simple dotted network pattern in

Fig. 3.23. Necklace Swinger, from the Adyton Fresco, Room 3, Xeste 3, Akrotiri (drawing S. P. Murray after Doumas 1992, pl. 101)

yellow enclosing yellow incurved lozenges. The wide band at the base of each tier, successively in yellow ochre, blue, and white, has a repeating, forked crisscross motif with variations in the fillers. The weightiness and deep colors of the overskirt give the impression of wool, while the diaphanous chemise seems more indicative of finely woven linen, or perhaps even silk (for which there is some evidence at this site).[82]

I have focused on the costume of the Necklace Swinger because her dress illustrates so well the capabilities of the Therans in the production of several varieties of luxury cloth, and because hers is the most decorative example of flounced overskirts represented in Theran fresco. The Veiled Girl in this scene (Fig. 3.24) and the two Crocus Gatherers from the east wall of the room above (Fig. 3.25) all wear shorter and somewhat simpler versions of this kilt, which repeat the forked crisscross motif in the broad tiers, and, in the latter scene, use a simpler version of the net pattern on the hips (a diagonal gridline with a dot or cross as filler). In other examples, when preserved on a large scale, these flounced overskirts do not include patterns, but utilize broad and narrow bands of contrasting color (see below).

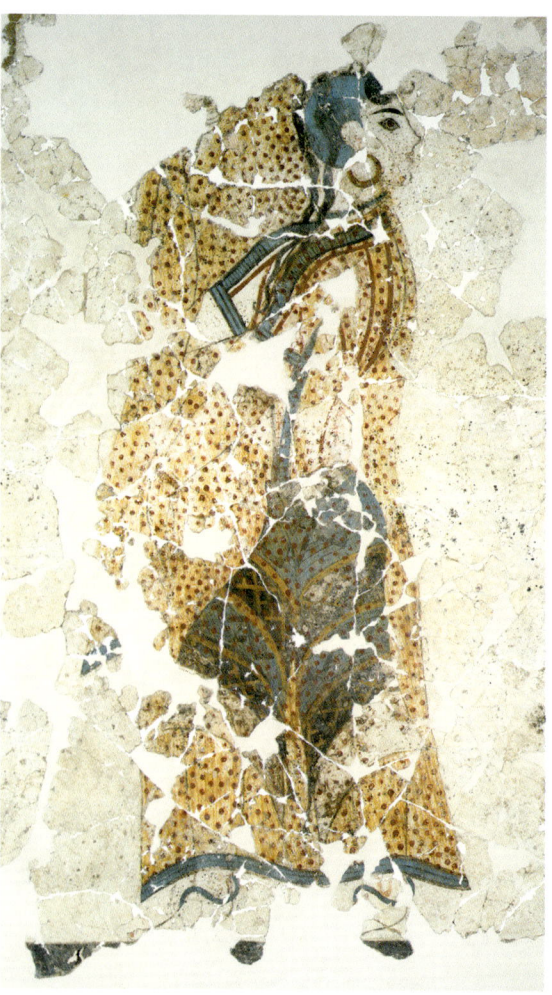

Fig. 3.24. *Veiled Girl, from the Adyton Fresco, Room 3, Xeste 3, Akrotiri (courtesy the Thera Foundation - Petros M. Nomikos)*

The Veiled Girl wears the more typical, opaque Minoan dress with thick trims (blue and black) and a tight bodice (in this case white with fugitive crocus blossoms and red piping), topped by a shorter flounced overskirt of only two tiers.[83] However, the display of diaphanous fabric continues here as well, and strikingly so, in the large swath of sheer yellow cloth that the girl holds out in front of her and drapes around her body. Hemmed with a blue band, the yellow fabric is sprinkled with an all-over pattern of red dots, perhaps indicating beads (although beads of this size might have been a bit weighty on the fine cloth).[84] Given the way she wears it, this appears to be a veil, although, considering the proximity of the shrine, she could also be displaying an offering of cloth or clothing suitable for a goddess.[85]

On the Wounded Woman, the seated figure who tends her foot, we see again a loose-fitting, sheer upper garment similar to that worn by the Necklace Swinger. This one too is depicted with a thin blue wash, but the pattern here is a dainty dotted net, and the long, orange sleeve tassels in this case clearly derive from the trims. It is not clear whether this garment functioned only as a blouse,[86] or whether it continued as

3. *Patterned Textiles as Costume in Aegean Art* 67

Fig. 3.25. Crocus Gatherers, from Room 3, Xeste 3, Akrotiri (courtesy the Thera Foundation – Petros M. Nomikos)

something longer, since the lower portions of the costume on this figure become a tumbled array of patterns and strips that are expressive of both her disheveled state and a very unique skirt. Rather than the traditional flounced overskirt, this woman wears a garment composed of vertical strips, alternately blue and white, each strip ending in its own angled tip. These strips hang loosely, although they may be joined in pairs, and they are attached to a wide blue belt with a design identified as a yo-yo pattern or a repeating row of crocus buds.[87] This design repeats as the vertical border band of what appears to be another skirt beneath, rendered in yellow ochre and patterned with a red dot network filled with red crosses. Whether this serves as a kind of liner for the outer skirt, or as a separate piece of clothing, is unclear; the color and heavier pattern seem to preclude it being matched with the upper garment as a dress. It has been suggested that the outer article represents a lappet skirt, a garment which, like the string skirt Barber describes, is worn to symbolize fertility and sexual readiness.[88] Although the underskirt makes this peekaboo effect less than suggestive, the lappet skirt may have carried an inherent symbolism about the role of the wearer. There are so few representations that may be interpreted as a lappet skirt that it is difficult to determine either its construction or its role.[89] What can be said about it here is that it occurs in a scene that highlights a variety of distinctive fabrics and female costumes in an activity associated with visiting an outdoor shrine. In this ritual context, bare feet, jewelry, delicate fabrics, and the finest clothes are *de rigueur* regardless of the rocky setting.

With regard to the men depicted in this area, there was a scene in a small adjoining space off the *polythyron/adyton* that was painted in three panels that represents boys of different ages, all nude, approaching a man with a white kilt and a large vessel.[90] Two of the boys carried vessels as well, but a third holds out before him a long strip of fabric, boldly decorated with a design of broad black and white stripes, filled with undulating lines in red and blue. This may represent an offering, or a garment that will be draped on the youth as a robing ceremony in an initiation rite.[91] Nearby in Xeste 4, fragments of a male processional scene, not yet fully published, show men wearing short kilts of a related, bold design (a thick, red running spiral on white, edged by black or blue borders) wound around their hips and tucked up under their belts.[92]

Crocus Gatherers and Goddess
The general consensus among scholars is that the frescoes of the *adyton* and those of Room 3 above it pertain to rites of passage in the maturation process of girls as they prepare to take on the duties of womanhood. The two young adults of the Adyton Fresco, with their luxuriant hair, full breasts, floaty transparent blouses and swinging lappets, are the picture of ripe femininity; but the focus in the scene above them (Fig. 3.21) is on adolescent girls and on the goddess of nature and fertility who will guide and succor them as they transition into adulthood.[93]

The girls, decked out in festive attire and large hoop earrings, gather crocus flowers in a rocky landscape (Fig. 3.25), and the stigmas (saffron) collected from them are presented to the seated goddess. All four of the girls in this composition wear chemises that fit tightly in the bodice (their breasts are still un- or underdeveloped), with sleeve tassels and the requisite trims in contrasting colors and simple patterns (striations, zigzags); the lower part of the chemise emerges below their flounced overskirts, where details indicate that the chemise is open all the way down the front and made of a sheer fabric that leaves the outline of the legs visible. These dresses are white, light blue, or yellow. Over their chemises, they each wear a flounced overskirt with wide bands of bold colors, tied at the waist by a cord. The costume of one of the Crocus Gatherers (Fig. 3.26), who appears by her shaved head to be the youngest, is somewhat more decorative than the others. Hers appears to be the only dress incorporating patterns (net and cross motif), and the sleeve tassels are replaced by more ornamental dangles, one of which appears again as a weight at the bottom of her flounced skirt.[94]

Fig. 3.26. Youngest Crocus Gatherer with a shaved head, from Room 3, Xeste 3, Akrotiri (courtesy the Thera Foundation – Petros M. Nomikos)

It is the goddess on the north wall (Fig. 3.27), however, who commands the most attention with her elaborately decorated attire, as we might expect. Here we have a figure who stands out as one of the most clearly recognizable images of a deity in Aegean art. Seated on an architectural platform, flanked by a tethered griffin, attended by a blue monkey giving an offering of saffron, surrounded by a landscape of crocus plants, and approached by girls on a smaller scale collecting crocus, she is our Great Goddess of nature. She is also *Potnia Theron*, the Mistress of Animals, whose attendant griffin illustrates that we are now in a supernatural realm. Her costume echoes the world over which she presides: she wears necklaces of silver and gold ducks and dragonflies, and a pale blue bodice covered with crocus blossoms that are repeated on its bright blue border bands. She has tucked a fragrant blossom behind her ear, above her huge earring. The hem band of her sleeve is patterned with a row of discs, rather like the beam ends on the skirt of the goddess/priestess of the Knossos Procession Fresco. Although it is poorly preserved, a white and black flounced overskirt trimmed with additional crocus flowers and blue bands highlights and animates the pose of her bending legs, with a bit of her sheer blue chemise peeking out below.[95] The Great Goddess sits high above everyone (human and beast), enthroned atop her dais and a stack of patterned cushions, colored in saffron yellow. The young maidens around her collect the fruits of Potnia's bounty and dress accordingly in the finery that honors the occasion.[96]

Why crocus, and why saffron? The crocus is a motif that dominates this painting, and it is also referenced in the Adyton Fresco and the Procession of Mature Women (discussed below). This has given rise to much discussion about its role in Theran society and the reasons for its collection here. In addition to the crocus being a fragrant flower often depicted in landscape scenes (including the fresco at Ayia Triada), this composition highlights the collection of the stigmas that are dried to produce saffron, a product that had many potential uses – as a multipurpose medicine, spice, perfume, cosmetic, artists' pigment, and dye.[97] Whatever symbolism the crocus and saffron in this scene may have carried relative to female initiation and maturation, I want to stress here its relation to textiles. The excavations at Akrotiri have produced strong evidence for the community's extensive textile industry; a primary use for saffron would have been as a dye that produced lovely yellow hues.[98] Textiles figure significantly in the fresco cycles in this building, among them yellow fabrics for two of the girls shown here and for the veil in the Adyton Fresco.[99] The girls gathering the saffron have reached the age at which they would have embarked seriously on their education in weaving, which would have begun with spinning and progressed to the loom.[100] The rites that were intended to induct them into the responsibilities of womanhood would have

Fig. 3.27. Seated Goddess, from Room 3, Xeste 3, Akrotiri (watercolor P. Rehak; courtesy J. G. Younger)

been associated not just with fertility and marriage, but also with domestic duties, and those would have been dominated by learning the skills necessary in the production of cloth.[101] The extravagant and, in some cases, unique textiles represented on the females in both frescoes may be seen not only as ritual regalia that speak of roles and status, but also as an illustration of the craftsmanship to which these initiates could aspire. In the next room, might we see a display of the advanced skills of their mothers?

Procession of Mature Women
In the adjoining hallway, four women form a procession, two on each wall (Fig. 3.28).[102] These women, who appear by their body types and bound hair to be a distinct adult group that is older than the others, wear garments that give further evidence of the variety and creativity of Theran elite attire. Again we see the use of light, filmy bodices or blouses with delicate washes of color and floral decoration (purple crocus blossoms, red lilies), but now with the addition of thick, fleecy mantles attached at one shoulder – in strong contrast to the light garments beneath. The mantles have puffy, irregular edges and wavy internal lines (Fig. 3.29), as if to denote their soft and fluffy texture.[103] They may represent the dyed fleece of sheep or a woven fabric that has a fleecy texture – in either case, these mantles signal the association of these women with wool and with the creation of luxury textiles. The wrapping of the women's hair in scarves or snoods highlights the further use of textiles, and they are distinguished by unique jewelry and numerous touches of murex purple on their garments, all of which add to the sense that these are women of privilege.[104]

The Lady of the Landscape presents the ultimate display of intricate textile fashion in this fresco (Fig. 3.30); her bell-shaped skirt, like the skirt of the Phylakopi "goddess," portrays a scene of rockwork and birds, in this case swallows.[105] In the Phylakopi scene, the goddess of nature is celebrated with an offering of cloth; in Xeste 3, the celebrants wear distinctive cloth (and fleece?) in her honor, and carry large bouquets of flowers, symbolic of the goddess's seasonal bounty. These are women whose high status in the community is demonstrated by the hours that they can devote

Fig. 3.28. Procession of Mature Women, from the upper-floor hallway entering Room 3, Xeste 3, Akrotiri (drawing A. Chapin after Vlachopoulos 2003, figs 22, 23)

to (or have others devote to) the creation of distinctive and luxurious garments that are suited to elite display in rituals and festivals, especially those that celebrate a goddess of nature and fertility.[106]

However, should we consider these women to represent matrons of Akrotiri, and their garments to represent the types of textiles that were actually made and worn by mortal women? That is a difficult question, and one which brings up one of the key problems in dealing with the Xeste 3 frescoes. We can imagine that the scenes reflect ritual practices, but are the images enriched with embellishments meant to signal the divine associations of the event? The figures contain so much detail that it is natural to consider the information as tangible and real, although at the same time there are clearly elements that warn us that we are in a liminal setting that merges with the supernatural: Potnia appears with her griffin, the Wounded Woman enacts a narrative that hints at mythic associations, and the matrons carry flowers that bloom in different seasons, putting us once again into that realm where all the bounties of nature occur simultaneously for the Great Goddess. The Mature Women, in spite of the communal nature of their activity, are quite individualized in their dress, their jewelry, and even in their facial features, but are they prestigious members of local ceremonies or characters typecast in a ritual storyline?[107] Are the images of women and girls in Xeste 3 reflections of Theran society or characters who have been costumed in a dramatic narrative?

Fig. 3.29. Rose Bearer, from the Procession of Mature Women, Room 3, Xeste 3, Akrotiri (courtesy the Thera Foundation – Petros M. Nomikos)

Fig. 3.30. Skirt of the Lady of the Landscape, from the Procession of Mature Women, Xeste 3, Akrotiri (drawing A. Chapin after Vlachopoulos 2003, fig. 22, with some restoration of the swallows)

Some of the evidence from Knossos may assist us here. Among the contemporary artifacts from the East Temple Repositories in the palace, the Snake Goddess (Fig. 3.3) wears the same piped bodice of the Veiled Girl, and the faience dress models (Fig. 3.4) feature elaborate, pictorial bell-shaped skirts with crocus plants as the central motif. The latter are facsimile garments offered to a deity, symbolic of the practice of gifting the goddess with real textiles and clothing, of which there must have been some extraordinarily lavish examples. The Procession Fresco, albeit a little later in date, depicts an event in which pageantry and visual impact are strong elements, and the fine textiles there are considered to be a showy display by the elite of the palace engaged in a traditional ceremony. We can reasonably surmise that elements of the palatial grandeur found at Knossos filtered into Cycladic societies, particularly when expressed through luxury goods such as jewelry and textiles, media through which ideas could be readily exchanged. At Akrotiri, these influences are filtered through the lens of local skills and capabilities to represent garments that are "showpieces" for prestige activities – perhaps a bit enhanced in grandeur for the enactment represented on the wall, but reflecting what was actually possible to produce with time, patience, imagination, and devotion.

The many frescoes of Xeste 3 have revealed, and are still revealing, extraordinary amounts of information, including aspects of gender-specific ritual, the depiction of pre-adults, an unequivocal image of the Great Goddess, and visual evidence for the production of many types of luxury textiles. In many ways, this building, although its primary function may have centered on cult and initiation rites, also represents a showcase of the textile arts. Its frescoes depict a festival dedicated to the collection of saffron (prominently used for dye), a rite for boys with a ceremonial display of cloth, a painted replica of a draped wall hanging (Fig. 4.6), and a cavalcade of the finest dresses of sheer linen or silk, fleecy mantles, and showy skirts – the creation of textiles is a current that hums throughout the decoration of this building. Even the highly ornamental wings of the griffin allude to textile motifs. The splendid costumes of the females in Xeste 3 are not only markers of their roles in these rituals, but they also serve as a display of Theran skills with the loom and needle. The creation of luxury textiles was itself a means of honoring the goddess through the crafts that she inspires – the human adaptation and utilization of the floral and faunal gifts provided by Potnia.

The House of the Ladies
While Xeste 3 displays a striking array of decorative garments, the women's attire depicted in the House of the Ladies (Figs 3.31–3.33) provides an interesting contrast.[108] Here too the women, identified as three extant females on two walls of Room 1b, are seemingly engaged in ritual activity, and they wear the typical Minoan outfit: a long dress with tight bodice and contrasting trims in the usual places, plus a flounced overskirt, secured with a cord. But here, we have no patterns in the flounced overskirts, no fancy flowers on bodices, no designs in the border bands. The fabrics represented are plain, varied only in the color of the cloth, the horizontal dotted lines on the lower portion of the dresses, the bands of contrasting colors, and some curious details in stitching and texture.

In the case of the Striding Lady (Fig. 3.31), the flounces of her overskirt are marked with unusual striations that lack the regularity, clarity, and color contrasts one would expect in traditional depictions of barred or striped flounces, and I believe the intention here is to show fringe or a fabric with a more unique texture, one that was either crinkled, finely pleated, or gathered.[109] The concise outline of the flounces tends to argue against a fringe, especially as compared to the more irregular fringe that is rendered along the open flap of the wraparound garment. The latter is one of the many details in these two scenes that have provided valuable insights into the design and construction of flounced overskirts: it appears to show the warp threads of the skirt trimmed as a decorative fringe and extending beyond the band that was attached to secure the exposed edge.[110]

Fig. 3.31. Striding Lady, from the House of the Ladies, Akrotiri (courtesy the Thera Foundation – Petros M. Nomikos)

Dotted horizontal lines on the dresses of two preserved ladies appear only on the lower section of the garment. I do not think this distinguishes a separate blouse and skirt, since the color of the cloth and the type of vertical trim remains consistent from top to bottom, but rather it may indicate some added stitching or the incorporation of a heavier thread into the weave of this part of the fabric, perhaps as reinforcement.[111] One dress has thick bands added at the bottom (a good way to rejuvenate a worn hem!), the other does not.

The Bending Lady's dress (Fig. 3.32) has some curious linear additions abutting the side seam on the bodice, depicted very lightly in yellow on (or within?) the white fabric. These wrap around the woman's narrow abdomen, perhaps with small loops and ties at the front,[112] and extend up into the armpit. The placement of these details does not indicate a separate garment applied overtop the dress or visible beneath it (both dresses are opaque), so perhaps it too is stitched or woven into the fabric – as a means of constricting the waist (like a 19th century corset) and reinforcing the armpit seam. Whether this is relevant to the Bending Lady's oversize breast is unknown to me, but the artist clearly took pains here to depict something subtle but specific.[113]

The Bending Lady carries an additional flounced overskirt, which she brings to another female who is only preserved in an isolated fragment that depicts her midsection. She wears a yellow chemise trimmed in blue and still unconfined by the overskirt that her companion holds. This fragment has been restored on the wall in a position that tilts her yellow-clad torso into a bending or leaning posture, and reconstructions have been published that depict her as seated.[114] However, the preferable placement of this figure should be upright and standing in profile, as I have previously proposed (Fig. 3.33). This profile, standing pose has recently been confirmed by Bernice Jones, who has detected a faint line detailing the curve of the figure's buttocks within the dress.[115] It appears from this detail that her dress is not the opaque fabric worn by her companions, a characteristic that may denote a difference in role or prestige (still no patterns, however!). In the fresco of the Crocus Gatherers, both the goddess and the girls wear sheer dresses, but in the Room of the Ladies the attire is considerably more austere and the differences may be meaningful. This third female in the scene, to whom the flounced skirt is offered, is clearly a focal point in the action. A robing ceremony seems to be enacted here, with the festal flounced skirt as a main element in the ritual. But who is the recipient of this textile? A goddess, a priestess, a young girl as an initiate, or someone else?[116]

The identity of the skirt's recipient is key to this scene, but too little information remains to apply iconographical markers to her. We do see the continued importance attached to textiles in the form of a distinct costume that carried special meaning – the significance of the flounced overskirt as a specific garment with ritual associations is stressed here,

Fig. 3.32. Bending Lady, from the House of the Ladies, Akrotiri (courtesy the Thera Foundation – Petros M. Nomikos)

Fig. 3.33. Presentation Scene, from the House of the Ladies, Akrotiri (restoration S. P. Murray)

Fig. 3.34. Women carrying festal skirts (left) on a seal impression from Ayia Triada and (right) on a seal from Knossos (drawing S. P. Murray)

corroborating its appearance on rings and seals as a cult item (as in Fig. 3.34).¹¹⁷ Whether the skirt functions here as an offering, as a vestment for a cultic role, or as the symbol marking completion of a rite of passage is difficult to ascertain, and the relative austerity of the garments makes it all the more puzzling. The plainer textiles in this scene may be due to the preferences of the artist or to a deliberate contrast in the ritual being performed, with changes in costume relevant to differences in social meaning, seasonal affiliation, or the degree of festivity/solemnity of the occasion.¹¹⁸ Here it seems that the form of the costume is primary, rather than its decorative possibilities (even the Bending Lady's sleeve tassel is barely noted). If it were not for Xeste 3, we would think that the Therans, by and large, made interesting garments, but used very plain and basic fabrics.¹¹⁹

Although the costumes of the women depicted in this room reveal little about the use of patterned cloth, above them the fresco explodes with textile references. The most ornamental elements in this scene are relegated to this curious zone above the figures, which features a bold rendering of a dotted net in red, joined at the interstices by blue incurved lozenges (or four-pointed stars), that is very much like the motif on the skirt of the Necklace Swinger. This entire decorative zone is bounded at the bottom by an undulating blue and black border, perhaps marking the bottom band of a large piece of fabric – such as a wall hanging, canopy, or tent.¹²⁰ If this is the case, an indoor activity may be indicated, in contrast to the more typical outdoor settings depicted in ritual scenes.

In general, Theran weavers do not seem to have had the same taste for dense rapport patterns and intricate edgings that we have seen expressed in prestige garments on Crete; rather their emphasis is on variations in weave, weight, and texture, and they seem to prefer using plain bands of contrasting colors for overskirts and trims.¹²¹ When they use patterns in their textiles, they tend to be either repetitive and less complex than those from Crete, as on the overskirts and bodice bands of the Adyton Fresco and the Crocus Gatherers, or extravagantly pictorial, as on the skirt of the Lady of the Landscape and the delicate floral cascades of some of the Xeste 3 bodices. The iconic role of the Minoan flounced skirt in gender-specific rituals is well attested here – a tradition that will filter into Mycenaean culture as well.

The Mycenaean mainland

In mainland Greece, the strong influence of Minoan culture on Helladic society reverberates through the art and iconography of the LH I–II periods, and in some subjects, primarily the representations of processions of women in festal attire at Thebes, Tiryns, and Pylos, the iconography of costume is tenacious until the close of the Mycenaean Age. The depiction of exuberant and complex Minoan-style fabric patterns will quickly fade, however, and the interest in decorative textiles will narrow in on the bands that edge bodices and skirts. By mid-LH IIIB (13th century BC), most fresco depictions of clothing will either show the use of rather plain textiles and simple tunics or a stylized and formulaic ritual dress.[122]

For the earlier periods, in which Minoan influence is strongest, we have only sparse fresco evidence, primarily at Mycenae and Thebes. In other media, however, such as on a selection of small-scale sculptures as well as in the miniaturized images engraved on seals and signet rings, the elaborate designs of flounced skirts and distinctively patterned fabrics are meticulously recorded. Some of these artifacts are so expressive of Minoan subjects, however, that they may be the work of Minoan artists. One such artifact is a gold signet ring from Mycenae (Fig. 3.35) that clearly illustrates the iconography associated with ritual offerings made to the Great Goddess.[123] Two women in the standard ritual attire of flounced skirts approach a seated female who wears a prominent necklace and an ornate skirt with a scale pattern (bodices are not indicated, probably due to the small scale). She is seated outdoors, on a rocky perch, in front of a large tree that canopies her – the goddess enthroned in nature. The offering bearers, in procession, have brought small bunches of flowers and the goddess (or priestess in her guise) has received the first gift (just as the goddesses of Xeste 3 and Phylakopi receive offerings). Two small attendants who flank the goddess may represent young girls, one reaching up to a tree, perhaps tugging at it, as is often shown in Minoan religious scenes, or plucking from it, in an activity of collection and offering reminiscent of the Xeste 3 girls.[124] The bezel of the ring is packed with additional religious symbols (such as the Minoan double axe), but, despite the minute scale and the crowded composition, the artist clearly considered the ornate skirts, with their dense patterns and arcing flounces, to be a priority in denoting the prominence of key figures in this ritual.

Fig. 3.35. Gold ring with presentation scene, Mycenae; National Archaeological Museum 992, Athens (drawing S. P. Murray)

In the Ivory Triad (Figs 3.36, 3.37), a small but intricately detailed sculpture from Mycenae, two crouching women with an attendant child are beautifully garbed in necklaces, elaborate flounced skirts, and a shared mantle.[125] On one skirt, the fabric shows the tricurved arch motif, while the other has a complex but indistinct design (perhaps including a four-petal motif). Flounces are rendered as rows of thick tufts (heavy pleats?). The mantle is covered with a dense, all-over pattern resembling beadwork, trimmed with a thick fringe.[126] Once again, the extravagant nature of their costumes (as well as the use of an exotic raw material to depict them) has led to the general assumption that these ladies are of very high status, presumably divine. The garments are depicted in strong relief, rather than light incision, giving an impression of highly textured textiles.[127] This use of relief may be a means of expressing the visual impact of these elaborate patterns in a medium where color could not do so; but it also reminds us that two-dimensional frescoes greatly restrict our knowledge about texture and weave. Depictions by Theran painters of sheer fabrics and fluffy mantles show attempts to work through these limitations.

In contrast, a gold ring from Tiryns (Fig. 3.38)[128] shows a processional theme in which the seated recipient is *not* shown with the flounced skirt and bare breasts of the Minoan tradition, but instead wears a long, closed tunic that is the mode of dress favored by Mycenaean men and women. This garment is similar to that worn by two groups of men in the Knossos Procession Fresco, perhaps reflecting an adaptation taken on in response to the Mycenaean domination of Knossos. The artist of the ring has delineated the vertical band at the front that dominates the decoration on such clothing, as well as the hem band, and has also tried to indicate with dashes either a texture or simple pattern in the fabric of the tunic. For the curious creatures that bring

Fig. 3.36. Ivory Triad, Mycenae; National Archaeological Museum 7711, Athens (National Archaeological Museum, Athens; photo W. M. Murray)

Fig. 3.37. Ivory Triad, Mycenae (back view); National Archaeological Museum 7711, Athens (National Archaeological Museum, Athens; photo A. Chapin)

Fig. 3.38. Gold ring with procession of genii, Tiryns; National Archaeological Museum 6208, Athens (after Rehak 1992, pl. xiib)

libation jugs to this woman (presumably a goddess due to the supernatural beings who serve her), the artist has adapted the network pattern with bead interstices to depict their imaginary backsides.[129] For the goddess, however, the omission of the long-established Minoan cult costume is notable – either the influence of Crete is waning or this is a (Mycenaean?) goddess with different attributes. We will see this contrast again in the frescoes of the Cult Center at Mycenae.

Mycenae

At Mycenae, scattered fresco fragments of elaborate garments indicate that there were once several paintings with the traditional Minoan theme of large-scale women, probably in a processional format.[130] A selection of these fragments illustrated in Figures 3.39 and 3.40 show a section of bodice in blue with a scale pattern, a blue skirt with a tricurved arch and papyrus motif and a black and white flounce, and the upper portion of a skirt with yellow interlocking crosses inset with red crosses, topped by a girdle with a running spiral band and an ivy chain.[131] These textile patterns all have close parallels with the kilts of the Knossos Procession Fresco and the costumes depicted in the shrine from Ayia Triada. Since these motifs will subsequently disappear from the Mycenaean repertoire in the representation of textiles, it would seem that the influence of Minoan intricately patterned fabrics lapsed by LH IIIB, to be replaced by simple scatter patterns (dots, dashes, crescents) or plain fabric with varied trims.

The Cult Center: Southwest Building
In the debris of the Southwest Building at Mycenae, an area closely associated with the nearby shrines of the LH IIIB period, assorted fresco fragments of female figures in differing scales were uncovered, indicating a number of scenes relating to processions and/or goddesses.¹³² A fragment of a white foot (half life-size) on a footstool or platform has been paired with the image of two similarly proportioned white hands holding a small female figure in a yellow tunic (Fig. 3.41). Included in the illustration is the restoration that I proposed in my doctoral dissertation after examining the original fragments, in which a seated figure (goddess or priestess) reaches to accept the miniature figure from another woman.¹³³ The fingers and positions of the two white hands represented are indicative of two individuals, and a scene of offering is reconstructed based on the iconography we have seen for seated women. The small figure, who has been interpreted as a figurine or a real girl,¹³⁴ wears a dress similar to that of the Ivory Triad child, but which also closely resembles the Mycenaean-style tunic on the Tiryns ring. The long tunic is closed at the front, with a vertical red band running down the front seam (and blending unnaturally into the sleeve trim), a red belt or tie at the waist, and a simple scatter pattern of clusters of triple red dots (perhaps representing beads, which could easily be applied in this manner).¹³⁵

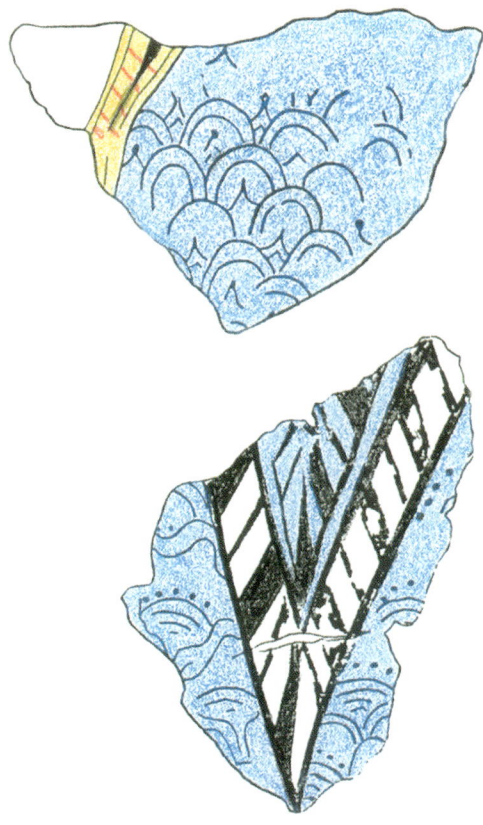

Fig. 3.39. Fresco fragments of female costumes, from Mycenae (drawing S. P. Murray after Lamb 1919–1921, pl. viii)

In the reconstruction of this scene, the garments of the larger females follow the iconographic tradition, and were adapted from a collection of associated fragments found in this area which depict portions of flounced skirts that contain barred flounces (in red, white, blue, and yellow) alternating with simple fabric motifs of repeated hooks or dot clusters.¹³⁶ The most elaborate of these (Fig. 3.42) is a skirt (or skirts) shown in overlapping

Fig. 3.40. Fresco fragment of female costume, from the Ramp House area, Mycenae (drawing M. C. Shaw)

Fig. 3.41. (above) Fragments and proposed reconstruction of fresco with presentation of miniature female figure, from the Southwest Building, Mycenae (drawing and reconstruction S. P. Murray)

Fig. 3.42. (right) Fresco fragments of skirt(s), from the Southwest Building, Mycenae (drawing S. P. Murray after Kritseli-Providi 1982, pl. Δ) sections: panels of yellow or blue, marked with four-dot clusters, alternate with multicolor, barred flounces, which are in turn topped by an edging band of white or yellow circles between blue borders.[137] The flounces give a light, ruffled impression, unlike the stiff and regulated renditions in most Mycenaean paintings, and the curved outline of the garment is reminiscent of the bent knees of seated figures in the festal skirt. Its scale, pose, and ritual character are suited to the recipient of the figurine, although there are some details that make this attribution tenuous.[138] Although the textile patterns here are not very complex in comparison to Minoan examples (and the figurine does not wear ritual attire), these flounced skirts give an impression of elaboration and richness, and still appear to follow the main design features and construction of Minoan prototypes.[139]

The Cult Center: The Room with the Shrine Fresco
Within the heart of the Cult Center, an extensive and enigmatic painting of three female figures was found largely *in situ* in Room 31 of the "Room with the Fresco Complex" (Fig. 3.43).[140] A block projecting from the frescoed wall presumably served as an altar and was painted along its upper side with bulls' horns ("horns of consecration") and a row of colored discs, motifs which find their origins in Minoan religious symbolism (such as that found in the Temple Fresco and the dress border in the Procession Fresco from Knossos, discussed above). Two large figures, about half life-size, are painted directly above the altar, and a third on a smaller scale appears beside it. They are thus directly associated with cult activity, and we can assume their prestige – but does their attire bear this out?

In the larger scene, the two females face each other, framed by yellow columns that indicate some kind of architectural setting. One wears the traditional Minoan costume for ritual occasions: a multicolored, flounced skirt and a tight bodice (red with blue trim, but much abraded), which no longer continues as a dress. The separate skirt is now rendered in a perfunctory manner as a "type" – the central division (or seam, in this case) and the downward arcs of contrasting flounces are maintained, but the outline of the skirt shows no projecting contours for the flounces. It is as if the entire skirt was pieced as a single A-line unit, perhaps due to an artist that was more familiar with readily visible Mycenaean long tunics than the fading traditions of Minoan female dress. The skirt conveys lingering impressions of contrasting colors and bold patterns, but these are reduced to simple vertical striations (using the black-on-blue, red-on-yellow formula) and an occasional white band with a wavy red line.

The other large figure wears a rather shapeless, light blue long robe with vertical rows of stripes, dots, and dashes (red and white), a prominent red and blue fringe running down the front (presumably the warp threads), and a line of red tassels or weights along the bottom. It has a heavy, woolen quality to it, rather as if the woman wearing it has wrapped a rug around herself. This image is quite unlike the other representations of high-status females we have seen, yet she is assumed to be a deity because of her size and placement in the scene.[141] She may be slightly larger than the other figure (at least in her feet), but what is their relationship? The woman on the right may represent a priestess in customary attire for performing ritual activities, or they may both be goddesses, one reflecting Minoan antecedents and the other a more characteristically Helladic deity.[142] The robed goddess holds in front of her a large yellow object identified as a sword, much in keeping with the more martial Mycenaeans. The robe conveys more details relative to

Fig. 3.43. Fresco of Room 31, Room with the Fresco Complex, Cult Center, Mycenae; Archaeological Museum of Mycenae (photo W. M. Murray)

weaving, and the design seems to express in a large textile format the Mycenaean taste for narrow decorative bands and stripes rather than Minoan-style fabrics with complex rapport or interlock patterns.[143] So, we may have here the meeting of two textile traditions, but also of two religious traditions, with goddesses who represent different spheres of power in the hybrid culture of the Mycenaeans. Between the goddesses hover two tiny, enigmatic male figures in silhouette, one in red above another in black, their arms extended toward the robed goddess – their roles in this scene remain a mystery.[144]

In the smaller scene abutting the altar, a female figure is shown with a sheaf of grain in each upraised hand. Although very little is preserved of her beyond her head and chest, it is clear that she wears a plain white tunic, a grey-blue mantle tied across one shoulder, and a hat (*polos*) that is usually associated with deities, priestesses, or sphinxes.[145] A bit of the hem of the mantle is preserved below, with pendant loops that may reflect the manipulation of the weft yarns of the textile. Vestiges of two upraised, yellow paws and a tufted tail indicate an animal companion, usually restored as a griffin, but more likely to be a lion due to the color.[146] Her small scale and her rather simple garments, which hint at the *chiton* and *himation* that will characterize post-Mycenaean Greece, may reduce her rank to priestess. The sealstone tied on her wrist denotes a level of status, but perhaps also the role of a cult official. The context and the animal companion, however, may identify her as a grain goddess, who perhaps has a less elevated status in this pantheon.[147] Thus, in this fresco we have three different modes of dress depicted, of which only one expresses the old, familiar signifier for prestige. It seems the Mycenaeans had some ideas of their own about what constituted status garments, although they would continue to display in their wall paintings a conventionalized Minoan iconography when depicting female processions of offering.

Thebes, Tiryns, and Pylos
Procession frescoes
The theme of presentation to (and collection for) a deity, which on Crete and in the Cyclades has such flexibility in interpretations, coalesces in the Mycenaean Age into a canonical formula of processional women marching formally in set costumes. Evidence for this theme has been detected at the major palatial centers on the mainland, and is best preserved at Thebes, Tiryns, and Pylos.[148] We can see in the earlier depiction of the Procession of Mature Women from Xeste 3 at Akrotiri (discussed above) some of the distinct elements that will be codified in the Mycenaean adaptation: a stately single file composed only of women, large-breasted and elegantly dressed (now invariably in flounced skirts, however), carrying floral offerings. Additional types of offerings may be paraded as well – *pyxides* (small boxes, for jewelry or cosmetics), vases, figurines, cloth, and perhaps jewelry, but flowers are prominent. In the mainland scenes, the recipient has not been preserved in recognizable form. Based on the iconography we have traced here, it should be a seated goddess/priestess, but in Mycenaean iconography the deity may have been implied solely by the formulaic line of women (as the presence of the goddess is implied in some Minoan nature scenes), or another symbol of the deity such as an altar could have been used. The dominance of floral offerings and the traditional Minoan attire of the processional women attest to a continuing tradition of honoring the goddess of nature. In keeping with this theme, the women are well dressed in variants of Minoan festal costume, but the depiction of this costume has become so stylized that the degree to which actual textiles are the models may be questionable. By comparing the three best-preserved scenes, however, we may get some clues.

The Procession of Women from Thebes (Fig. 3.44) appears from its stylistic characteristics to be the earliest in this group, probably LH II.[149] This scene survives primarily in fragments of the midsections of at least nine women, with some moving to the left, but most are moving to the right. They originally decorated a small room in the Kadmeia, where they may have marched along both side walls and converged on an unknown goal on the smaller north wall (rather like

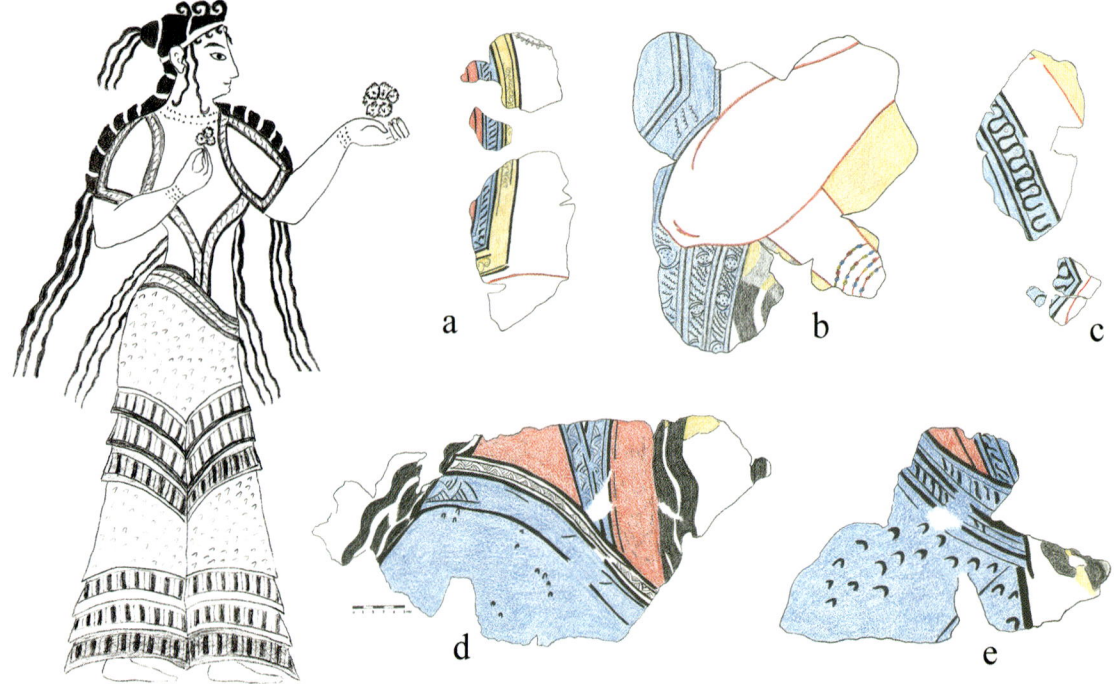

Fig. 3.44. Selection of fresco fragments and reconstruction of a figure, Procession of Women, Thebes (fragment b: drawing S. P. Murray; fragments a, c, d, e, and reconstruction: S. P. Murray after Reusch 1956, pls 2–4, 7, 15)

the mural from Ayia Triada).[150] The line of life-size figures was placed about one meter above the base of the wall, therefore above the viewer, a position that would have increased the imposing and stately effect of the ceremony.[151] The women wear the typical tight, open bodice, shown both frontally and in profile views, and a long, flounced skirt with a sloping girdle at the hipline. Although poorly preserved, the skirts seem consistently to be of blue fabric with a simple design of small black crescents (Fig. 3.44d, e). While the flounces are still multicolored, the color changes do not occur within each strip of flounce, but vary only from row to row (white flounces marked with red bars, blue or yellow flounces with black bars), representing a garment that relies more on construction than weaving expertise for its effect.

The decorative intricacies of this costume are concentrated on the edgings of the bodice – this is where the artist has focused on the careful execution of a variety of patterns with differing degrees of complexity. While the main fabric of the bodices is plain (red or blue), the motifs in the multiple border bands, in addition to the usual thin and thick black bars, include rope pattern, ivy or papyrus chain (Fig. 3.44a), scalloped lines, reversed crescents, dotted rosettes (Fig. 3.44b), hatched chevrons or adder-mark (Fig. 3.44d), and a sinuous running-S pattern (Fig. 3.44c). Some bodices are less rich in their decoration than others, perhaps indicating a hierarchy of figures in the composition, and some trims match the overall color of the bodice, rather than producing the impression of strong contrasts so typical of Minoan ritual costume. There is a delicacy and decorative versatility in these trims, but, other than these variations, the women must have seemed rather repetitive in their costume. Nonetheless, the general tradition of fancy dress, abundantly accessorized with bracelets, necklaces, and hair ornaments, worn by elite women for a ritual occasion was still expressed, and its effect would have been felt by the viewers of this fresco. However, the garments here are taking on the appearance of a ceremonial uniform, and the emphasis is more on the variety of the offerings than on the display of textile extravagance. The display of beautiful and intricate fabrics that was so prominent in the Knossos Procession Fresco is not a priority here, but we can see in the Mycenaean taste for ornamental edgings a practical side: the production of intricate textiles was relegated more efficiently and quickly to the band loom.

At Tiryns and Pylos, scenes of processional women followed much the same layout as at Thebes, with life-size figures in festal attire arrayed in single file at a raised level; but the costumes show a reduction and simplification of motifs.[152] Both paintings were associated with palaces, but they were found in nearby dumps, probably the result of redecoration projects, and so the architectural context is uncertain. In the Tiryns procession (Fig. 3.45), the women are consistently depicted in a left or right profile (no frontal views), indicating two streams in the procession. All seem

Fig. 3.45. Selection of fragments and reconstruction of a composite figure, Procession of Women, Tiryns; National Archaeological Museum, Athens (figure: drawing S. P. Murray after Rodenwaldt 1912, pl. viii; fragments: drawings S. P. Murray, photo A. Chapin courtesy National Archaeological Museum, Athens)

to be in similar carrying poses, with an exaggerated shoulder position, overlarge breast, and elaborately dressed hair. All bodices are red, with borders that are still ornamental but more limited in variety: only the central band is patterned, and only one of three motifs (interlocking leaves, flame, or circles with central dots) is used, while all examples are flanked by the barred bands (blue with black, yellow with red bars) that are habitual in this period. The skirt fabrics (blue or yellow) have simple all-over patterns with either a basic scale motif or a net of rippled lines; and the flounces, once again shown with the more traditional alternating blocks of color (red, yellow, blue, and white), and some decorative borders, now have minimal outer contours. The skirt shows only slight peaks in its outline, and seems more akin to the example we have seen from the Room with the Fresco at Mycenae (Fig. 3.43), which is roughly contemporary. The redundant nature of the poses and costumes, varied only in the items carried, the small variations in skirt colors, and the central band of the edgings, turns the whole figure into a repeated pattern in a rather static processional march. The Minoan festal costume is maintained in its general visual effect, but the details of its construction are mutating, and this is apparent in the Pylos procession as well.

At Pylos, the interest in ornamental trims is further reduced to a single border band decorated with a meager rippled line, and a skirt in which the "flounces" look like variants of the bodice bands (Fig. 3.46). There is no longer any recognition of a separate overskirt, split at the front, like those we saw illustrated on Crete and Thera, but now it appears to have blended or evolved into a bell skirt with arcing panels pieced into it. There is still a vivid variation in colors, and some use of pattern (crescents, vertical ripples), but the impression here is that the painter is following an artistic formula that may have little parallel with textiles. One detail does give a glimmer that some real, observed textile motifs may still be inserted into the scheme: a skirt panel in red with a net pattern ornamented with white dots at the interstices – a wonderful echo of the Knossian "Lady in Red" (Fig. 3.2) from two centuries earlier! For the most part, however, the garments and the oddly posed figures, crowded stiffly into the limited space of a panel, look as if they have been reduced to a series of component parts with perfunctory decors that are easily rendered with a brush.

Among the gifts carried in the Tiryns and Pylos processions, there are some possible references to cloth as an offering. From Tiryns, an intriguing fragment shows a hand grasping what seems to be a dedication of blue cloth with fine black striations, very similar to that held by the goddess in the Phylakopi fresco (Fig. 3.20), and perhaps including a figurine as well.[153] In the Pylos procession, flowers are the most frequent offerings, but one woman (Fig. 3.46, left) carries what appears to be a bit of cloth cut off by the edge of the wall.[154] Some enigmatic fragments from the Thebes procession may also represent textiles, perhaps as offerings of a patterned fabric (Fig. 3.47), or as elements of a structure, but they are too fragmentary for a reliable interpretation.[155]

In the Mycenaean processional scenes, especially those of LH IIIB, the stylization of the figures and their costumes is distancing them from reality – the basic action of formal procession with gifts is conveyed here, but the garments represent a faded Minoan style which may at this point have more to do with lingering artistic iconography or religious tradition than with the actual prestige attire of Mycenaean ladies. Formality is an expected trait in a stately processional rite filled with ceremonial trappings, but these mannequins have become stiff and remote, a quality accentuated by their elevated position and the use of heavy black outlines. One

3. Patterned Textiles as Costume in Aegean Art

Fig. 3.46. Two women in the Procession of Women, from the Northwest Slope Plaster Dump, Pylos (watercolor P. de Jong, courtesy the Department of Classics, University of Cincinnati)

Fig. 3.47. Fresco fragments of fabric(?), from the Procession of Women, Thebes (drawing S. P. Murray after Reusch 1956, pl. 13)

sees in these scenes a borrowed heritage of Minoan religious customs, but not a display of Mycenaean women parading their skill at the loom. While elite women may have enacted these rites in some form, it is more difficult to imagine whether these variations on the festal costume reflect what they actually wore, or wove.

Mycenaean dress beyond the strictures of Minoan traditions seems to have put little emphasis on ornate fabrics other than decorative bands. We may get a more convincing picture of the kinds of textiles they made for people in elite positions from two other examples of frescoes from Pylos. In an isolated group of fragments from the fresco dump (Fig. 3.48), we find evidence

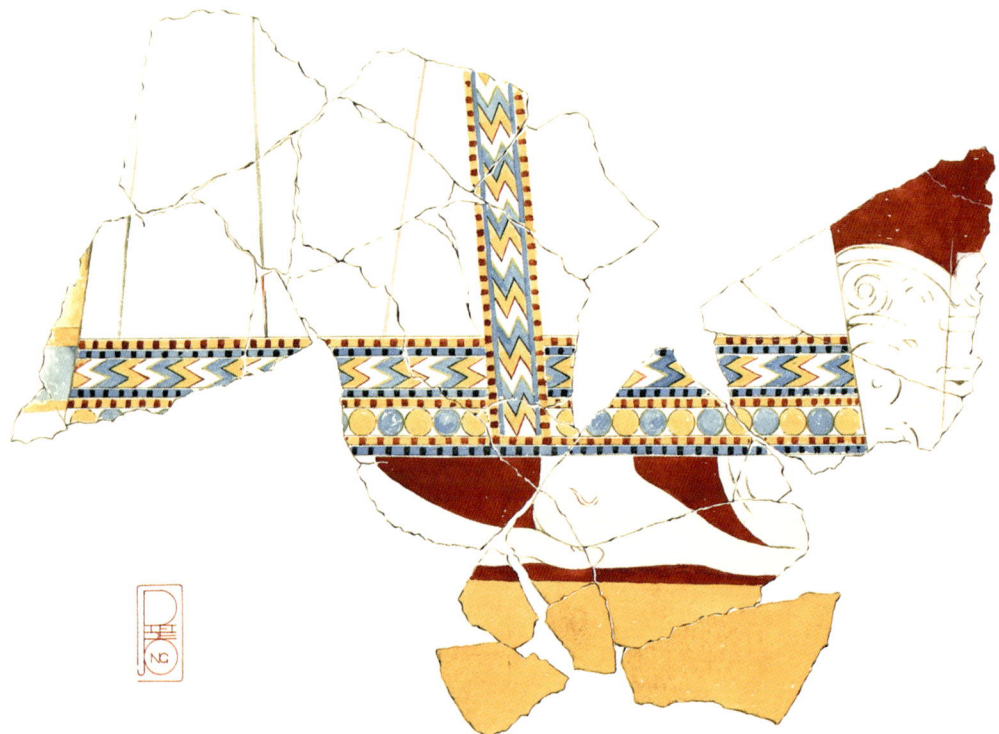

Fig. 3.48. Fresco fragment of dress border, from the Northwest Slope Plaster Dump, Pylos (watercolor P. de Jong, courtesy the Department of Classics, University of Cincinnati)

that the production of colorful and highly ornamental edgings remained popular for certain prestige clothing.[156] The bottom of a white, A-line garment is trimmed with a vertical band and horizontal hem border in an intricate and elegant design of blue, yellow, and white nested chevrons above a row of colored discs (beam ends), reminiscent of the goddess/priestess from the Knossos Procession Fresco. It probably depicts an extravagant version of the long tunic favored by the Mycenaeans and used by both men and women, more often shown in a plain color, with plain bands in the typical places (on sleeves, hems, and down the front). Men are shown wearing fancier versions of these long tunics in a small-scale scene from the vestibule to the throne room at Pylos, which seems to include at least one woman and several men in various modes of dress advancing toward a structure (shrine?).[157] Given the nature of the scene, and the movement of the figures toward the actual doorway to the throne room, the representation of high-status members of the court can be assumed. There seems to be an effort here to note, nominally, the more impressive appearance of the tunics worn by elite men for this special occasion: stiff, white fabrics with scatter patterns of dots, ψ-motif (Fig. 3.49), or dotted rosettes.[158]

Fig. 3.49. Man in long tunic, Vestibule Fresco, Pylos (drawing S. P. Murray after Lang 1969, pl. 120)

The "sacred knot"

A curious item in the wardrobe of ritual garments called the "sacred knot" gives us a striking example of the importance of textiles as emblems that carried meaning in their own right. It appears to have been fashioned from a long, narrow strip of cloth, sometimes with a rippled, striped, or plaid pattern, which was knotted so as to create a prominent loop that projected upward, while its two fringed ends hung below. It is sometimes depicted at the nape of the neck of a female figure (Figs 3.50, 3.51 right), but it is shown infrequently, which indicates that it was something more specialized than the typical ritual costume. It also appears independently in sculptural form (Fig. 3.51 left) or represented in glyptic examples, often in association with known religious symbols.[159] It is thus interpreted as a textile with sacred meaning that, when worn, identified a goddess or high-level priestess, and when shown on its own, it symbolized that goddess. It would presumably have figured in robing rituals, in which the fabric would be brought to the goddess (ostensibly a priestess surrogate or a statue), tied, and attached in the canonical spot.[160] The blue fabric given to the goddess in the Phylakopi fresco (Fig. 3.20) and carried along with a figurine in the Tiryns Procession of Women may be connected to this

Fig. 3.50. La Parisienne, fresco detail from Knossos (photo A. Chapin. Archaeological Museum of Heraklion, Hellenic Ministry of Culture and Sports – Archaeological Receipts Fund)

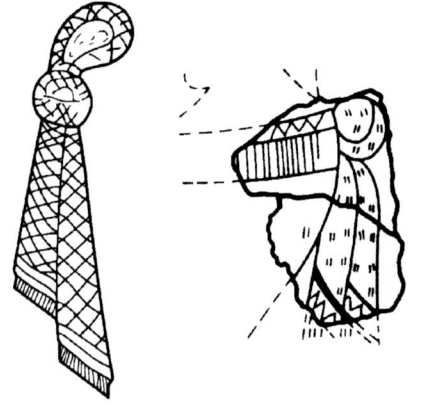

Fig. 3.51. Representations of sacred knots: (left) ivory relief, Southeast House, Knossos; (right) fresco fragment, Mycenae (after PT, fig. 15.14)

process.[161] Although the ornately patterned and sewn garments worn in ceremonial events were a means of displaying the prestige of the participants and of showing off the expertise of weavers, the sacred knot provides a compelling expression of the divine significance that could be given to a simple, unconstructed piece of nicely woven cloth.

Conclusion

"Clothes make the man" (or the woman!). Inherent in this saying is our perception that garments can be indicators of status, profession, income, or culture. In the modern world, even the simple T-shirt can be a vehicle for expression, albeit through the medium of printing on fabric rather than the nature of the fabric itself. In preindustrialized societies, the fabrication of cloth and the construction of garments was a long and laborious process, and clothing that displayed richly colored or patterned fabrics and complex designs marked one as "special" – be it in social, religious, political, or economic spheres. Whether it is the ability to follow fashion trends, to import exotic materials, or to command/devote considerable man- (or more likely woman-) hours to the creation of a garment, "fancy" bespeaks prestige. Thus the gaily colored, patterned, flounced, and boldly trimmed dresses and skirts of women and girls in the frescoes are likely to represent an elite social group as well as, in the images discussed above, an event that transcends the mundane. The abundance of jewelry depicted on many of these figures reinforces the impression. Where a context is discernible, these upscale representations occur in scenes associated with aspects of ritual and ceremony, and the most richly dressed figure is assumed to represent the most prestigious participant – a goddess, or a priestess enacting the role of goddess.

On Crete, we have seen in the garments of the Pseira and Ayia Triada goddesses a *tour de force* of visual illusion and intricacy that demonstrates the high level of textile artistry that Minoans were capable of producing for prestige garments in LM I.

The kilts of the Procession Fresco at Knossos indicate that the penchant for luxury fabrics with rich rapport patterns continued into the LM II period, and that such fabrics were also a venue for elite display on men, at least in ceremonial contexts. From Thera, we see a great variety of fabrics, from light and sheer chemises and veils to fleecy mantles to skirts that seem like murals of their own; and we learn details about how these textiles were used to express elite roles, gendered rites, and social identity. The Mycenaeans, while increasing their political dominance in the Aegean, adopted the iconography of prestige from the more luxurious Minoans as a means of displaying the trappings of their success. But the people of the mainland were, by nature and industry, less interested in garments made with intricate fabrics that required specialist weavers, and more inclined to define "fancy" through the efficient production and application of decorative trims.

Some garments in the paintings from Crete, Melos, and Thera seem in their richness and complexity to test our belief in their veracity, but Aegean peoples had the fibers, dyes, and technology to make such textiles, provided they had the luxury of time and the motivation for such extravagant display. It is reasonable to question whether deities may have been represented in "supernatural" costumes that were perceived as being beyond the capabilities of humans, but the fact that we often have trouble distinguishing who is a goddess among the females in a scene tends to argue against that distinction. It is more plausible to suggest that the most extravagant garments in the frescoes represent aspirations more than fantasies. They represent, as Iris Tzachili has expressed, "the measure of the desires of a society which recognizes itself" there.[162] Both cloth and finished garments were given as offerings to deities, and these had to be either of the finest quality or of a distinctive ritual type, such as the sacred knot. Elaborate articles like the faience model dresses with crocus emblems (Fig. 3.4) or the flounced skirts carried on seal impressions (Fig. 3.34) demonstrate that the goddess was gifted with elegant textiles, and wore them as a mark of her prestige and power. Women enacting the role of the goddess, or in emulation of her as devotees, honored her and reinforced their own elite roles by wearing special ritual attire. Girls, passing from puberty into adolescence, enacted roles that linked them more closely, through costume, reverence, and service, with the goddess who brought fertility and protection.

This intricate and highly decorative attire required not only the substantial creative talents of weavers with skill, patience, and plentiful materials, but those with the luxury of time – namely women whose high rank in society allowed them the leisure to develop their expertise or who could employ others to do so. The status of such women would be demonstrated in the wearing and gifting of the fine textiles they wrought. For the women of Thera, we may perhaps see a local textile tradition that excelled at creating cloth with varied weights and textures, as well as fine embroidery; whereas the elite weavers of Crete loved ornate patterns and plenty of them. Like Penelope and Helen in Homer's tales, these women expressed both their talents and their prestige in the finest products of the loom, long perished, and displayed them in ceremonial activities, long past, but splendidly illustrated in paint and plaster on the walls of their shrines, homes, and palaces – an enduring testimony to the vital role played by the textile arts in social and religious custom in the Aegean Late Bronze Age.

Acknowledgement

My gratitude is extended to Anne P. Chapin for the invitation to contribute this chapter, for her much-valued advice, her generosity in the use of many of her own illustrations, and her diligence and skill in organizing this volume. Maria Shaw has served not only as a frequent resource on the subject matter of this chapter, but she has also set the standard for sound scholarship, meticulous study, and precise illustration. Susan Lupack is to be commended for her editing talents in reviewing the manuscript. My thanks as well to John Younger, Michael Nelson and Elizabeth Barber for the use of their drawings and photographs. Mark Porlides and Philip Davis, my teaching assistants in the Department of History at the University of South Florida, steadfastly took on greater duties, allowing me to focus more time on this project when it was most needed. William M. Murray, as always, has been my sounding board, my advisor, and my constant support.

Notes

1. Elizabeth Barber's *Prehistoric Textiles: The Development of Cloth in the Neolithic and Bronze Ages* (hereafter *PT*) presents a collection of much of the key evidence, see particularly pp. 314, 317, 322. See also Marcar 2004 on design features and the textile origins of the patterns reproduced in frescoes.
2. Of the colors used in frescoes, white and black reflect the natural colors of wool, while red, blue, yellow (saffron yellow or yellow ochre), and murex purple are related to the dyes that were available to be used on fabrics. The brown shades of wool are not depicted, however. The information provided by two-dimensional frescoes about texture and weight is limited, but there are attempts by painters to convey these characteristics in some of the garments discussed, and the fit and movement of fabrics is often expressed.
3. The ritual use of cloth both as costume and offerings in the paintings of the Neopalatial Period has been summarized in Donahue 2006.
4. For further discussions of typologies of Aegean clothing, see Chapin 2008, 57–67 and Crowley 2012; the latter considers some of the examples of flounced garments that I discuss here to be pants rather than skirts. Lillethun (2003) makes an analysis of bodice types, and Televantou (1982) details the Theran garments. See also the articles by B. Jones in the bibliography, which treat the Minoan chemise (termed by Jones as a Minoan *chiton* or *heanos*), flounced skirts, and other specific garments. Trnka (2007) summarizes clothing types within a chronological context.
5. The variations are presented by Televantou (1982, 119–123, fig. 3).
6. B. Jones 2003, 443.
7. See further B. Jones 2001; 2003; Lillethun 2003, 468.
8. The primary publication on the excavations conducted by Arthur Evans at Knossos is the multivolume *Palace of Minos* (hereafter *PM*). The Neopalatial structure reflects a rebuilding of the palace in MM IIIB (after destruction by earthquake), followed by a series of renovations and redecorations in the LM periods. Significant damage occurs at the end of LM IB, followed by Mycenaean control of the palace in LM II–IIIA.
9. Knossos is the only palace that contains evidence thus far of extensive figural frescoes, while curiously there are several examples from nonpalatial contexts. See further Immerwahr 1990, 181–185; Rehak 1997, 164; Shaw 1997; 1998, 68. For the dating of the frescoes found in the palace area, see Immerwahr 1990, 170–179; Hood 2005.
10. Lady in Red: Cameron 1971; Peterson 1981a, 28–29, 174 (cat. no. 1); bodice pattern in *PM* II, fig. 457a. Loincloth: red dots and partial triangles on the main cloth, and a black border dotted with white and yellow; *PM* II, 751; Cameron 1978, 588; Peterson 1981a, 27–28, 177 (cat. no. 9). Cameron ascribed this fragment and some others to a life-sized male procession, which may have included "offerings of lotus flowers," that he restored to the Grand Staircase.
11. *PM* I, 501–506; *PM* III, 440–442.

12 The dress models included a hole for suspension and display. For the association of such garments with rituals of costuming the goddess or her priestess, see nn. 72, 116.
13 Seager 1910, 32–34, pl. V. The reliefs decorated a room in what is generally designated as a house. See the updated study in Betancourt and Davaras 1998.
14 Shaw 1998.
15 A relief fragment of an embossed band from Knossos, perhaps representing a belt, shows similar designs in white dots on a red field. See *PM* III, 37; Shaw 1998, 69; Hood 2005, 58–60.
16 Shaw 1998, 59, 65 notes indications that flounces and girdles on the costumes of these figures were rendered in relief, but not the painted patterns.
17 *PT,* 317–313. See also Shaw 1998, 69.
18 Betancourt 2007, 187–188 suggests that this design reflects a "knotted-net" overlaid atop the blue fabric of the skirt. The motif of a beaded net connected by decorative "clasps" also appears in the fresco of the Room of the Ladies on Thera (Fig. 3.32). A similar tricurved lozenge chain in relief, "bound" with yellow ring-like "clasps" and enclosing a flat field of large rosettes, appears in a fresco from Xeste 3 that seems to represent a hanging textile display (Fig. 4.6). For these examples see Chapter 4.
19 Shaw (1998, 64) notes the fineness of the grid of tiny 7 mm wide squares. See Chapter 7.
20 Shaw (1998, 69) discusses an alternate motif used in the museum reconstruction.
21 Shaw 1998, 59 (A2–A5).
22 Shaw (1998, 74) considers them to be "possibly equally luxurious in their outfits" and states that it is "impossible to determine their relative status."
23 Shaw 1998, 75. The bibliography on the seated goddess is extensive, including many discussions about the goddess in the Xeste 3 fresco at Akrotiri (discussed in this chapter) and her appearance on seals and rings (for a synopsis of the iconography see Peterson 1981a, 123–135; Niemeier 1989, 173–175).
24 *CMS* I, no. 17; Athens, National Archaeological Museum 992.
25 The Xeste 3 goddess also displays a hierarchy of scale, although this is generally not an iconographic feature in this period (see also Shaw 1998, 75).
26 The "goddess" in this reconstruction is seated on the right, which has also been interpreted as a common element in this theme (Niemeier 1989, 173–174; Shaw 1998, 75).
27 Shaw (1998, 71) discusses the textile origin for rapport patterns on LM IB pottery. See also Marcar 2004.
28 *PT,* 314–322. Barber sees in this time period on Crete a notable growth in the production of the "most difficult and time-consuming types, the most complex …designs" based on "extremely elaborate build-ups of fairly simple and familiar elements into repetitive strips, chains, grids, and rapports" (*PT,* 319–320).
29 The frescoes from Ayia Triada have been published in detail by Militello 1998. For additional bibliography, see Immerwahr 1990, 180 (AT no. 1); and Rehak 1997.
30 Militello (1998, 99) gives the room dimensions as 1.71 × 2.35 m. The fourth wall gave entry into a larger *polythyron*.
31 Immerwahr (1990, 180) describes the setting as the shrine of a mountaintop sanctuary. On the platform, see Rehak (1997, 167–171), who notes that this architectural feature, and the placement of this scene opposite the doorway, combine to form significant markers of the status of this woman, making it "reasonable to designate her a goddess . . . rather than a human being or priestess."
32 Militello 1998, 107, 259–261. On interlock patterns, see *PT,* 328; Shaw (2003, 185; and Chapter 7) details the use of a grid in laying out this complex design.
33 Shaw (2000, 58) notes that the black, four-pronged filling ornament within the blue quatrefoils is omitted in some restorations, but appears in that of Gilliéron (see Militello 1998, pls D, E for a comparison). The difficulty that artists have had in recreating this garment in reconstructions is testimony to the complexity of the pattern.
34 Militello 1998, 107; a faint blue strip also marks the hem band of one sleeve, indicating that at least one arm of the figure was raised.
35 Militello 1998, 253–259, 279–282 (with corresponding bibliography); see also Rehak 1997, 168. Chapin (2004, 58) strikes a note of caution, advising that while the identification of this figure as a goddess has its merits,

"her divinity is not wholly conclusive" since she is "neither enthroned nor accompanied by supernatural attendants" like some more definitive examples.

36 Militello (1998, 101) describes the fabric as light yellow with bluish streaking, and suggests that it was originally light blue; B. Jones (2007, 151) considers it to be white. While the garment depicted has been described by some as pants, fragments of a sleeve in a matching fabric indicate a full-length dress. See the discussion in Militello 1998, 259.

37 Fragment V2 (Militello 1998, 116-118, pl. Fb) shows another small segment of the dress fabric with the decorated band, which Militello interpreted as part of the sleeve of the other arm, leading him to restore the figure's torso in a frontal view. B. Jones (2007, 151–158) has offered an alternative reconstruction (and recreation) in which the decorative band of fragment V2 is placed vertically along the abdomen of a figure in profile, marking the front of the bodice. The front bands of the bodice would ostensibly continue down the front of the dress (merging into a single band if the skirt portion is seamed closed as it is according to the design by Jones), but this presents an inconsistency since the band shown between the figure's knees does not show comparable decoration. It matches only in color and bears no additional enhancement. However, a profile pose facing the east wall would better link her to the goddess.

38 Gold ring (Fig. 3.12): Ashmolean Museum 1919.56 (*CMS* VI, 278); see also *PM* II, 842. The new reconstruction of the adorant at a baetyl by Militello replaces earlier interpretations in which the woman was thought to be picking crocuses, as reconstructed by Cameron 1987, fig. 10. For the iconography and function of baetyls, see Niemeier 1989, 174–177 and Warren 1990. When women are depicted on seals and rings performing this ritualistic action (accompanied by other religious iconography), the female enactors lack the iconic flounced skirt, in contrast to the other females in the scenes (as is the case in the fresco). There seems to be a specific idiom for this ritual "story." A seal impression from Ayia Triada (*CMS* II.6, 4) depicting a woman wearing a similar Minoan chemise and kneeling at a baetyl, forms an interesting parallel with the fresco – could the painting be reflected in local glyptic production, or could both reflect a local ritual enactment? See further Niemeier 1989, 177; Crowley 2012, LIV 62.

39 See Militello (1998, 258–262) for a full discussion of these textile motifs and their appearance in other contexts, with additional bibliography. See also *PT*, 315–320; Shaw 2000, 52–63.

40 As discussed by Barber (*PT*, 317–324), who notes as well the use of a similar four-petal motif on the kilt hem of a male figure on the pommel of a sword from Mallia (*PT*, fig. 15.5). For the "Little Priestess," see n. 119.

41 See Chapter 8: A painting on the ceiling of the tomb of Amenemhet, a scribe during the reign of Thutmose III (c. 1476 BC), displays the same pattern as that of the skirt on the goddess, with additional color highlights (*PT*, 338-340, color pl. 3; Chapter 8, Fig. 8.15).

42 Chania: Shaw 1998, 68; Mycenae: Peterson 1981a, 194–195, nos 62–63, from the area of the Ramp House; Pylos: see below, Chapter 7, and Lang 1969, 186, pl. R, from the palace.

43 Rehak 1996, 43–44. On the Minoan cloak, see Crowley 2012. For an illustration of the Harvester Vase, see Marinatos 1984, fig. 39.

44 The grid is visible in Immerwahr 1990, pl. 18.

45 In the case of the overlapping legs of the adorant, the artist has tried to separate them visually by the use of the colored band that trims the front of the dress.

46 If this figure was intended to represent a priestess rather than the goddess, then her pose might indicate a ritual action meant to summon the presence of the deity. However, if the reconstruction of the adorant clutching a baetyl to summon the goddess is correct, then it is preferable to interpret this as the goddess's epiphany.

47 Whether the villa was entirely a private residence is debatable, but the small size and internal location of Room 14 gives it an exclusive character. The room faces a *polythyron*, a Minoan architectural feature that is sometimes associated with areas of religious character, as in Xeste 3 at Akrotiri. A low platform is built into the floor at the east end of the room, immediately below the central scene, and this has led some to interpret the room as a sleeping chamber (cited in Rehak 1997, 165, n. 15). However, it may also be read as a focusing device (Rehak 1997, 164), directing attention to the goddess and giving the illusion of elevating the most sacred aspect of the scene (especially if she is arriving in a ritual of epiphany). This platform may also allude, on a smaller scale, to the riser on which a goddess sits in scenes depicted on seals and rings (see Niemeier 1989, 172–174; Rehak 1997, 170–171), and notably in the fresco of Xeste 3.

48 Akrotiri: Doumas 1992, pl. 2 (Papyrus or Sea Daffodils in the House of the Ladies), pls. 66–68 (Spring Fresco). Knossos: Warren (2005, 131–148) links a series of fresco fragments from the North Building at Knossos with the Ayia Triada fresco because both depict flowers (including crocus), a life-size woman (or women) in flounced attire, and some form of architectural construction.

49 Stürmer (2001, 73) labels these small rooms (particularly those at Ayia Triada and Akrotiri) as *"Naturkulträume,"* spaces with intensive murals of nature, directly or indirectly associated with cult. He suggests they were preparation and meditation rooms for a priestess (and storage space for sacred equipment). Chapin (2004, 56–59) views the hybridization of plant species in this and other frescoes, as well as the simultaneous depiction of blooms of different seasons, as a means of intimating a supernatural landscape, infused with divine power and fertility (as also Marinatos 1984, 92). She interprets such landscapes not only as wealthy displays of luxury, but also as symbolic messages that "the elite class was entitled to its intimate connection with divine power" (61). For further discussion of plants in Minoan art, see Chapin 2004, n. 46 and associated bibliography.

50 Shaw (1998, 56, 68) suggests close associations "of a religious or theocratic nature" between Pseira and Knossos in the use of plaster reliefs of human figures. See also Shaw's discussion of the role of Knossos in the transmission of themes and conventions (1997, 484, 491 with additional references). Rehak (1997, 163–175) sees in the combined imagery of the goddess with wild animals, rocky terrain, crocus plants, and architecture the dissemination from Knossos of a theocratic message linking the palace with the peak sanctuary. More broadly, Chapin (2004, 61) links the ability of the ruling class to commission expensive landscape paintings infused with religious symbolism as a means to communicate that "divine power fully supported the Minoan elite."

51 Ladies in Blue: a few very disconnected fragments indicate at least three elegant ladies with multiple necklaces and patterned yellow bodices trimmed with blue borders containing spiral motifs (*PM* I, 544–547); Hood (2005, 79) considers this fresco to date to "probably MM IIIB;" Immerwahr (1990, 172) considers it to be MM IIIB–LM IA. Corridor of the Procession: numerous burned fragments (Fig. 3.13, left) representing textile motifs and jewelry were excavated beneath the pavement of the LM II passage containing the Procession Fresco (*PM* II, 680–681; he notes remnants of plant motifs as well); Hood (2005, 66) and Immerwahr (1990, 172) consider them to be contemporary with the Ladies in Blue. Northwest Fresco Heaps: highly decorative fragments showing details of animals (griffin, sphinx, birds, bucrania) and intricate patterns (Fig. 3.14) that were thought by Evans to represent the textiles of dresses (*PM* III, 37–42; *PT*, 320–321; Hood 2005, 58–60, pl. 9) and a multilayered girdle, the latter in relief (*PM* III, 37–45; Hood 2005, pl. 9); all dated generally MM IIIB–LM IA. An isolated fragment showing the shoulder and hair of a large-scale woman (Fig. 3.13, right) was found near the Fresco Heap (*PM* II, 682, fig. 431; Peterson 1981a, 27, cat. no. 8; Hood 2005, 58). For many of these examples, see also Donahue 2006, 12–13 and Shaw 1998, 67–70, who considers the Ladies in Blue and Lady in Red to be the equivalent in "scale, stature, and richness of dress" to the Ayia Triada "goddess and her female companion"(69). Shaw (1997, 489–491) and Donahue (2006, 13–14) catalogue the sparse evidence for representations of large-scale women at additional sites on Crete.

52 *PM* II, 704–736; the fresco is discussed in detail, with a catalogue of the fragments, in my dissertation (Peterson 1981a, 29–41, cat. nos 2–6). See also Immerwahr 1990, 85–90, 174–175 (Kn no. 22); Wilson 2009, 35–81.

53 For a discussion of the episodic composition of the painting, see Boulotis 1987, 148–150.

54 Group A, found *in situ* on the east wall: *PM* II, 719–736; Peterson 1981a, 174, cat. no. 2. Boulotis (1987, 148, fig. 5) illustrates the group from the west wall, which he designates as "Group D." Evans makes only passing mention of it (*PM* II, 684, 720). Evans considered the closely packed groups of men in long robes as priests/musicians, but there is little room for instruments and, as pointed out by Wilson (2009, 50–51), the close formation "suggests persons of lesser importance," perhaps officials or acolytes associated with the religious bureaucracy of the palace.

55 Flounces on these figures are placed as horizontal trim at the hem and as V-shaped decoration above, and are barred in blue, red, and white (Peterson 1981a, 174, cat. no. 2; Boulotis 1987, 148, n. 18). The woman's skirt in Group A, which is better preserved, shows an ogival or papyrus pattern on the blue fabric between the flounces (see also Chapter 7, and Shaw 2000, 53, fig. 2A).

56 *PM* II, 590–596, 604, figs 368, 371; Hägg (1987, 131–132) indicates that these artifacts had fallen from a large room in the northwestern part of the *piano nobile*, beyond the staircase toward which the Procession Fresco led. Another fresco fragment of this frieze motif includes the blue/black and red/yellow bands above and below (*PM* II, 604, fig. 377).

57 As discussed by Davis 1987.

58 *PM* II, 724. Shaw (1998, 74) notes the architectonic features of her dress, which she considers to be suggestive of a divinity. This female figure lacks some of the features that would help identify her as a goddess, such as a seated pose or attendant animals, so a high-status role as the priestess, robed and feted as the goddess surrogate, might be preferable (Peterson 1981a, 32; Wilson 2009, 54). On the identity of this figure, see the bibliography in Boulotis 1987, 152–153.

59 Boulotis (1987, 150) suggests that it is the fringe of a sacred knot that will be arranged on the goddess/priestess as part of her ritual attire; Evans (*PM* II, 724) proposed that it might be a veil. While these streaks or ribbons are not characteristic of a veil, they are also thick and irregular in comparison to other representations of fringe (e.g. on the Harvester Vase or the flounced skirts depicted in the Room of the Ladies scenes from Thera [Figs 3.31, 3.32]). Given the clustering of men around her, a scene involving the robing of the deity would be plausible at this stage of the ceremonial progression and would ostensibly include textiles and jewelry. Linear B texts that may indicate the use of offerings of cloth anointed with oil (Nosch and Perna 2001, 477) might be considered with respect to this vignette as well.

60 Boulotis 1987, 147; also referenced in Peterson 1981a, 32–33 and noted as a garment with religious associations, perhaps for "officiants of cult." Crowley (2012, 235) associates this garment with the puzzling "hide apron" worn by some male figures on seals, which she considers to be leather. The markings – dashes and F-shapes – on the skirts depicted on the sarcophagus are thought to be an artistic convention for indicating fur or fleece (thus the "tail"), but the Knossian version is fabric, as is a robe with similar markings on a male figure in a fresco from Pylos (Lang 1969, pl. 120, 7 H 5). We might consider whether the "hide skirts" on the sarcophagus are instead meant to indicate garments with a heavier weave and texture, whose antecedents were animal hides or fleeces. It is an interesting parallel that the sarcophagus illustrates a range of costumes similar to those of the Procession Fresco – kilts, long tunics, hide skirts (as noted by Rehak 1996, 45), but no flounced skirts. For a full treatment of the Ayia Triada Sarcophagus, see Long 1974.

61 For a discussion of the Minoan kilt and other male attire, see Rehak 1996. Concerning the items the men carried, those extant are large vessels, seemingly of rich materials like silver, gold, copper, and colorfully veined stone (see *PM* II, 705, 722–725, fig. 451; Peterson 1981a, 34–36, 241, n. 20; Boulotis 1987, 150, with bibliography in nn. 24–27). Some, like the long, conical rhyton held by the Cupbearer, have ritual significance. However, since so few items have been identified from the many that would have been depicted, I am hesitant to categorize the objects carried. Related scenes from seals and rings, or from processional frescoes found on Mainland Greece, show additional types of gifts as offerings, some of which are discussed in this chapter (see also Peterson 1981a, 111–119; Wilson 2009).

62 The pendant nets are also discernible in white on the pair of men behind the goddess/priestess in Group B, indicating that they wore kilts (Peterson 1981a, 34). This decorative addition would have been a hindrance in daily wear, but here they emphasize the fancy-dress occasion. The jingle of the finials in the formal procession would have added an aural enhancement to the visual effect that would have further distinguished the role of these members of the ceremony. The effect of these brief but ornamental costumes "in action" struck me when seeing them realized in a 2011 production of the Minoan-themed "Idomeneo" by the San Jose Opera, for which I served with Anne Chapin as archaeological consultant.

63 The Cupbearer's kilt might be considered a "fat" quatrefoil motif, since the four arms are curved. Each arm contains a blue dotted rosette, with a red rosette placed in the center. The kilt was painted yellow ochre, but a light wash of red gives it an orange cast. The interlocking pattern on the other kilt is a quatrefoil set within a cruciform motif, with a solid black lozenge at the interstices and another in outline in the center of each light blue cross. The scale-pattern kilt is in blue and red on a yellow or white ground. For a full description of the figures, see Peterson 1981a, 34–41, cat. nos 4–5; for discussion of the patterns, see Chapter 7, and Shaw 2000. A kilt with an interlocking "anvil" design and quatrefoil border is represented on a male figure in repoussé on a MM II gold sword hilt from Mallia (see *PT*, fig. 15.5; Rehak 1996, 42–43).

64 Shaw 2000, 56; 2003, 185. See also Marcar 2004, 234. The thin gold ornaments, perhaps from the shroud, were found in the chamber tomb of a woman on the north slope of the Areopagus (Immerwahr 1973, 7–11, fig. 16), dated c. 1400 BC. This design seems to be related to the waz-lily, which has out-turned leaves. Variants of this ivy-like motif are seen in paintings, perhaps of textiles, from Xeste 3 (Vlachopoulos 2008, 494, figs 41.47–41.50).

65 Peterson 1981a, 34, 175, cat. nos 3–4.

66 Rehak (1996, 44–45) and Hood (2005, 66) summarize the state of the date debate: Hood opts for LM II while Immerwahr (1990, 175) settles on LM II/IIIA; see also the detailed analysis of the evidence in Marcar 2004. Shaw (2000, 62) favors a date of LM IB–LM II based on the elaborate textile patterns represented and their appearance on male garments (fancy fabrics tend not to reflect Mycenaean tastes in menswear). Boulotis (1987, 155) suggests that mural painters, or their students, who had worked in the LM IB period might still have been available to work in the same vein in a LM II redecoration of the palace. Wilson (2009, 56–59), examining the fresco from the viewpoint of gendered activities, sees in the ratio of preserved participants (5 women, approximately 28 men) a diminishing of previous female dominance in ritual and a rise in the power of men, who now appear in a variety of garments and roles that are indicators of rank and status. If we ascribe a LM II date for the creation of this scene, this may reflect the merging of Minoan religious custom with the masculine authority of the new Mycenaean ruling class.

67 For a representative sample of interpretations of this scene, with attendant references, see Boulotis 1987, 153–155; Cameron 1987; Marinatos 1987. Cameron viewed the procession as part of a grand, multistage festival to the Great Goddess of nature, who is enacted by a woman as "goddess-impersonator." Wilson (2009, 77) suggests that the scene reinforced the prominence of the palace in religious matters (and its role in maintaining the social order) by presenting a grandiose picture of proper ritual behavior.

68 We can get a glimpse of the effect that these vibrantly patterned kilts had on observers in a series of 18th Dynasty Egyptian tomb paintings representing the men of Keftiu (the Egyptian term for Aegean peoples), particularly in the tombs of Rekhmire (Fig. 8.1) and Menkheperreseneb (Fig. 8.8). See Chapter 8; Peterson 1981a, 138–147; Rehak 1996, 35–39.

69 Atkinson et al. 1904, 70–74; Morgan 2007, 380–389.

70 Morgan 2007, 384.

71 PT, 320–322. For bibliography on this topic, see Chapin 2008 and Blakolmer 2012a, 330 n. 39.

72 Morgan 2007, 386 with bibliography; see also Donahue 2006, 42–43. Morgan (1990, 260–264) draws parallels between this scene and those from Ayia Triada and Thera (Xeste 3 and the House of the Ladies) in terms of illustrating phases in the preparation and presentation of ritual cloth or clothing.

73 Morgan 2007, 372–380, 386–389. The lily blossoms, in white on a red background, were found slightly to the south of the others, presumably from an adjoining room on the upper floor. Fragments indicating several blue monkeys and a frieze of spirals, found in Trench ΠS, came from a room abutting the Pillar Crypt, whose access at the upper level is undetermined, but which Morgan considers to be part of a unified cycle of paintings in the Pillar Crypt area. For the role of monkeys in connection with the goddess, see Marinatos 1987a.

74 Morgan 2007, 388–389. Morgan also draws parallels between a number of fresco cycles in Cycladic contexts (as well as at Ayia Triada) that juxtapose and relate scenes of pure nature with scenes of "culture" that depict ritual activity (see also Marinatos 1984, 87–89, fig. 59; Morgan 1990, 264).

75 The scholarship on this site is too vast to include here, but for a selection of publications, see the bibliography in this volume and Vlachopoulos 2008, as well as the proceedings published in *TAW* I–III and *The Wall Paintings of Thera* (Sherratt (ed.) 2000). The primary excavation reports are by Marinatos in *Thera I-VII*; the frescoes are presented collectively in Doumas 1992. I have attempted to include in the notes those publications that incorporate substantial bibliographical information, and apologize for any omissions in citations.

76 For a selected bibliography on Aegean costume and its role in cult, see in particular Chapin 2008 (esp. n. 27); and also Laffineur 2000; B. Jones 2003 (esp. n. 46); Lillethun 2003; Murray 2004; Donahue 2006; Crowley 2012. Televantou (1982) presents a detailed catalogue and discussion of the female clothing and textile motifs in the Thera frescoes.

77 Tzachili (1990; 2005; 2007) suggests that textile manufacture may have been one of Thera's main trade products; she notes the particularly high concentration of loom weights in the West House, which she proposes as a workshop. Tzachili stresses the collection of saffron for dye, as do Douskos (1980) and Morgan (1990, 262). On the evidence for the production of murex purple, see the latter as well as Aloupi *et al.* 1990. Additional evidence is cited in Vlachopoulos 2012, 40, n. 66.

78 *Thera VI*; Doumas 1992, 126–175; Vlachopoulos 2007, 2008, 2012; and this volume.

79 Vlachopoulos (2007, 109) notes that an olive tree is shown between the horns (see pl. xxviia). See also Marinatos 1984, 63–64, 73–84, figs 43, 53.

80 Discussions of the age markers used in the Theran paintings may be found, with additional bibliography, in Davis 1986; Doumas 2000; Chapin 2002; Rehak 2002; 2007; Vlachopoulos 2007, n. 26.

81 Televantou 1982, 117; Rehak 2004, 86–90. Vlachopoulos (2007, 113–114) notes the use of murex purple for selected elements of the female garments and crocus blossoms in this scene as well as in those on the upper floor; this pigment is quite fugitive and is not visible in illustrations. Murex shells found around the site verify its use (see Aloupi *et al.* 1990). On the evidence for murex purple dye in the Aegean, see *PT*, 228–229. Porter (2000, 623) has identified the garland draped across the Necklace Swinger's chest as clusters of saffron stigmas.

82 Panagiotakopulu 2000 notes the discovery in the House of the Ladies of a silk moth cocoon in a context denoting domestic use. For other discussions about silk as a fiber in the Aegean, see Burke 2012; Van Damme 2012. On the production of flax on Thera, see Tzachili 2007, 193.

83 A similar bodice design appears on the Snake Goddess from Knossos (Fig. 3.3). The red bands that cross the bodice may be a means of decorating or masking seams in a pieced garment, which would have enhanced the fit. See B. Jones 2001, 259–262; Lillethun 2003, 468.

84 B. Jones (2003, 442–443) recreated the veil with carnelian beads. The use of beads in relation to garments has been strikingly illustrated by the discovery of 40,000 beads in a LH III tomb at Dendra, some still visibly arranged in a five-color chevron pattern, perhaps as decoration on a leather or cloth belt (*PT*, 172, 312).

85 Although the view that this is a veil is generally accepted, Donahue (2006, 54–57) has suggested that the item may represent a luxury gift for the deity of the shrine, carefully displayed, in keeping with evidence that fine textiles and garments are standard cult offerings.

86 See Chapin 2008, 65.

87 Yo-yo: *PT*, 317; crocus buds: Rehak 2002, 41; 2004, 89. See also Televantou (1982, 118, 124) for a description of the skirt.

88 Barber 1994, 42–70; Chapin 2000, 23; 2008, 66. Scholars who have interpreted this figure and her wounded foot as symbolic of marital initiation include Davis 1986; Chapin 2002; Vlachopoulos 2007. For alternate views, see Marinatos 1984, 73–84; *PT*, 317; Rehak 2002, 42; 2004, 94.

89 Crowley (2012, 233, no. 12) illustrates an example on a signet ring.

90 Doumas 1992, 130, 146–151.

91 Doumas 1992, 130, pl. 109. Vlachopoulos (2007, 110) interprets this scene and the Adyton Fresco as episodes related to a marriage rite, in which textiles, whether kilt or veil, are central elements (see also Davis 1986; Koehl 2001). Murray (2004, 115) and Donahue (2006, 55–56) consider a cloth offering as a possibility.

92 Doumas 1992, 176–177, pl. 138; Rehak (1996, 47–48) notes the relationship of these earlier kilts to those of more intricate textile design in the Knossos Procession Fresco.

93 Sources on this topic are many, but see in particular Marinatos 1984, 60–72; Chapin 2002; Rehak 2002, 2004, 2007.

94 Lillethun 2003 undertook a recreation of the bodice of the girl with the shaved head, and found that the version that produced the best fit, considering both the image and actual ease of movement, used linen and a bias cut.

95 See Televantou (1982, 116) for a description of the skirt of the goddess and her fig. 1 for a schematic illustration of all of the flounced skirts (and their patterns) in this scene and the Adyton Fresco.

96 For the interpretation of these girls as acolytes in service to Potnia, see Rehak 2007.

97 Rehak 2002, 48–50; 2004; Marinatos 1984, 65; 1987, 130–132. Douskos (1980, 120) considers Thera as a special source of saffron, which held great economic value as an export.

98 See n. 77, above. Morgan (1990, 262) also considers saffron's key role as a commodity in their economy to have been in the production of dyed cloth.
99 Parallels have been drawn by a number of scholars regarding Classical references to saffron yellow cloth for the Panathenaic robe given to Athena and the garments of girl acolytes of Artemis in the sanctuary at Brauron, as well as other associations. Note especially Davis 1986, 403; Marinatos 1987, 132; *PT*, 518; Rehak 2004, 93; Vlachopoulos 2007, 113. For additional parallels between the Xeste 3 frescoes and later Greek culture, see Vlachopoulos 2007.
100 Barber (1994, 88) suggests that this would occur at age 10–12. Chapin (2002, 16) estimates the ages of these girls at approximately 11–13 years based on body development; Rehak (2007, 209) suggests that they are 8–10 years old.
101 Cameron (1978, 582) perhaps implies this in his reference to the ritual as a preparation "for the domestic duties of womanhood," but he does not elaborate.
102 *Thera VII*, 22–28, 32–38; Doumas 1992, 168–171; Vlachopoulos 2007, 114. The processional scene decorated a corridor that led to the service stairs (Room 8).
103 See also Chapin 2008, 66–67.
104 Chapin 2008, 77. For a detailed discussion of the jewelry, see Shank 2012.
105 Vlachopoulos 2007, 114; he describes as well a fragmentary fifth woman, possibly seated, in the decoration of this room. She appears to wear a garment decorated with a marine theme, including flying fish, thereby supplying another intriguing link with the frescoes of the Phylakopi shrine. Chapin (2008, 56–57, 69–78) notes that the extravagant costume of the Lady of the Landscape cannot in this case be ascribed to a goddess as would normally be the assumption. The swallows echo the motif in the nearby Spring Fresco (Doumas 1992, 99–107). In another plausible example of figured textile, Shaw (1978) has interpreted a badly abraded and indistinct fresco fragment from Katsamba (Crete) as an embroidered belt with a line of birds among reeds and rocks, shown above a bit of dotted-scale pattern in a larger format, which perhaps indicates the skirt of a large female figure of LM IA date.
106 The position of this composition of richly dressed women in a subsidiary location of the building has caused some puzzlement about the role of these matrons in relation to the youthful celebration of the goddess in the next room, but the large bunches of flowers they carry put them into the role of offering bearers, and their maturity may signal the continuity of ritual traditions through the generations, which, though communal, still allow for individuality among the elites. For further discussion on this topic see Chapin 2008, 72–78; Vlachopoulos 2008, 493; Shank 2012, 564. The flowers, as noted by Chapin, represent blooms of different seasons.
107 Vlachopoulos (2012) has observed that jewelry excavated at the site does not parallel the design and richness of that shown in the frescoes, and he sounds a note of caution about the iconography of extravagance in these scenes. Barber (*PT*, 321) also suggests that textiles with elaborate pictorial scenes might be artistic invention for divine garb, although she considers them nonetheless to be quite achievable.
108 *Thera V*, 11–15, 38–41; Peterson 1981b; Marinatos 1984, 97–105; Doumas 1992, 33–43; Murray 2004.
109 Televantou 1982, 116 (fringe); Immerwahr 1990, 58 (pleats); Murray 2004, 106.
110 Televantou 1982, 114; B. Jones 2012.
111 This detail is indicated as well in the companion to the Snake Goddess from Knossos (see *PM* I, 505–506, fig. 382; Chapin 2008, figs 2–5 [7a]).
112 As noted by B. Jones 2012.
113 Marinatos (1984, 101) has suggested that she is lactating, but whether this design is related to postchildbirth paraphernalia is not apparent.
114 Marinatos 1984, fig. 71.
115 B. Jones 2012.
116 Marinatos (1984, 104) suggests a priestess; Morgan (1990, 261–262) a goddess. B. Jones (2012) has argued for a young initiate on a smaller scale (she would have to be at a higher level). In Murray 2004, I have considered all of these possibilities and also the possibility that the figure represented a statue (see also Morgan 2005a, 37). Morgan (1990, 261, 264) connects this scene to the Crocus Gatherers and to the Phylakopi fresco as representative of three stages in a ritual robing cycle: 1) collection of saffron for dye, 2) production and/

or presentation of cloth, and 3) presentation of the sewn garment (see Murray 2004, 113; Morgan 2005a, n. 26 for further bibliography and comparanda on robing rituals).
117 For discussion and additional examples of this theme on rings and seals, see Murray 2004, 113–14.
118 Murray 2004, 122–123.
119 An exception is the boldly patterned bodice of the "Little Priestess" from the West House (Doumas 1992, pls 24–25), which, although only composed of two colors, displays a wonderful play on visual illusion in its rapport of white four-petaled flowers on blue reverting to blue four-pointed stars encircled in white (see *PT*, 317, 329).
120 Immerwahr 1990, 49; Shaw 1998, 69; Murray 2004, 125. See n. 18, above for pattern parallels with the Pseira fresco. For the relation to the knotted-net motif, see Betancourt 2007, 187.
121 See also Tzachili 1990, 387–388; and *PT*, 316–320, in which Barber posits that the regional differences may relate to the fact that the more elaborate designs were more affordable in areas close to the palace centers.
122 Barber (*PT*, 325) notes that even the edgings become "increasingly plain" in LH IIIB, and that "elaborate cloth and clothing, if it persisted, …may have been only for deities and royalty and/or only for state occasions." She ascribes the earlier Mycenaean predilection for ornamental band designs to be a heritage from nomadic ancestors, for whom the conveniently portable band loom could more easily and quickly produce the fanciest results (Barber 1997, 517).
123 *CMS* I, no. 17 (Athens, National Archaeological Museum 992), found in the Ramp House, dating possibly to LH II (Peterson 1981a, 257, n. 11). Niemeier (1989, 173, fig. 4:1) places this ring in the category of seated goddesses with adorants and notes the problems with determining whether it is Minoan or Mycenaean.
124 Rehak (2007) does not include this ring in his discussion of girls as acolytes, although this seems to be a very likely interpretation of their role in this scene. See Murray 2004, 117–118.
125 Poursat 1977, 20–21, no. 49 (Athens, National Archaeological Museum 7711). The sculpture is generally placed in the LH IIIA period, perhaps because it was found in the palace, but it is more characteristic of the Minoan-inspired style of LH I–II. Rehak (2007, 221) considers it an heirloom and notes Barber's opinion, based on textile patterns, that it is Neopalatial. The child is probably a girl wearing a simple version of the dress of the Crocus Gatherers (Rehak 2007, 221; B. Jones 2009, 318).
126 B. Jones (2003, 442–443) and Rehak (2007, 220) consider the mantle to be beaded, although the design is so dense that one must wonder if wearing a shawl composed almost entirely of beads would serve any purpose other than ornate display. Could this be a tufted weave?
127 An ivory female figurine from Prosymna also depicts a Minoan-style skirt with a striking emphasis on texture: the main fabric appears heavily ribbed and the bottom of the skirt is trimmed with individual rosettes that look like thick appliqués. See Konstantinidi-Syvridi 2012.
128 *CMS* I, no. 179. See Rehak (1992, 47, n. 85) for extended bibliography.
129 These hybrid creatures, usually called Minoan *genii* or demons, appear in a number of contexts, including fresco fragments from Mycenae and Pylos (discussed by Rehak 1992, 47–48, with citations). Always shown standing upright, in some versions it has features of a lion (as on the ring) or a donkey, and in all cases has this unusual, variegated element down its entire back, with a distinct dip at the bottom and a wavy front edge. In some instances, it resembles crocodile hide, but here it gives the impression of something shell-like or heavily ornamented. Note on this ring as well the use of the triglyph and rosette frieze beneath the ground line (see n. 56, above).
130 Peterson 1981a, 58–63; Immerwahr 1990, 114, 117, 190 (My no. 1c). Those featured here came from the Ramp House area, and probably date to LH II–IIIA.
131 Blue fabrics: Peterson 1981a, 62–63, 194–195 (cat. nos 62, 63), figs 50, 51. Skirt with cruciform interlock: Peterson 1981a, 62–63, 199 (cat. no. 84), fig. 59; Shaw 2010, 321, fig. 30.3. Barber (*PT*, 322–324, fig. 15.10) considers the latter fresco fragment to be contemporary with the Knossos Procession based on the interlocking pattern. The red crosses are barely discernible now – a visible dot marks the center and arms of each cross.
132 Peterson 1981a, 201–205 (cat. nos 91–103); Kritseli-Providi 1982, 37–53, pls 4–11; Immerwahr 1990, 119–120, 191 (My nos 3–5); B. Jones 2009.
133 Peterson 1981a, 65, 202 (no. 92), fig. 65. B. Jones (2009) favors a reconstruction with only a single seated figure holding the offering, although she considers both alternatives.

134 Rehak (2007, 222, fig. 11.13), considers the small figure to be a "living girl" based on the "naturalistic" representation and the similarities he sees with the Veiled Girl from Thera, but he is incorrect in restoring her with the partially shaved head of a Theran adolescent (see also B. Jones 2009, 313). Small details on the poorly preserved head show fillets in red and white that are consistent with arranging hair (Peterson 1981a, 202, no. 92). B. Jones (2009, 317–318) and Wilson (2009, 267) also regard the figure as a real girl. Figurine or idol: Peterson 1981a, 202; Kritseli-Providi 1982, 41–42; Laffineur 2001, 387; Morgan 2005b, 165. Laffineur (2001) also notes the Linear B references to statues carried in processions. I do not consider the image as a living girl because of its disproportionately small size and the method by which she is held with a thumb placed under her skirt, which, while an odd method of grasping any figure, would seem particularly peculiar if a real being is represented. The figure may be symbolic of gifting a child, but I do not think it is meant to reproduce one. See also Immerwahr (1990, 119, 191 [My no. 4]), who variously refers to this figure as a statuette, a doll, and a facsimile denoting the dedication of a child.

135 Beading has been suggested by B. Jones (2009, 318–321), who considers this to be a Theran-style dress, open at the chest, but this is not congruent with the representation. Rehak (2007, 222) compares the yellow garment with red dots to the veil in the Adyton Fresco.

136 B. Jones (2009, 322–336) has ascribed numerous small skirt fragments with red, white, and yellow flounces to the nearly life-size figure of the "Mykenaia" from this area, whose upper body is well preserved. Her bodice (with breasts covered) is plain yellow with undecorated red and white bands, but yellow sections of the skirt fabric have an all-over decoration of fine red hooks, perhaps indicative of texture rather than pattern. See Peterson 1981a, 193–194 (no. 59); Kritseli-Providi 1982, 37–40. This figure wears and holds an abundance of jewelry, but in this case her status is not signaled by extravagant textiles.

137 Peterson 1981a, 203 (no. 95); Kritseli-Providi 1982, 49–50 (B 20–21), pl. Δ, pl. 9. B. Jones (2009, 310–312, 321–322, fig. 4) rearranges the fragments in a manner that is more congruent with a seated figure, and provides several reconstructions that incorporate them into the offering scene; the rows of circles that she views as attached discs look more like the repeating patterns produced on a band loom.

138 Since it reinforces my reconstruction, I would like to be convinced by the case B. Jones makes for this being the skirt of the goddess in this scene, but there is a troublesome detail that led me previously to reject this interpretation: the color change of the skirt from a yellow panel in the upper portion to a blue panel in the lower portion does not remain consistent from one leg to the other, which is extremely unlikely in Aegean iconography. This anomaly led me to wonder if two overlapping women might be indicated in this fragment or perhaps a pair of skirts that are displayed independently, unworn. In my reconstruction, the seat for the goddess is adapted from some additional fragments found in the same location (Kritseli-Providi 1982, 43–44, B4, 5, pl. 7).

139 A recently restored female processional scene from Tiryns includes among the carried items a small female figure, one arm similarly raised, but dressed in a flounced skirt and adorned with beads or dotted ribbons in her hair. She is grasped around the waist (one preserved foot dangles free) by a woman in a full-length tunic very like the one on the Tiryns ring (see Papadimitriou, Thaler and Maran 2015); my thanks to Bernice Jones for bringing this painting to my attention.

140 The fresco, discovered in 1968, had been whitewashed before the building was destroyed in LH IIIB2 and was not visible during the shrine's final period of use. The initial publication of the Cult Center was by Taylour (1969). For more thorough discussion and bibliography on this fresco, see Peterson 1981a, 64–65,190–191 (no. 53), 251, nn. 110–113; Marinatos 1988; Rehak 1992; Morgan 2005b (comprehensive bibliography on the excavation of the area is in her n. 6). Morgan (2005b, 162–163) dates the fresco to early LH IIIB, since the shrine went out of use by mid-LH IIIB, around 1230 BC. Immerwahr (1990, 191, My no. 6) dates it to mid-LH IIIB.

141 The closest clothing parallel is the fringed mantle of the so-called Little Priestess from the West House at Akrotiri (Doumas 1992, pl. 24), which is monochromatic saffron yellow except for two rather plain, horizontal zones of decoration at the bottom and midsection (a wide blue band edged by a striated white border). The long tunic shown on the Tiryns ring (Fig. 3.38) is similar in its basic shape and its decoration of vertical dashes (it is not identical, contra Rehak [1992, 47], as the Tiryns tunic is not fringed, a detail which may indicate that the garment in the fresco is not seamed at the front).

142 I have previously suggested (Peterson 1981a, 65) that the woman in Minoan garb is a priestess making an offering to the robed goddess, but I am now less convinced about their identities. Marinatos (1988, 247), Rehak (1992, 58), and Morgan (2005b, 168) identify both of them as goddesses, but Morgan notes that, although their clothing differentiates them, it "does not permit us to speculate on the divine versus mortal status of these women," and it is their positions in the scene that give them balanced status.

143 It has parallels with the vertically striped kilts of some of the Keftiu men from the Tomb of Rekhmire in Egypt, which Barber relates to the Mycenaeans (see Chapter 8; and *PT*, 333–338; Barber 1997, 517).

144 I have suggested previously (Peterson 1981a, 65) that they may represent votive figurines; Rehak (1992, 48–49) suggests votaries of reduced status and scale; Marinatos (1988, 248) and Immerwahr (1990, 121) consider figurines as a possibility, but suggest that they are souls due to the hovering position (rejected by Laffineur 2001, 389–390). Morgan (2005b, 169) has noted the options, and wonders if the proximity of the tiny figures to the sword might allude to soul-like ancestors of a warrior aristocracy who honor a warlike protector-goddess.

145 One of the female officiants in the sacrificial scene on the Ayia Triada sarcophagus wears a similar hat, as does a large figure known as the "White Goddess" from the palace at Pylos (Lang 1969, 83–85, 49 H nws). For additional examples, see Rehak 1992, 51–52. The mantle has some similarities to those worn by the "Little Priestess" (see n. 119, above) and the Mature Women from Xeste 3.

146 Lion: Peterson 1981a, 191; Rehak 1992, 55–57, reconstruction pl. xviiia. Griffin: Marinatos 1988, 246–247.

147 Rehak (1992, 52) notes that sealstones were gifts to deities, who therefore may be depicted wearing them; he concludes (57–58) that all of the figures in the scene are deities of differing ranks (see also his n. 118 for a bibliography of earlier interpretations). Marinatos (1988, 246–247) concludes that this is a priestess bringing offerings to the altar; Morgan (2005b, 167–168) considers the costume to be indicative of a priestess, but notes the confusion created by the presence of a lion or griffin; she also remarks on the discovery of a Linear B tablet at Mycenae that refers to a deity called *potnia siton*, "Mistress of the Grains." Also see Laffineur 2000, 389; Wilson 2009, 249–250.

148 The major publications on each of these frescoes are Reusch 1956 (Thebes), Rodenwaldt 1912 (Tiryns), and Lang 1969 (Pylos). They are also discussed extensively in dissertations on processional paintings by Peterson 1981a and Wilson 2009.

149 LH II: Reusch 1956, 41–47; LH IIIA: Immerwahr 1990, 201; first half of 14th century BC: Peterson 1981a, 58.

150 Room N: *c*. 5.35 × 2.7 m. The amount of wall space would have allowed up to 12 figures (Peterson 1981a, 46) and, with this decoration, it may have served as a private shrine. Wilson (2009, 149–156) interprets the scene as having two focal points and, therefore, as possibly showing two different ritual actions.

151 The Procession Fresco at Knossos puts the viewer on nearly the same plane as the figures, resulting in a more inclusive effect, while the placement of the Thebes procession would have conveyed a more remote, exclusive character (Peterson 1981a, 46). Minoan scenes have a greater tendency to draw the viewer in, whereas Mycenaean paintings generally give the impression of making announcements.

152 Both are dated to LH IIIB, in the 13th century BC. Tiryns Procession: Rodenwaldt 1912, 66–90; Pylos Procession of Women: Lang 1969, 52–60, 86–91 (51 H nws–53 H nws), 217–219, pls 34–40, E, O. Both frescoes: Peterson 1981a, 69–86 (nos 104–138, 143–145); Wilson 2009, 163–232.

153 Rodenwaldt 1912, 87, pl. X; Boulotis 1979, 59–67.

154 Wilson 2009, 170–175 puzzles over the minor role and undistinguished appearance of this bit of cloth and speculates whether the association here with flowers may relate to Linear B tablets at Pylos linking textiles with perfumed oil, and perhaps with the offering of anointed cloth (see here n. 59). The flowers held up to the nose in this scene might be a means of symbolizing their use in perfumes.

155 Peterson 1981a, 55–56, 187–188, 248, n. 81 suggests that these items may represent textiles. Wilson (2009, 145–155) reconstructs the former (Fig. 3.47) as a large swath of patterned cloth held by one of the processional women, but proposes that some of the other fragments formed a painted or inlaid block serving as an altar or seat.

156 Lang 1969, 59–60, 85 (no. 50Hnws), pls 31, D, N; Peterson 1981a, 81–83, 223 (no. 147). The figure, about half life-size, is female due to the white feet. Lang terms the figure as a priestess and proposes that she faced a life-size, seated goddess whose head is preserved (49Hnws).

157 Lang 1969, 38–40, 64–68 (5H5–15H5), pls 3–11, 119–120; Peterson 1981a, 84–86, 224–227 (nos. 148–158). Wilson (2009, 185) suggests that the varying scales and costumes denote people of differing ranks and occupations, with the hefty men in long robes giving the impression of well-fed officials.
158 Scatter pattern: *PT*, 325. One fragment (Lang 1969, pl. 7, no. 10 H 5) shows horizontal dashes on either side of a vertical band that is wider than those on the other tunics. It is reminiscent of the skirt design of the Bending Lady from Thera (Fig. 3.32), but no details of the flesh are preserved to identify the figure's gender.
159 See examples in *PT*, 327–328; Lenuzza 2012, with bibliography in nn. 17, 18. The most well-known renditions are the figure of "La Parisienne" (Fig. 3.50) from the Camp Stool Fresco at Knossos (Immerwahr 1990, 95, pl. 44; *PM* IV, 2, 379–396), an ivory knot from the Southeast House at Knossos (Fig. 3.51 left), and faience sculptures from the Shaft Graves at Mycenae (all in *PT*, fig. 15.14). Crowley (2012, 231–232) discusses the sacred knot and associated clothing in glyptic.
160 Boulotis (1987, 150) interprets the item in front of the goddess/priestess in the Knossos Procession Fresco (discussed above) as a length of fringed fabric that will be arranged on her as a sacred knot. The appearance of the flounced skirt as a carried garment with religious significance has been noted above in the Room of the Ladies fresco and on related images on rings and seals (Fig. 3.34).
161 See n. 153 on the Tiryns fragment. Lenuzza (2012, 258–260) associates the Phylakopi fresco with one of the stages in a ceremony that involves the presentation of a blue "sacred stole," which is then draped or tied on the priestess as part of her transformation into goddess.
162 Tzachili 1990, 380. One is reminded of the myth of unfortunate Arachne, who considered her skill at the loom to be comparable to that of a goddess.

4

Palace and Household Textiles in Aegean Bronze Age Art

Maria C. Shaw and Anne P. Chapin

Populations around the world, and throughout the centuries, have utilized textiles in "household" settings; whether that household was a mud brick house, villa, or a palace, textiles were part of the inhabitants' surroundings. Textiles serve as window curtains and wall hangings, as rugs and ceiling coverings; they are used to "close" doorways and divide interior spaces; they are spread over beds as sheets, blankets, and pillows; they are laid over tables; and folded or knotted into carriers, sacks, and bags. Made into tents, textiles can even serve as temporary or portable dwellings. In short, household textiles serve a host of domestic functions. Unfortunately, they are also made of organic materials that easily degrade, and only rarely do bits and pieces survive from prehistoric Aegean archaeological contexts. And whereas images of textiles in the form of clothing are relatively common in Aegean art (see Chapter 3), household textiles are more difficult to recognize in the pictorial record, and hunting for them could be viewed as a losing proposition. Yet the archaeological evidence demonstrates that the textile arts developed slowly over millennia, from the first Paleolithic production of string and cord to Neolithic innovations in spinning, weaving, and sewing.[1] Many of the tools, technologies, and techniques developed millennia ago for producing textiles are still used today, often by people of diverse cultural and geographical backgrounds (see Chapter 2). As such, the textile arts constitute a living tradition, conservative by nature, which suggests that when an Aegean fresco looks rather like a household textile recognizable from later eras, it could, in fact, have been inspired by a type that was actually made in the Bronze Age.[2]

This chapter, then, surveys surviving artistic evidence in an effort to identify what these lost fabrics looked like, how they were used, and what their significance may have been for the people who created and used them. The most useful artistic renditions are the frescoes and painted stucco reliefs, which were produced at relatively large scales and thus provide detailed information on the textiles they imitate. If we categorize the images in terms of the ways cloth was displayed while in use, there would be three main categories: (1) cloth hung vertically over wall surfaces (as wall coverings); (2) cloth used as ceiling coverings; and (3) textiles used as floor coverings or carpets. These are the three categories that will be considered in this chapter, but cloth can also be attached to moveable screens and used as mobile space dividers, which are known as *paravents* or screens. Similarly, cloth can be suspended like a curtain, sometimes from the ceiling. In each of these proposed household functions, the cloth material is mobile and its use is flexible. Additionally, mats and similar items made of fibrous materials such as straw, wicker, and other fibers, could have functioned as floor or ceiling coverings. The question of whether the ancient representation refers to cloth or matting can perhaps be determined by the character of the design – unless it is a case of artistic hybridization, which would prevent one from making an exact identification.

The catalogue below discusses wall paintings and painted plaster reliefs identified as possible imitations of household textiles in each of the three categories (wall, ceiling, and floor). Shield frescoes, which may reproduce some elements made of cloth, are also reviewed. The material is presented chronologically and geographically, starting with frescoes from sites in Crete, followed by those from contemporaneous Aegean sites, and then by artworks discovered in Mycenaean contexts that are dated to the later phases of the Bronze Age. The chronological sequence used in this presentation has the added benefit of helping us to note any potential developments, or imitations, in the artistic representations.

Wall hangings

These frescoes appear to imitate wall hangings made of textiles (either wholly or in part) and "hung" (that is, painted) on walls. They appear in the Neopalatial period on Crete as decoration in villas and palaces, and on the Cycladic islands at Akrotiri on Thera as house embellishment. Later, in the Mycenaean era, frescoes depicting wall hangings are identified on the Greek mainland at the palatial centers of Mycenae and Pylos. Even though actual wall hangings do not survive from prehistoric Greece, the existence of these frescoes suggests that elaborately patterned cloth was widely used as sophisticated interior decor in a variety of elite structures throughout the Late Bronze Age. These frescoes appear in both domestic and ritual settings, which also implies a wide range of use for the textiles they imitate. Finally, given the time and expense required to produce large and elaborate cloth wall hangings, it is likely that the paintings sometimes functioned as cheaper imitations of the real thing.

Paintings of textile hangings also appear as wall and ceiling decoration in ancient Egypt (discussed in Chapter 8). They are generally believed to imitate Aegean cloth that was imported by the Egyptian elite and used to decorate real structures.

LM I Wall Hanging Fresco, from the Royal Villa at Ayia Triada, Crete

This painting is reconstructed in Figure 4.1 on the basis of plaster fragments that were found scattered in Court 11 and Room 13 of the so called Royal Villa at Ayia Triada,[3] not far from Room 14, which is known for its famous fresco showing richly clad women in a lush landscape (Fig. 3.8).[4] The fresco's textile pattern is marked by a blue background against which spreads a network of lozenges outlined by white lines. Decorative zigzag white lines are set vertically within each lozenge in alternate rows, while the next row is filled with red star-like motifs set within white lozenges. A thin, undulating white band separates this upper patterned zone from one that is painted solid red. As restored, intriguing details are placed along the upper border: a thin black band that serves as the upper frame, and an "arcade pattern," which is similar to the images of the textile hangers used in the Mycenaean era to suspend figure-eight shields from walls (see, for example, Fig. 4.20, below). The painted details in this fresco probably represent hangers and suggest that it depicts a wall hanging of patterned cloth. According to this reading, its undulating hem is bordered by a white band that separates it from a solid red area, which may represent the overall color of the wall against which the cloth was suspended. We can speculate that the fresco imitates a wall hanging used as house decoration or a window curtain. Still another option (though less attractive, given the fresco's placement on a wall), is that the painting imitates cloth suspended vertically from a ceiling, hanging nearly down to the floor, to serve as a divider of a large architectural space.

LM I Pattern Fresco with Waz-Lilies, from the villa at Epano Zakros, Crete

This fresco is restored from fragments discovered in a rural "villa" situated in the environs of the Minoan palace at Kato Zakros (Fig. 4.2).[5] The structure was excavated in the 1960s by Nicholas Platon, but remains incompletely published.[6] The fresco fragments preserve a lively net (or diaper) pattern of diagonally crossing blue and black barred lines, with the intersections marked by blue rosettes

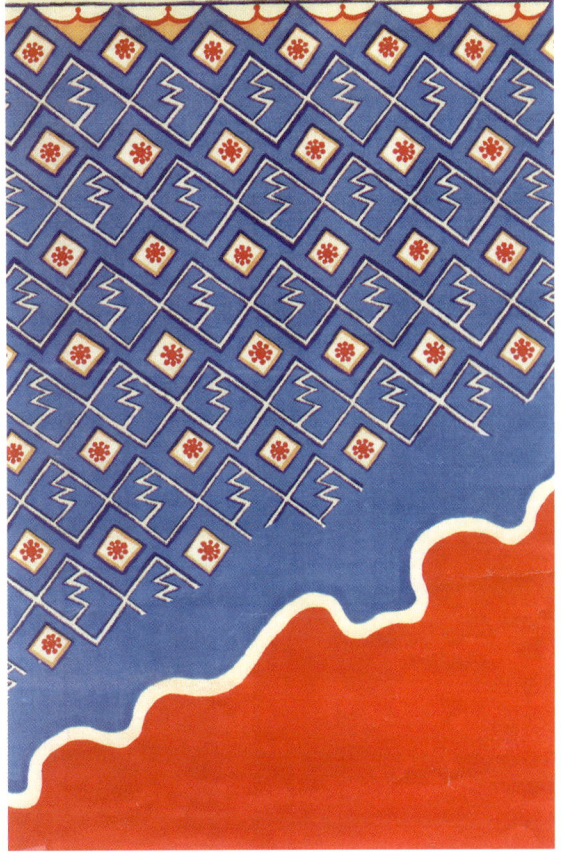

Fig. 4.1. Fresco of a Wall Hanging, from the Royal Villa, Ayia Triada, Crete (restoration M. C. Shaw and G. Bianco)

drawn in black, with red centers, and a single red *waz*-lily painted inside each lozenge. The *waz*-lily is an Egyptianizing motif formed by the stylized combination of lily and papyrus flowers. A chain of red lily flowers frames the network. Thin bands of blue, each outlined in black and embellished with a thin wavy black line, supply borders to the various parts of the composition.

Pending formal publication, interpretation of the composition remains tentative.[7] If the fresco is understood as representational rather than merely decorative, then the slanting lines could suggest a wooden trellis formed of diagonally arranged wooden slats painted blue and black, and fixed by clasps in the shape of rosettes at their intersections to form an attractive lattice pattern. However, the net pattern finds numerous parallels in depictions of patterned textiles used for clothing (e.g., the Lady in Red from Knossos, Fig. 3.2), and it is possible that the design preserves a bit of a decorated garment similar to those painted in the Procession Fresco at Knossos.[8] The blue and black barred motif is paralleled by the edging of the relief rosette fresco from Xeste 3 (see below, Fig. 4.6), and the lily chain suggests the influence of jewelry design.[9] It is possible that this composition was inspired, at least in part, by patterned textiles, perhaps in the form of woven wall hangings or quilts.

Fig. 4.2. Fresco from the Minoan villa at Epano Zakros, Crete (restoration courtesy L. Platon)

Since the fragments were found in a basement, it is not known which room the fresco once decorated in the villa. Evi Saliaka suggests it originally decorated a floor or a ceiling,[10] but it seems doubtful that the decoration belonged to a floor, given the delicate nature of the painted design and the exposure to wear that one would expect in such a context. According to Lefteris Platon, whose father, Nicholas Platon, discovered the fresco in his excavations of the building, the piece is more likely to have decorated a wall on the upper floor, given the small scale of the iconographic motifs, which require closer scrutiny – perhaps a window or a doorframe.[11] Future inspection of the reverse (back) side of the plaster fragments would reveal the surface to which it was applied.

Fresco of Interlocking Circles, from Building 1, Galatas, Crete

This new restoration of a fresco from Building 1, a large and well-built structure situated about 100 m southeast of the Minoan palace at Galatas on Crete, is made from fragments discovered in 1992 and published by Giorgos Rethemiotakis as a red lozenge net design.[12] This restoration, which remains tentative pending publication of the building and its contents, explores the idea that the fragments belong to a design of interlocking circles with the intersections marked by red dots (Fig. 4.3). The resulting geometric pattern is an optical illusion, however, and can also be viewed as repeating four-petaled flowers and as a rapport design of curving lozenges (four-pointed stars) rendered in double outline. The multiple readings of this fresco's design reflect the playfulness of much Minoan decorative art.[13]

The Galatas fresco's interlocking circle/four-petaled flower rapport pattern, as a painted motif, may have its origins in "Woven Style" Protopalatial (MM IB) polychrome pottery decoration,[14] but a closer stylistic parallel can be found in the depiction of patterned cloth worn by the young woman of the Priestess Fresco from the West House at Akrotiri, dating to LM IA (Fig. 7.3).[15] The curving lozenges also recall similar motifs in the flounced skirt of the Necklace Swinger of Xeste 3 (Fig. 3.23) and the four-pointed stars of the red-dotted net pattern from the House of the Ladies (discussed below as a fresco depicting a wall hanging; Fig. 3.32), both from Akrotiri. Even more closely related may be the textile pattern of the kilted bearer (No. 22) in the Procession Fresco from Knossos (Figs 3.18, 7.10, 9.2) in which a lozenge design similarly painted in outline is framed by interlocking rapport patterns of quatrefoils and crosses. These artistic parallels emphasize the textile associations of both the lozenge pattern and the interlocking circle/four-petaled flower motifs, and raise the possibility that the fresco from Galatas depicts a textile – possibly a wall hanging, based on the rather large scale of the textile design and its lack of figural elements.

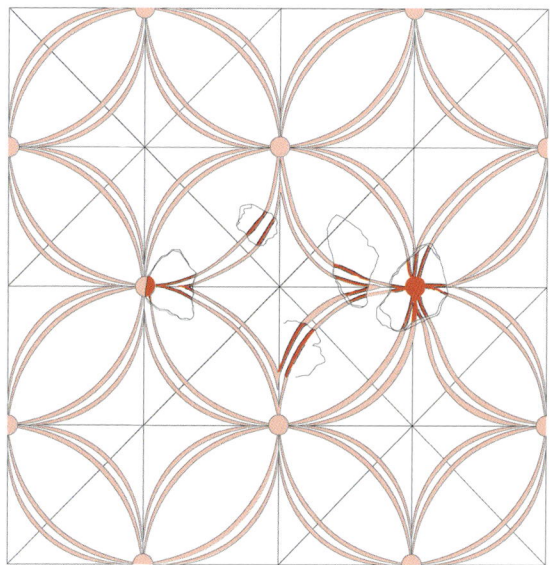

Fig. 4.3. Fresco of Interlocking Circles, from Building 1, Galatas, Crete (restoration G. Bianco and A. Chapin)

Rethemiotakis initially dated the fresco fragments to MM IIIB/LM IA,[16] and indeed, renewed excavation of the building in 2008 confirms that Building 1 was destroyed by earthquake in LM IA.[17] But before this, the structure enjoyed a long architectural life. Additional fragments of the fresco, together with fragments of another composition depicting twigs and leaves, were found in a small corridor that had been packed with hundreds of MM IIIA vessels and more than 20 circular, portable hearths. Rethemiotakis suggests that Building 1 suffered a major destruction at the end of MM IIIA (as did the early Galatas palace), and, in the course of rebuilding, this small corridor had been filled, its doorway blocked, and the space sealed, never to be used again. This closed archaeological context indicates that this fresco and the pictorial fragment discovered in the Galatas palace in another secure MM IIIA context, are among the earliest examples of pictorial fresco painting yet discovered in the Aegean. Further, the ceremonial functions of the pottery associated with the Fresco of Interlocking Circles suggest that Building 1 played host to important ritual events before its MM IIIA destruction. The building's LM IA ceramics, which reportedly comprise cups, jugs, and amphorae used for drinking and storage of liquids, together with other finds, indicate a continued ritual use in the Neopalatial era.[18]

MM III–LM IA Swastika and Rosette Fresco, from Poros-Katsambas, Crete

This tentative restoration of a swastika and rosette is based on a fresco fragment discovered in a Neopalatial house in Poros-Katsamba, the Minoan harbor town of Knossos.[19] The fragment, which is on display in the Herakleion Museum, preserves a repeating pattern of black swastika designs painted on blue squares, alternating with red rosettes on white squares (Fig. 4.4). The swastika

Fig. 4.4. Swastika and Rosette Fresco, from Poros-Katsambas, Crete (restoration G. Bianco and A. Chapin)

designs point counterclockwise, and alternating arms extend to the black borders of the square. The rosettes are made from central red circles surrounded by a ring of red dots just touching the central circle's perimeter. An indeterminate area of black can be seen on one edge of the fragment; whether or not it constitutes part of an undulating border of black cannot be determined here.

The Fresco of a Wall Hanging from the Royal Villa at Ayia Triada (Fig. 4.1) similarly features a large-scale repeating design with red rosettes set in white squares and zigzag lines on blue (though at Ayia Triada, these do not seem to be swastikas). The Rosette Fresco from Room 9 of Xeste 3, Akrotiri, features large-scale petal rosettes set in groups of four in white lozenges; this fresco too probably depicts a wall hanging (discussed below, Fig. 4.6). Rosettes are found as textile patterns in the costume of Lady A at Pseira (Figs 3.5, 7.4), and similar red circles are placed in a net pattern that decorates the flounced skirt of the youngest Saffron Gatherer of Xeste 3 (Fig. 3.26).[20] Although swastikas are not common as spotlighted motifs in Aegean representations of textiles, similar designs can be detected if one looks at the *frames* of four-pointed cross interlock designs, such as those that are found on the kilt of one of the offering-bearers in the Procession Fresco from Knossos (Figs 3.18, 9.3), and on a fresco fragment of a woman's skirt from Mycenae (Fig. 3.40).[21] The large scale of the Swastika and Rosette Fresco, however, makes it unlikely that it represents a piece of human clothing; rather, a wall hanging seems likely. The placement of each swastika and rosette motif within its own individual square panel, moreover, recalls historical and contemporary quilting practices, and raises the possibility that the Minoans also sewed quilts from smaller pieces of cloth and added the central motifs as embroidery or appliqué designs.

MM III–LM IA *Quatrefoil Fresco, from Poros-Katsambas, Crete*

The Quatrefoil Fresco, from a Neopalatial house in Poros-Katsambas, is tentatively restored from a fresco fragment published by Nota Dimopoulou-Rethemiotaki and placed on view in the Herakleion Museum (Fig. 4.5).[22] The fragment depicts approximately half of a large blue quatrefoil thickly outlined in black on a white ground. The interior of one arm of the quatrefoil is painted with thin, black, parallel lines; its exterior is embellished with a line of carefully painted, small red dots reminiscent of the bead-like, red-dotted net pattern in the wall hanging painted above the women in the House of the Ladies at Akrotiri (discussed below; Fig. 3.32). The other quatrefoil arm preserves no similar decoration.

Similar to many Minoan frescoes depicting intricate geometric designs (and as discussed in Chapter 7), the Quatrefoil Fresco was planned and executed with the assistance of thin guidelines lightly incised into the plaster while still wet, before the process of painting had begun. A thin, light incised line is just visible along the line of the red dots (and it runs through the quatrefoil motif); a second incised line marks the quatrefoil's central axis. Presumably additional lines were incised into the wet plaster beyond the surviving fragment to extend the geometric pattern in all directions, but these are not preserved.

Quatrefoils are found in the MM II Protopalatial floor fresco from the Phaistos palace (discussed below as a possible representation of a rug or a mat; Fig. 4.14) and on the LH IIIB Mycenaean megaron floor fresco at Pylos, where it appears as a textile motif interspaced with other textile and rockery designs (see Chapter 5, especially Fig. 5.5). In Neopalatial wall painting, images of quatrefoils are identifiable as complex textile patterns on garments worn by human figures. The "Goddess" of Ayia Triada wears a flounced skirt featuring a vibrant red, blue, black, and white pattern of interlocking quatrefoils (Figs 3.9, 3.10, 7.6). The kilt of the famous Cupbearer of Knossos (Figs 3.19, 7.11) features an interlocking pattern of orange quatrefoils ornamented with dot rosettes and outlined in black, and one of the offering bearers (No. 22) in the Procession Fresco wears a kilt similarly decorated with interlocking quatrefoils, though these are blue with black outlines and set within interlocking crosses (Figs 3.18, 7.10). The four-pointed flowers of the *trompe l'oeil* textile design on the Priestess Fresco from the West House at Akrotiri (Fig. 7.3)

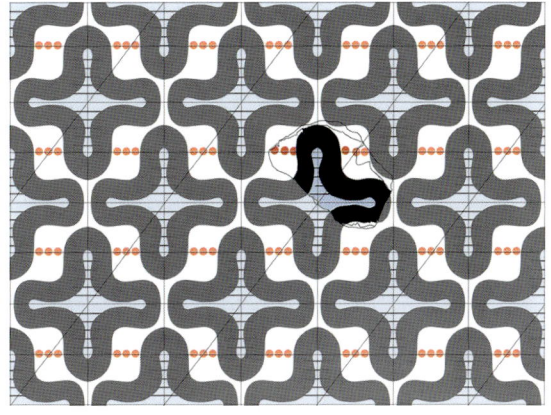

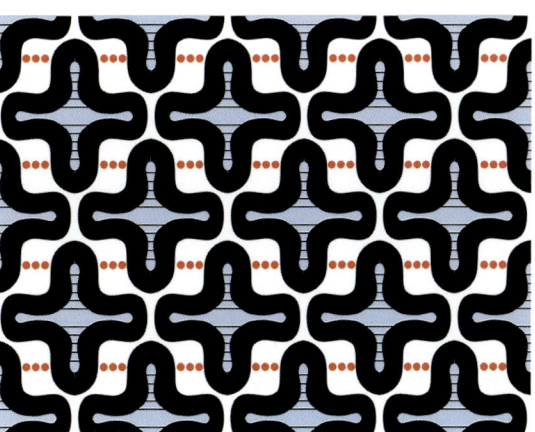

Fig. 4.5. Quatrefoil Fresco, from Poros-Katsambas, Crete (restoration G. Bianco and A. Chapin)

are broadly similar in shape to the Minoan quatrefoils, and also work as a rapport (repeating) design.

While the Poros fragment preserves no visible trace of additional quatrefoil motifs, the use of incised guidelines strongly suggests that the design was repeated at regularly spaced intervals. The chains of small red circles also indicates greater spacing between the quatrefoils than is found in the comparanda. Figure 4.5 accordingly restores a gridded design in which the quatrefoils alternate with lines of red circles to create a lively and bold geometric pattern. The strong textile associations of the quatrefoil motif, together with its large scale, suggest that this fresco could depict a wall hanging. Alternately, it could belong to a life-size human figure.

LC I Fresco of Rosettes, from Room 9 in Xeste 3, Akrotiri, Thera

This ornamental painting spread across two walls of a second-story room in Xeste 3 (Fig. 4.6).[23] In addition to this decorative painting, the second story was painted with large, psychedelic spiral designs extending across much of the floor space. In the lower two floors, Xeste 3 was endowed with a plethora of figural paintings depicting youths, mostly female but some male, apparently engaged in the performance of rituals related to the transition from childhood to adulthood (discussed in Chapter 3). Such rites of passage were commonly celebrated in early societies, as, to a degree, they have been since. Rather expectedly, cloth appears to have played an important role in such occasions, both as part of the dress of the participants as well as the theatrical setting within which the activities took place.

The painting of Room 9 is marked by a tricurved ribbon network made of three horizontal rows of interlinked lozenges. Their contours are defined by double undulating white bands modeled in low relief and painted as if passing through yellow rings rendered, interestingly, in foreshortened views. Quoting Spyridon Marinatos, Christos Doumas thought the relief bands were inspired by ivory carving,[24] but it seems more likely that the ancient painter tried to illustrate some kind of textile, perhaps a quilt, and that the raised bands likely depict padded cylinders of cloth that were twisted and curved to create the undulating pattern.[25] Enclosed within each lozenge is a group of four rosettes. The petals of most are rendered in blue – potentially representing silver medallions – but some are painted yellow, possibly representing ornaments made of gold or gold foil. Noteworthy too is the festoon-barred border that appears at the top edge of the composition. This decorative pattern looks forward to the barred bands seen in later Aegean representations of costume, where narrow, decorative barred bands were used to reinforce the seams and hems of elaborate costumes (e.g., Fig. 3.45). Here, the barred band seems to imitate the reinforced edges of a wall hanging. There does not seem to be a similar indication of a hem at the bottom of the fresco, though this may well be a matter of preservation. A wall hanging in this case makes the most plausible interpretation for the representation.

LC I Red-Dotted Net Pattern, from Room 1 of the House of the Ladies, Akrotiri, Thera

This painting extended across portions of the south and north(?) walls of Room 1 in the House of the Ladies. The composition is best known for its figural decoration showing three female figures engaged in a costuming ritual (discussed in Chapter 3, this volume), but above the figures is an eye-catching red-dotted net pattern with four-pointed stars (lozenges) painted blue and black at

4. *Palace and Household Textiles in Aegean Bronze Age Art* 113

Fig. 4.6. Fresco of Rosettes, from Room 9 in Xeste 3, Akrotiri, Thera (courtesy the Thera Foundation – Petros M. Nomikos)

the interstices (Figs 3.31–3.32).[26] This network design was laid out with the assistance of lightly incised diagonal lines marking the path of the red dots, just visible in published photographs.[27] It is demarcated by colored bands: just below ceiling level, horizontal border bands are painted red, yellow, black, and white; below, undulating black and blue bands follow the fresco's figural contours. This arrangement of compositional motifs suggests that the women are meant to be seen by the viewer as moving under, or perhaps in front of, the red-dotted net pattern, and that the pattern represents a decorative wall hanging.

Accordingly, the net pattern has been identified either as a painting of a wall hanging (either woven or embroidered) or as a painting of a beaded net with metallic stars.[28] Without the physical evidence of surviving wall hangings or beaded nets, it is difficult to determine which scenario might more closely reflect Theran interior decor, but Elizabeth Barber, an expert weaver herself, questions how such a pattern could have been woven.[29] If the fresco represents an embroidered hanging, then perhaps it was attached at its upper and lower ends to cloth bands and gathered up along its lower hem, as implied by its undulating bottom edge. If, however, it represents a net of threaded red beads, then their red color could suggest carnelian or some other kind of red-colored semiprecious stone, and the blue color of the stars could refer to metallic (silver?) clasps at the intersections of the threads, to help keep them in place. Philip Betancourt sees the knotted net as a sacred symbol which appears in Neopalatial art, both as a costume pattern (as on the seated lady from Pseira, Figs 3.5–3.6) and as an object draped over (sacrificial?) bull figurines.[30] Fritz Blakolmer draws connections between the netted stars of the House of the Ladies and symbolic depictions of a celestial realm in the context of ritual performance. It seems likely that the wall hanging – painted or actual – transformed the space into a special setting for ceremony.[31]

LH IIIB Quadruple Ivy Leaf Fresco, from Corridor M, Area B, of the Southwest Building, Mycenae

The fresco restored in Figure 4.7 is developed from fragments published by Ioanna Kritseli-Providi and depicts a repeating motif of four-stemmed ivy leaves, each radiating from small concentric white circles outlined in black.[32] The leaves are heart-shaped and painted red or yellow; they have fan-shaped umbel elements which are connected to stems and colored yellow and white. All elements of the pattern are outlined in black against a blue ground. At regular intervals between the radial ivy groups are additional concentric white circles. An undulating white band outlined in black provides a lower border to the ivy pattern; only blue ground is preserved on the far side of the band. The reconstruction of Figure 4.7, which is published here for the first time, provides a suggestion as to how the composition might once have appeared; the fresco's repeating pattern and undulating lower border suggest that it likely depicted a hanging patterned textile.

Ivy leaves offer an interesting iconographic choice for either a painting or a wall hanging. "Sacral ivy" was identified by Sir Arthur Evans as a cultic motif in Minoan Crete, where it appears in the decoration of a wide variety of objects, including painted pottery, jewelry, and even a carved stone column from Knossos.[33] In fresco, the motif is best known as a "living plant" in the Monkey and Blue Bird Fresco, found in the House of the Frescoes at Knossos.[34] It appears as a stylized motif in the frieze running above the Antelopes and the Boxing Boys in Room 1 of Building Beta at Akrotiri.[35] Ray Porter identifies the plant of inspiration (in nature) as a climbing, bush-like evergreen vine, *Hedera helix* L.[36]

The Mycenaean fresco fragments were found in Corridor M, in Area B of the Southwest Building of the Cult Center at Mycenae. This location is near the find spot of the "Mykenaia," the famous fresco of a woman holding necklaces now restored not as a seated figure, but as a standing figure walking to the left.[37] Both the restored Quadruple Ivy Leaf Fresco and the Mykenaia share a similar blue background, which opens the door to the possibility that they once belonged to the same composition. This possibility is represented by the restoration in Figure 4.8, in which the Mykenaia is shown with an ivy-patterned wall hanging above her. Though the fresco design thus created is consistent with other frescoes depicting women under undulating (textile?) designs, such as that found in Room 1 of the House of the Ladies at Akrotiri (Fig. 3.32), it remains a tentative association only.

LH III Rockery Fresco, from Area Γ of the Cult Center at Mycenae

More than 100 fragments of this painting were found in Area Γ, some distance away from the main part of the Cult Center, in a fill that was dumped during the LH IIIC period. The rendering is clearly inferior, judging at least from the part preserved, where we see a conglomeration of crudely rendered multicolored rocks painted red, yellow, and blue, with wavy markings acting as the veins.[38] A selection of the fragments were photographed, drawn, and restored as a Rockery Fresco, presented here for the first time (Fig. 4.9).[39] This mass is contoured at the bottom by an undulating white band, which

Fig. 4.7. Quadruple Ivy Leaf Fresco, from the Southwest Building, Mycenae (restoration M. C. Shaw and G. Bianco)

recalls the curving border bands of the other representations of household textiles reviewed in this chapter. In this instance, the regularity of the white border suggests that the rock pattern refers not to nature, but to a design on cloth below which a solid blue area begins. Within the latter is where one might expect a figural scene (now lost), perhaps of women painted on a relatively large scale. Judging from the fresco's stylized character, it may depict a wall hanging in the form of a padded or stuffed patchwork quilt with the rocks rendered in relief.

LH III fresco with a male figure and a possible wall hanging, from the Northwest Slope Plaster Dump, Pylos

This composition, catalogued by Mabel Lang as 57 H nws, is represented by seven joining fragments that preserve part of a man's red-skinned face in left profile and a painted background divided into distinct decorative zones: a lower area of blue, an intermediary wavy white band outlined in black, and above, paneled decoration imitating textile patterns and perhaps variegated stone (Fig. 4.10).[40] Decoration inspired by textiles is preserved by bits of a net pattern painted in black on tan and framed by a wavy tan line outlined in black. The textile origin of this motif is clear. It appears (though rather infrequently) on textiles used for clothing in Mycenaean art, as seen on the skirt of a female processional figure from Pylos (Fig. 3.48).[41] The pattern is often found in the Neopalatial period when parallels for the net pattern are numerous and include the textiles worn by the Lady in Red from Knossos (Fig. 3.2) and the Necklace Swinger from Xeste 3 at Akrotiri

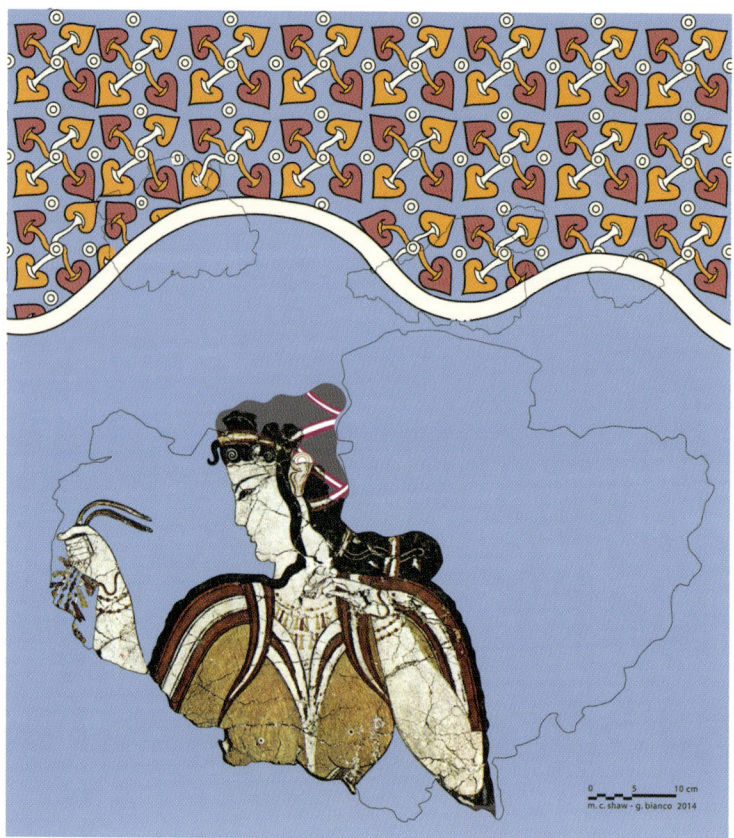

Fig. 4.8. Tentative restoration of the Quadruple Ivy Leaf Fresco with the Mykenaia, from the Southwest Building, Mycenae (restoration M. C. Shaw and G. Bianco)

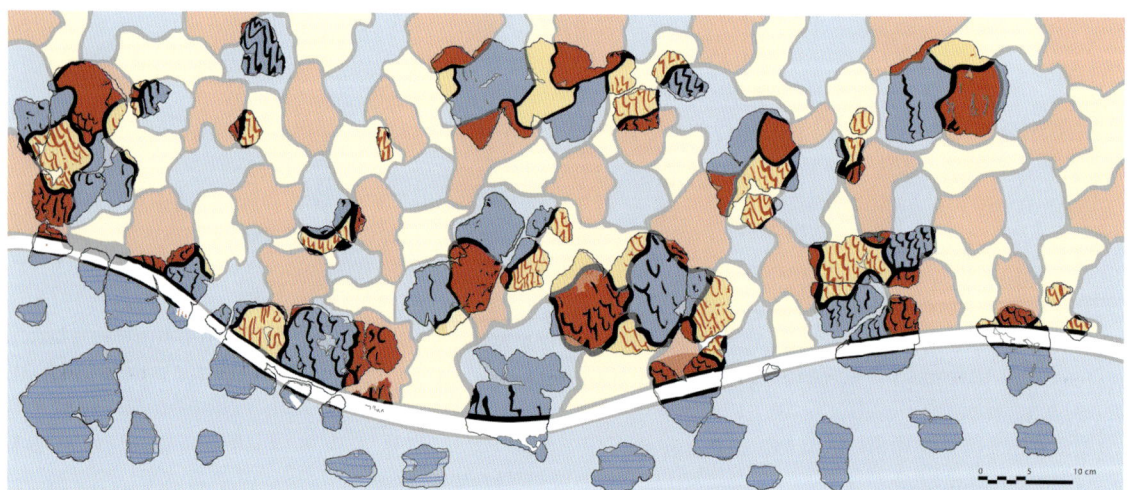

Fig. 4.9. Rockery Fresco, from Area Γ, Cult Center, Mycenae (restoration G. Bianco, A. Chapin and M. C. Shaw)

(Fig. 3.23). It seems possible that in the Mycenaean period, the net pattern's associations shifted somewhat away from clothing toward architectural embellishment.

In 57 H nws, the net pattern appears beside a panel featuring horizontal wavy (ripple) lines painted in red and white against a blue background. This pattern reminded Lang of a variegated stone dado design, as seen on Pylos 14 D nws (Fig. 5.9) and she identified it as imitation stonework.[42] But the ripple design, too, may have been influenced by textiles, considering the many flounced garments of the Mycenaean era painted with wavy linear patterns, such as those worn by the processing ladies of Pylos 51 H nws (Fig. 3.46) and the "Staff Bearer" in the Room with the Fresco of Mycenae's Cult Center (Fig. 3.43). Both patterns of this fragment, then, have textile associations, and it seems that the fresco of 57 H nws could depict a male figure moving below (or behind?) a cloth wall hanging – one that could have been made by sewing together cloth panels of different patterns, rather like a patchwork quilt. According to this reading, the white border could describe the lower edge of the wall hanging, which undulates just like the other possible wall hangings discussed above.

Ceiling coverings

Ceiling frescoes, like the ceilings and roofs of ancient buildings, rarely survive today, and, since textiles used in household contexts are not preserved, information on this subject is limited. Fortunately, imitations of Aegean textiles are preserved on some painted ceilings in Egyptian palaces and elite tombs of Middle and New Kingdom date, which suggests that such textiles were exported from the Aegean to Egypt (discussed in Chapter 8), and two Aegean frescoes may preserve evidence for cloth ceiling decoration.

LM IA Relief Fresco of Quadruple Spirals, from the palace at Knossos

This fresco is best known from watercolors and drawings of the extant plaster fragments and a restoration (Fig. 4.11) made in 1902 by Theodore Fyfe, an artist and architect employed by Sir Arthur Evans during his excavation of the Minoan palace at Knossos.[43] These depictions show white running spirals rendered in relief, with the round central area within each spiral rendered flat and painted with a rosette. Additional rosettes painted red and yellow, and outlined in black against

Fig. 4.10. Fresco with a male figure and a possible wall hanging, from the Northwest Slope Plaster Dump, Pylos (watercolor P. de Jong. Pylos II, pl. N, courtesy the Department of Classics, University of Cincinnati)

a blue ground, adorn the spaces between the relief spirals. Blue quatrefoil medallions, each centrally decorated with a large yellow, red, and black rosette, are spaced at regular intervals over the spiral design. The use of relief here conveys well the impression of a quilt, perhaps enhanced by cording or stuffing as seen in the medieval Italian *trapunto* quilting technique. Such manipulations of fabric are well known from historical traditions of sewing and quilting,[44] but can only be surmised for the Aegean Bronze Age from the fresco evidence. If the fresco actually refers to quilting, then it indicates that such items of household cloth were in use in Minoan times. That quilting could have continued into Mycenaean times is suggested by the next catalogue entry, as well as the previously discussed examples from Mycenae and Pylos (Figs 4.9, 4.10).

Fig. 4.11. *Relief Fresco of Quadruple Spirals, from the palace at Knossos (watercolor T. Fyfe. Ashmolean Museum, University of Oxford. PM III, pl. xv)*

LH IIIB carved decoration on the limestone ceiling of the side chamber of the Treasury of Minyas, Orchomenos

Though not a fresco, this carved ceiling from a Mycenaean tholos tomb called the "Treasury of Minyas" is included in the catalogue due to its being very similar to relief frescoes, both in terms of scale and iconography (Fig. 4.12).[45] The drawing illustrated here (Fig. 4.13) provides a good idea of the decoration: the central panel has a repeating pattern of interlocking spirals and fan-shaped papyrus flowers with projecting "buds" and framing calyx motifs. This is bordered by a double frieze of rosettes. The outer perimeter of the relief decoration resumes the interconnected spiral/papyrus decoration. This ceiling helped to inspire Evans's placement of the Knossos Quadruple Spiral Fresco (Fig. 4.11) to a ceiling.

The repetitive character of the Orchomenos ceiling decoration is evocative of a patterned textile hung on a ceiling. If interpreted as an imitation of a large cloth, then it may have been made from individual cloth panels that were stitched together. The bands decorated by carved rosettes act as frames. Of interest is the fact that the rosettes in this decoration are rendered in a somewhat higher relief, which might suggest that they imitate padded quilting, as was also proposed for the relief frescoes of Xeste 3 (Fig. 4.6) and the palace at Knossos (Fig. 4.11). Alternately, the rosettes could imitate metal attachments in gold or silver foil, sewn onto cloth, rather than woven, as was the case with the gold roundels found by Heinrich Schliemann at Mycenae in Shaft Graves III and V.[46] As discussed in Chapter 8, the interlocking spiral pattern was imitated in painted ceilings of New Kingdom date in Egypt. Such analogies offer evidence for artistic interconnections between the Aegean and Egypt, and quite possibly, for the export of fancy Aegean woven products for elite Egyptian consumers.

Fig. 4.12. Carved decoration on the limestone ceiling of the side chamber of the Treasury of Minyas at Orchomenos (photo A. Chapin)

Fig. 4.13. Relief decoration of the side chamber ceiling in the Treasury of Minyas at Orchomenos (drawing M. C. Shaw and G. Bianco)

Floor coverings and carpets

Aegean buildings were sometimes painted with floor frescoes characterized by patterned or pictorial decoration.[47] Floor frescoes of the pictorial type, such as the Marine Floor Fresco from the LM IIIA shrine at Ayia Triada,[48] remain relatively scarce, but floor frescoes with decoration imitating costly materials, such as stone slabbing or woven materials, are somewhat better known.[49] This chapter focuses on the latter and includes possible imitations of woven, braided, or knotted carpets made of textile materials, and mats plaited from straw, grasses, reeds, or other materials.

In order to be considered a floor fresco possibly imitating carpets or mats, two criteria must be met: (1) the painting must be identifiable as floor decoration (and not wall or ceiling decoration); and (2) the design must imitate patterns associated with weaving or plaiting.

Frescoes of this class are identifiable in Minoan (Protopalatial and Neopalatial) and Mycenaean (LH III) contexts. The more complex geometric patterns – quatrefoils, meanders, and labyrinth designs – of the Minoan floors do not seem to imitate the natural veining of stone slabs, so it is likely that these floors imitate woven materials, such as rugs or floor mats. The Mycenaean painted floors, however, are more open to interpretation, and are considered separately in Chapter 5.

MM II painted plaster floor with quatrefoil design, from Room LIV, palace at Phaistos

This plaster floor is painted with brown rectangles around brown quatrefoils, in the *incavo* technique (Fig. 4.14).[50] In *incavo* painting, the surface of the still-damp plaster was cut away to form grooves, into which another color was inlaid. Sara A. Immerwahr observes that the technique may reflect Egyptian influence, and was often used for floors, presumably to preserve the pattern and maintain its fresh color despite wear and tear from people walking over it.[51] The pattern of quatrefoils and rectangles evokes rug decoration or a mat pattern.[52] The space it once decorated had a built-in, bed-like platform coated with unpainted plaster, which suggests that the room could have functioned as a bedroom.

MM III painted plaster floor fragments with meander design, from Chalara, at Phaistos

Like the quatrefoil floor fragments, these are also discussed by Pietro Militello, who confirms their identification as a floor fresco dated to the MM III period (Fig. 4.15).[53] These fragments bear a rectilinear meander-like design, painted brownish red in the *incavo* technique, which also evokes a rug design. The meander design can be compared to the Labyrinth Fresco from Knossos, discussed below, and the Bull and Maze Fresco from Tell el-Dab'a, in northern Egypt (Fig. 7.14).[54] The suggestion is that this floor fresco imitates real rugs that were once woven and used in Bronze Age Crete.

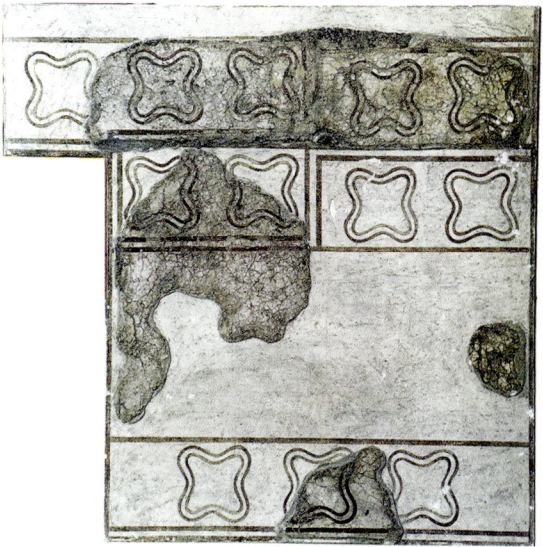

Fig. 4.14. Plaster floor fragments painted with quatrefoils, found in Room LIV, palace at Phaistos (Militello 2001, col. pl. A, top)

The MM III Labyrinth Fresco, from Corridor 94, palace at Knossos

The fresco fragments, painted in the *incavo* technique, preserve bits of a labyrinth pattern painted in dark red (Fig. 4.16). They were found in a corridor running north–south (Corridor 94) a short distance to the east of the Hall of the Double Axes, near other fresco fragments painted with veining to imitate a dado of gypsum or alabaster slabs. Though Evans imagined a corridor with walls painted with the labyrinth design above the imitation stone dado,[55] the use of the *incavo* technique means

Fig. 4.15. Plaster floor fragments painted with a meander design, from an unspecified context at Phaistos (Militello 2001, col. pl. A, bottom)

that the Labyrinth "Fresco" must now be recognized as belonging to a painted plaster floor (Fig. 4.17).[56] The appearance of a similar labyrinthine pattern on the ground of a bull-leaping scene in the Aegeanizing Taureador Fresco from Tell el-Dab'a (Figs 7.14, 7.15), offers an interesting iconographic parallel.[57] In this composition, the motif gives the appearance of patterned floor decoration, one rather evocative of a patterned rug or mat.

MM IIIA Zebra Fresco, from the north side of the Royal Road, Knossos

This fresco, which remains formally unpublished, is nevertheless known from excavation reports, photographs, and a restoration by Mark Cameron (Fig. 4.18).[58] Fragments corresponding to its distinctive design have been identified from various contexts, but reportedly, 18 of these were identified in the Royal Road Excavations, sealed below an MM IIIB floor. The fragments preserve undulating black and white bands framing irregular spaces that resemble lopsided quatrefoils, each painted yellow with spots of red ringed by black outline. The black and white bands bring to mind zebra stripes, hence the name of the fresco, while the spots of yellow suggest the hide of a leopard or other spotted cat. The fact that the *incavo* technique was used for the yellow spaces suggested to Cameron that the composition should be identified as a floor fresco, and that it was "directly inspired by Minoan 'mosaiko' or crazy-pavings in stone even though the representational significance of this composition lies rather with rugs or carpets made from animal pelts."[59] The composition is included in this survey of evidence for household floor coverings because of the suggestion that it imitates animal skins – presumably exotic imports from Africa of zebras and either leopards or cheetahs – that could have been sewn together or imitated through textile design.

4. Palace and Household Textiles in Aegean Bronze Age Art 123

Fig. 4.16. Fragments of the Labyrinth Fresco, from the palace at Knossos (M. A. S. Cameron, courtesy the University of Western Ontario, Canada)

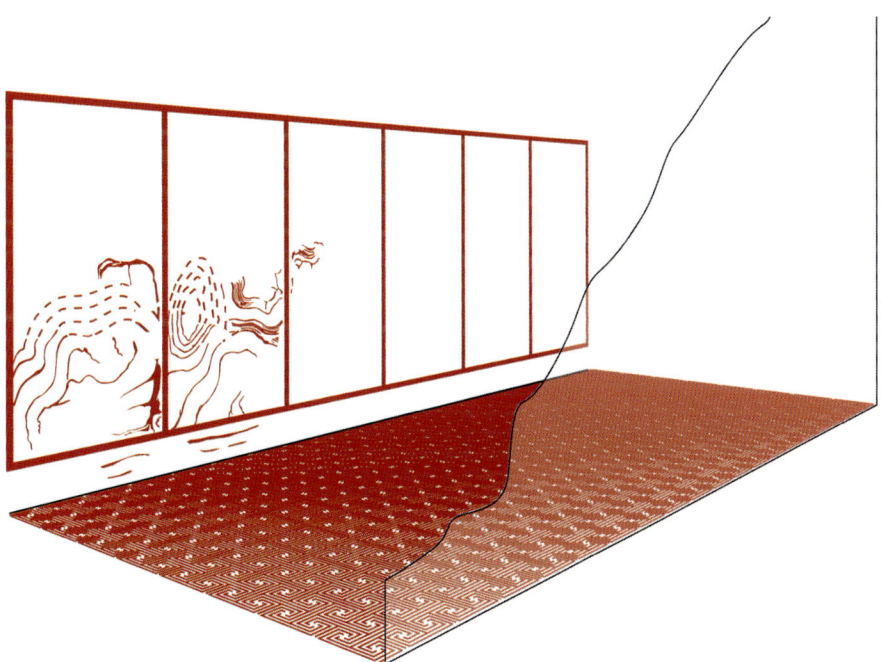

Fig. 4.17. The Labyrinth Fresco restored to the floor of Corridor 54 of the palace at Knossos (restoration M. C. Shaw and G. Bianco)

Aspects of the pattern preserved in the fragment pictured in Figure 4.18 suggest a certain regularity of design despite the apparent irregularity of the individual details. That is, if one regards the Zebra Fresco as a repeating pattern of uneven, red-spotted yellow quatrefoils, then it can be observed in the fragment of Figure 4.18 that the upper and lower points of the quatrefoils approach a horizontal axis of symmetry about which the quatrefoils are distributed. This axis of symmetry is indicated in the restoration drawing of Figure 4.19 as a horizontal dotted line. When extended as a repeating pattern, the reliance of the Zebra Fresco on a predetermined grid becomes quite evident and indicates that the composition required considerable planning before its execution. Interestingly, the fragments themselves do not preserve direct evidence of the grid: published photographs do not show impressed or incised grid lines, nor do they indicate painted sketch lines. Nevertheless, this evidence of design suggests that the composition itself was laid out according to a grid. For further investigation of the artist's grid in Aegean art, see Chapter 7.

Shield frescoes

Figure-eight shields – a signature component of Bronze Age armor – are known from artwork in a wide variety of media (e.g., painting; stone, ivory, and metal relief; and glyptic – that is, seals, sealings, and signet rings). Details of their depiction indicate that they were constructed of oxhides stretched over wooden frames bent into a characteristic figure-eight shape. One set of (ceremonial?) shields from Pylos, however, appears to have been made of textile materials rather than oxhide, and a shield fresco from Mycenae suggests that decorative supports used to hang shields on walls incorporated textile elements. Accordingly, this review surveys these shield frescoes for their evidence pertaining to textiles.

Fig. 4.18. Zebra Fresco, from the north side of the Royal Road, Knossos (drawing G. Bianco)

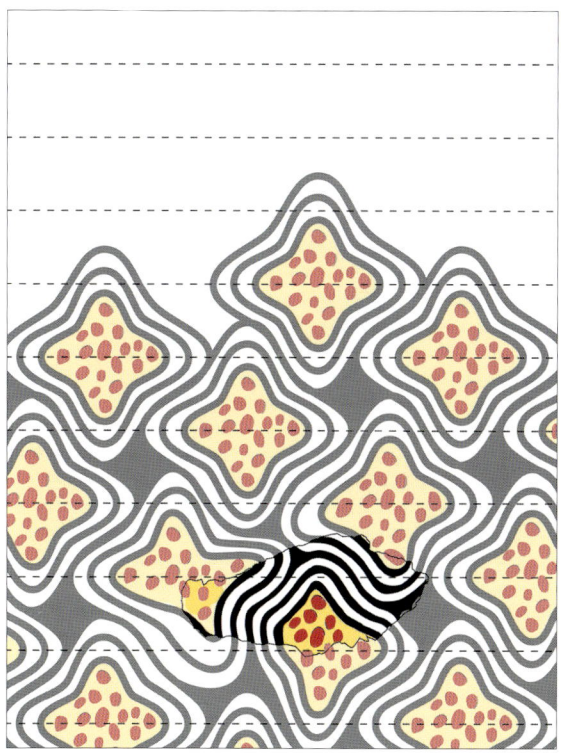 

Fig. 4.19. *Zebra Fresco restored as repeating pattern (restoration M. C. Shaw and G. Bianco)*

Fig. 4.20. *Figure-Eight Shield Fresco, from the Southwest Building at Mycenae (National Archaeological Museum, Athens; photo A. Chapin)*

LH IIIB Figure-Eight Shield Fresco, from the Southwest Building at Mycenae

This fresco preserves three figure-eight shields in relatively complete condition, and fragments of other shields, painted as if hung on a wall ornamented additionally with a spiral frieze.[60] The shields are realistically shown as having been made of oxhide, as in the single example illustrated here (Fig. 4.20). It is included in this survey of hanging textiles because the rendering of details suggests that cloth was possibly used as part of the suspension device – the round disk shown above each shield.[61] Assuming that the fresco faithfully depicts the actual object, the disk could have been made from a plank of wood, with small holes pierced through it to allow for a round piece of cloth of the same diameter to be secured to it. The edges of the cloth seem to be marked by festoon stitching, which would have kept the cloth from unraveling. The small central circle in the disk likely signifies the head of a nail hammered onto the wall to keep the disk in place, and the wavy lines linking the round hanger to the shield probably represent the strings used to suspend the shield. Their lower ends must have been attached to some metal bracket at the rear part of each shield.

LH IIIB Cloth Shield Fresco, from Stairway 54 in the palace at Pylos

Mabel Lang identified this fresco as "Rosettes with Streamers" (13 F 54) and published a color restoration by Piet de Jong incorporating some of the 40 reported fragments.[62] Renewed study

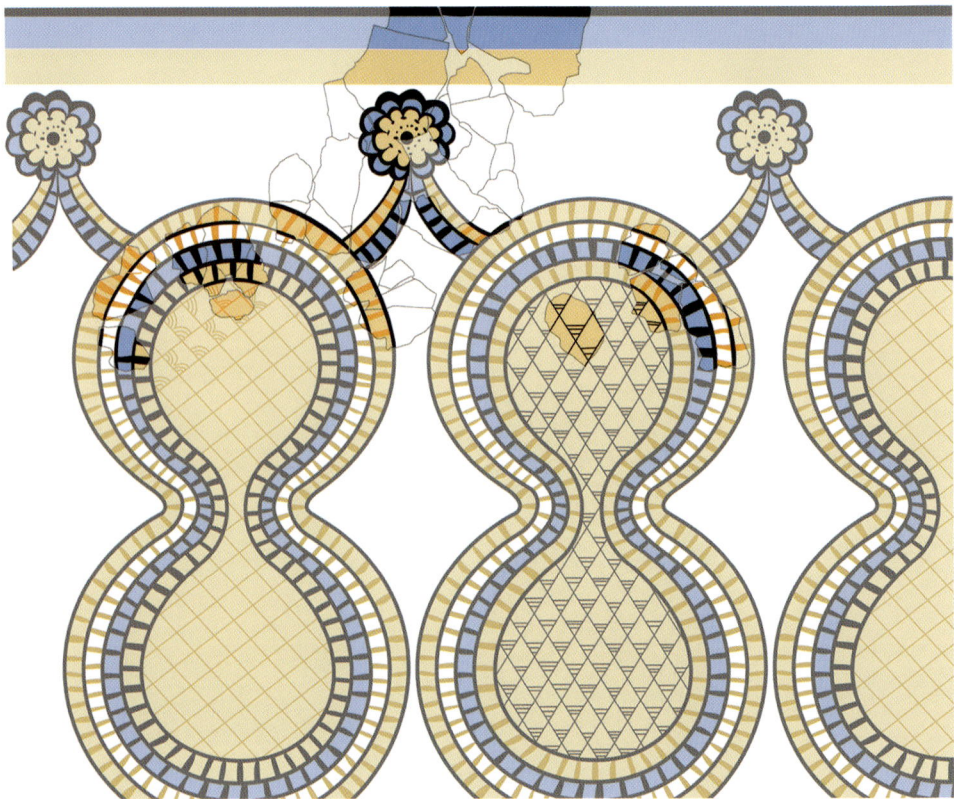

Fig. 4.21. Cloth Shield Fresco, from the palace of Pylos (restoration M. C. Shaw and G. Bianco)

of the composition, however, suggests that the fresco more likely depicts a frieze of figure-eight shields (Fig. 4.21).[63] Interestingly, these shields are shown in the painting as if constructed of cloth, with cloth also covering the heads of the nails or pegs on which the shields were suspended. Variations exist in the kinds of decoration chosen for each shield, using two alternating combinations of colors: blue bands crossed by black bars and yellow bands by red bars. Such alternating colors are commonly used in the frames of paneled compositions in Mycenaean art, where intentionally or not, they create the impression of the panels being part of a large woven piece of cloth.

Details from the fresco and other artistic depictions of figure-eight shields allow for speculation on the shields' manufacture. A cloth shield was likely made on a wooden form or frame that had been bent into its characteristic figure-eight shape, perhaps after being soaked in hot water or steamed. Then, a piece of cloth could be cut, stretched, and secured onto the shield frame. It is possible, perhaps even likely, that layers of cloth were laminated together (as layers of oxhide may have been laminated) in order to give the shield greater strength and durability.[64] The piece of cloth at the center bears a net pattern in the shape of lozenges or diamonds and appears to be laid on the frame in a smooth, flat manner; that of the barred bands surrounding it was potentially, or even likely, padded to allow for such shields to appear more substantial.

Artistic and archaeological evidence suggests that real figure-eight shields made of oxhide and wood were used for warfare during the Shaft Grave period, *c.* 1600–1400 BC, but that they had probably gone out of use before the LH IIIB period, *c.* 1300–1200 BC, when the Shield Frieze at Pylos was painted. This suggests that the LH IIIB frescoes of figure-eight shields, including those from the Southwest Building at Mycenae (Fig. 4.20) and the palace at Tiryns,[65] were archaicizing in subject matter and probably produced for symbolic or ideological purposes. It is possible that figure-eight shields were associated with a Mycenaean war goddess, perhaps akin to a Bronze Age Athena or Hera,[66] but definitive proof is lacking. It is possible that real versions of the painted cloth shields, which would have been unsuited for warfare in any era, were carried in ceremonial occasions or parades in the later Mycenaean periods. Interestingly, the space this fresco decorated, Stairway 54, can be identified as belonging to soldiers' barracks.[67] If so, then perhaps the cloth shields preserved the memory of real shields used by earlier generations; they had become a sign of the Mycenaeans' cultural identity, having been transformed into a ceremonial attribute of palatial power.

Discussion

This accumulation of artistic evidence for lost palatial and household textiles indicates that elaborate wall hangings, ceiling textiles, and floor coverings likely existed in both the Minoan and Mycenaean eras, and that they served as elegant, luxurious room decorations, as well as indicators of elite social status. In the Minoan period, frescoes such as the Wall Hanging Fresco from Ayia Triada (Fig. 4.1), the Fresco of Interlocking Circles from Galatas (Fig. 4.3), and the Quatrefoil Fresco from Poros-Katsambas (Fig. 4.5) all feature complex decorative patterns and suggest that actual wall hangings demanded significant skill at the loom. The Swastika and Rosette Fresco from Poros-Katsambas (Fig. 4.4) and the Fresco of Rosettes from Xeste 3 (Fig. 4.6) indicate further that wall hangings incorporated a variety of sewing techniques, including quilting or cording, embroidery, or appliqué in cloth or metal. The red-dotted net pattern from the House of the Ladies (Fig. 3.32) even implies the existence of wall hangings made from nonfibrous materials, such as beads of carnelian and stars of silver. The investment of time and material resources required to make actual wall hangings must have been considerable and underscores the significance of these textiles – and the frescoes that imitate them – as luxury goods demonstrating the wealth and social position of those whose homes they adorned.

These messages of luxury, wealth, and elite identity reinforce other signs of rank and social status culturally coded as palatial elements in the architectural plans of these nonpalatial buildings – e.g., the lustral basin of Xeste 3, the light well of the House of the Ladies, and the *polythyra* (pier-and-door partitions) of Ayia Triada and Xeste 3.[68] These spaces, so characteristic of elite Minoan architecture, cannot be assigned a single purpose. Rather, a certain flexibility of function and diversity of use seems likely in which rooms dedicated to domestic life seamlessly yielded to spaces suited for cult and ritual performance.[69] The textiles, as revealed by the frescoes, may have supported an adaptable use of space by offering a simple means through which rooms could be transformed into sacred environments appropriate for religious rites or ceremonies.

The Mycenaean frescoes that appear to imitate wall hangings suggest both continuity of tradition and a certain standardization of the iconographic type. The Quadruple Ivy Leaf Fresco

(Fig. 4.7), the Rockery Fresco (Fig. 4.9), and the possible wall hanging fresco from Pylos (Fig. 4.10) all incorporate repetitive, textile-like patterns that end with an undulating white wavy band (a hem?) similar to that preserved in the Wall Hanging Fresco from Ayia Triada (Fig. 4.1). But rather than decorating houses and villas, as in the Neopalatial period, these frescoes are associated with palaces – that is, with the Mycenaean centers of religious, political, and economic power. At Mycenae, the Quadruple Ivy Leaf and the Rockery Frescoes were found in the area of the Cult Center, a portion of the citadel dedicated to religious ritual. This find context implies continuity in the function of these frescoes as backdrops to religious ritual, and in the use of textiles to delineate (or, "mark off") cultic settings.[70]

Frescoes imitating ceiling coverings are too few in number to put forward any comprehensive hypothesis for their use, but the palatial context of the Quadruple Spiral Fresco from Knossos (Fig. 4.11) and the elite funerary context of the Orchomenos ceiling (Fig. 4.12) suggest that they imitate elegantly designed ceiling coverings commissioned by the wealthiest and most powerful Bronze Age patrons. Similarly, the floor fresco survey highlights the palatial associations of the paintings (Phaistos, Figs 4.14, 4.15; Knossos, Fig. 4.16; and the Mycenaean palaces discussed in Chapter 5). It is also interesting to note that the interest in repetitive, labyrinthine designs evoke associations with later Greek myths of the labyrinth at Knossos.

And finally, the shield frescoes of the Mycenaean era (e.g., the Figure-Eight Shield Fresco from Mycenae, Fig. 4.20) depict shields characteristically made of oxhide and incorporate textiles only for their cloth-covered suspension devices. But one fresco from Pylos features figure-eight shields made of cloth (Fig. 4.21) – in real life, they would have been useless in battle, but perhaps they were thought to be suitable for ritual and ceremony. Being modeled after old-fashioned (antique?) armaments, their display, whether on the walls of the palace or as real objects carried in Mycenaean pageants, parades, or ceremonies, would have kept the memory of actual figure-eight shields alive, and with it, a sense of how significant the early Mycenaean warrior culture must have been for the Mycenaean states at the time when they were first emerging as cultural and political powers in the Aegean. Such ancestral shields, rendered nonfunctional in their cloth forms, demonstrate how textiles – both real and fictive –were used by Mycenaean Greeks, like the Minoans before them, to define and reinforce their own sense of history and cultural identity.

This review of possible artistic representations of palatial and household textiles presents evidence for the existence of actual wall hangings, ceiling and floor coverings, and even cloth shields that were manufactured and displayed during the Aegean Bronze Age. This survey also offers insight into the many roles that textiles played in Minoan and Mycenaean cultures. Underlying all is the sense that these luxury items were displayed in households as prestige goods that demonstrated the wealth and social position of those whose homes and palaces they adorned. And, given the elaborate designs and the extensive investment in time and raw materials that must have been required to create these luxury textiles, it is possible that some frescoes served as less expensive substitutes. As such, household textiles and the frescoes they inspired appear to have played significant roles in the cultural self-definition of both the Minoans and Mycenaeans.

Notes

1. *PT*, 9–162.
2. Blakolmer 1994.
3. For details of the reconstruction, see Shaw and Laxton 2002. See also Militello 1998. Thanks are due to Pietro Militello, who facilitated Shaw's study of the fragments.
4. See Chapter 3.
5. L. Platon 2002, 154–155, pl. xlviiib; Saliaka 2008.
6. N. Platon 1964, 163–167; 1965, 216–224. The plan of the villa is only partially preserved. For overviews of the building and its contents, see Driessen and Macdonald 1997, 234–235; L. Platon 2002, 153–155.
7. For recent discussion of the fresco program, see Vlachopoulos, Platon, and Chrysikopoulou 2011.
8. L. Platon 2002, 154.
9. See, for instance, the gold necklaces from the Zapher Papoura graves near Knossos and the Kalyvia cemetery near Phaistos (Marinatos and Hirmer 1960, pl. 120, left), or the lily necklace on the Priest-King Fresco (*PM* II, 2, frontispiece).
10. Saliaka 2008.
11. L. Platon 2002, 145–156, pls xliv–xlviii; pers. comm. 2010; Vlachopoulos, Platon and Chrysikopoulou 2011, 440–441, 442.
12. Rethemiotakis 2002, 57, 60–61, pl. xviia.
13. On visual illusions in complex rapport textile patterns, see *PT*, 317, 328–329, figs 15.2, 15.15.
14. MacGillivray 2007, 111–112, fig. 4.5.
15. Doumas 1992, pls 24–25.
16. Rethemiotakis 2002, 57.
17. See http://chronique.efa.gr/index.php/fiches/voir/785/.
18. See http://chronique.efa.gr/index.php/fiches/voir/785/.
19. On the excavations, see Dimopoulou-Rethemiotakis 1993.
20. Doumas 1992, pl. 120.
21. For the association of interlocking crosses with swastikas, see *PT*, 328; Barber 1998, 14.
22. Dimopoulou-Rethemiotaki 2004, 375–376, fig. 31.27.
23. Doumas 1992, 174–175; pls 136–137. Maria Shaw thanks Andreas Vlachopoulos for providing pictorial details not easily discernible from illustrations of the painting (pers. comm. 2002).
24. Doumas 1992, 131, citing *Thera* VII, 27.
25. Raised, padded, and stuffed quilting techniques are well known from historical times but can only be surmised for the Aegean Bronze Age from artistic evidence. On manipulating fabric to create structure and relief, see Wolff 1996.
26. *Thera* V, 39–40; Doumas 1992, 35, pls 6–8.
27. Doumas 1992, pl. 8.
28. For the various views, see Immerwahr 1990, 49; Shaw 1993, 679; 1998, 69; Blakolmer 1994, 22; 1996; 2012a; True 2000, 353–355; Murray 2004, 125, and Chapter 3, n120.
29. *PT*, 329.
30. Betancourt 2007b, 185–189.
31. Blakolmer 1994, 22–27.
32. Kritseli-Providi 1982, 63–64, pl. 19, col. pl. Z.
33. *PM* II, 2, 478–489.
34. Cameron 1968.
35. Doumas 1992, pls 78, 79, 83.
36. Porter 2000, 625.
37. Kritseli-Providi 1982, 37–40, pls 4–5, col. pl. Γ; B. Jones 2009; see Chapter 3.
38. Kritseli-Providi 1982, 76, pl. 25.
39. The authors sincerely thank Dr. Katerina Kostanti, Curator of the Prehistoric Collection at the National Archaeological Museum in Athens, for permission to photograph and study this fresco. The museum work

was conducted in July 2014 by Anne Chapin, and the fragments were drawn and restored by Giuliana Bianco in consultation with Anne Chapin and Maria Shaw.
40   *Pylos* II, 92–93, pls 42, 117, D, N.
41   In Mycenaean art, the net pattern appears more often on dado designs in association with stylized rockery or stone patterns, as in Pylos 14 D nws, discussed in Chapter 5 (Fig. 5.9).
42   *Pylos* II, 92–93, 173–174, pl. Q. If the pattern imitates stonework rather than textiles, then the combination of motifs (textile and stone) interestingly parallels the decoration of the painted plaster floors in Mycenaean palaces, discussed by Emily Egan in Chapter 5.
43   *PM* III, 30–31, pl. xv; Kaiser 1976, 270, fig. 417; Immerwahr 1990, 63, 142, 178, fig. 39c.
44   On cording and stuffed quilting techniques, see Wolff 1996; on *trapunto*, see Morgan and Mosteller 1977.
45   Schliemann 1881; Marinatos and Hirmer 1960, 165–166, pls 160, 161; Papazoglou-Manioudaki 1990.
46   Schliemann 1880, 165–173, figs 239–252; Karo 1930–1933, 44–47, pls xxviii–xxix; Demakopoulou 1990, 284–285.
47   Hirsch 1977a. On undecorated floors, see J. W. Shaw 2009, 147–152. For a detailed study of plaster floors (some painted) at Kommos, see M. C. Shaw 2006.
48   Militello 1998, 321–335, pls 11b, 12b, 13.
49   Minoan *tarazza* floor plasters, which have high lime content and are mixed with small water-worn pebbles, are not included in this survey as they do not imitate textile designs. On *tarazza* floors, see J. W. Shaw 2009, 149–150.
50   Levi 1976, 85, pls xxiv, lxxxva; Immerwahr 1990, 22, 183, fig. 6c; Militello 2001, 45–46, 146–148, col. pl. a, top (F 54).
51   Immerwahr 1990, 206–207, n12.
52   Levi 1976, 85–86; Blakolmer 1994, 4; Militello 2001, 146–147.
53   Militello 2001, 146, 148–149, col. pl. A, bottom (F CHZ). See also Levi 1976, pl.lxxxvb; Immerwahr 1990, 22, 183, fig. d.
54   Militello 2001, 149; Bietak, Marinatos and Palivou 2007.
55   *PM* I, 356–358, fig. 256.
56   Shaw 2012a.
57   Bietak, Marinatos and Palivou 2007, 47–50.
58   *ArchDelt* 1961–1962; Cameron 1975, vol. I, 726, vol. iii, 124, 131, 155, 187; Evely 1999, 248.
59   Cameron 1975, vol. iii, 187.
60   Kritseli-Providi 1982, B-32–46, 54–62, pls E, ΣΤ, 12–17.
61   *Contra* George Mylonas (Kritseli-Providi 1982, 56–57, referenced by Immerwahr 1990, 140), who believes that the rosette functions not as a hanging device but as a symbolic head, which transforms the shield into a palladion.
62   *Pylos* II, 153, pls 86, Q (13 F 54).
63   Shaw 2012b.
64   Recent experiments with laminated linen have successfully reproduced the *linothorax*, the lightweight cuirass of Macedonian armies. See Aldrete, Bartell and Aldrete 2013.
65   *Tiryns* II, 34–40, no. 40, pl. v. See, also *PM* III, 304–306, fig. 197.
66   See, for example, Rehak 1984; 1999; O'Brien 1993, 126–130.
67   Shaw 2012b.
68   The plan of the villa at Epano Zakros is only partially preserved and cannot be assessed here.
69   On the architecture of Neopalatial palaces and towns, see McEnroe 2010, 93–116; for Akrotiri, see Palyvou 2005.
70   The fresco with a male figure and a possible wall hanging (57 H nws) was found in a dump next to the Pylos palace, so its palatial context seems assured even if it cannot be restored to a specific room.

5

Textile and Stone Patterns in the Painted Floors of the Mycenaean Palaces[1]

Emily C. Egan

For over a century, painted plaster floors have been considered among the most iconic features of Mycenaean palatial megara.[2] Discovered first by Heinrich Schliemann and Wilhelm Dörpfeld at Tiryns in the early 1880s, and subsequently by Christos Tsountas at Mycenae, and by Carl Blegen and Marion Rawson at the Palace of Nestor at Pylos, these floors are remarkably consistent in their design. Each is laid out as a grid, with its squares embellished with brightly painted patterns and/or figural elements. On account of these formal similarities, considerable debate has ensued over how these paintings should be interpreted, and more specifically over what type of floor covering they were meant to represent – textiles or cut stone? This chapter will revisit this question as it pertains to the well-known, yet enigmatic, painted floors from the megaron at Pylos. A review of the extant evidence of the floors themselves, alongside various comparanda and new data, will demonstrate that these floor paintings, rather than replicating textiles or stone, were designed to emulate *both* materials simultaneously, creating fantastic hybrid surfaces intended both to impress and to instruct the ancient viewer.

Situating the study of Mycenaean painted floors

In 1966, the painted floors of the megaron at Pylos, with its two anterooms and the main Throne Room (Fig. 5.1, Rooms 4–6), located in the southwestern Greek Peloponnese, were published by Blegen and Rawson in the first volume of their series, *The Palace of Nestor*.[3] In their text, the excavators offered detailed descriptions of each floor's gridded layout and decoration accompanied by watercolor illustrations by artist Piet de Jong. The three floors' grids were marked out by pairs of impressed parallel lines spaced roughly 5 cm apart, and painted red. The decoration of the squares (Fig. 5.2) comprised a total of ten different abstract patterns painted in hues of red, yellow, blue, black, and white. The patterns included tricurved arches, scattered circles, scales, parallel zigzags, interlocked quatrefoils, diagonal or transverse parallel wavy lines, sets of two or four radiating concentric arcs, and cross-hatching. A single figural element, a red-brown octopus, was situated in a square near the room's northeast wall, opposite a floor cutting identified by Blegen and Rawson as the placement for a throne.

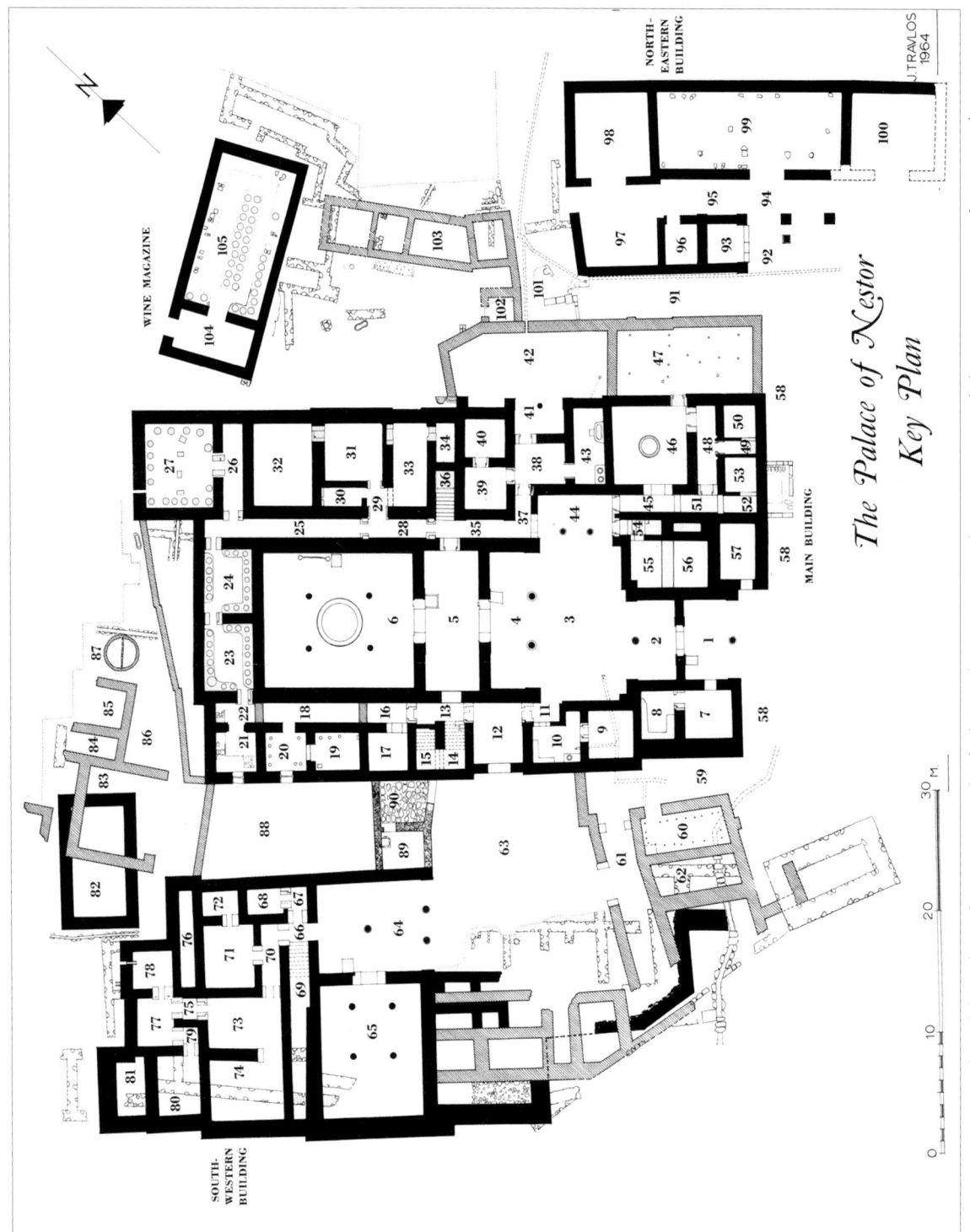

Fig. 5.1. Plan of the Palace of Nestor. J. Travlos (Pylos II, pl. 143, courtesy the Department of Classics, University of Cincinnati)

5. *Textile and Stone Patterns in the Painted Floors of the Mycenaean Palaces* 133

Fig. 5.3. Restored section of Arc Dado (1 D 64) from the Palace of Nestor (P. de Jong, color version of Pylos II, pl. 140, courtesy the Department of Classics, University of Cincinnati and the American School of Classical Studies at Athens; photo A. and J. Stephens)

Fig. 5.2. The painted floors of the Pylos megaron (E. C. Egan, after P. de Jong, with elements courtesy the Department of Classics, University of Cincinnati and the Trustees of the American School of Classical Studies at Athens)

In 1969, a first attempt to interpret these patterns was made by Mabel Lang, who, in her examination of the palace's painted dadoes, identified similar abstract designs as representations of different types of stone (Fig. 5.3).[4] Lang's conclusions, which built upon similar theories developed by Winifred Lamb in the 1920s for the floors at Mycenae,[5] was advanced in the 1970s by Ethel Hirsch, who argued in her doctoral thesis that the gridded and painted floors of the Mycenaean palaces were meant specifically to imitate Minoan stone-slab pavements.[6]

Although anticipated by Lamb and Lang, Hirsch's argument represented the first major refutation of the alternative (and widely held) opinion that Mycenaean gridded floors were meant to represent pieced carpets.[7] The initial observations on this latter subject were made by Dörpfeld, who in 1885 described the incised grids that he found in the large megaron at Tiryns as reminiscent of a "carpet pattern."[8] While it is unclear whether Dörpfeld meant to say that the incised grid simulated an actual carpet or whether he was simply referring to an overall decorative effect, the idea that gridded Mycenaean floors represented carpets stuck, and was subsequently adopted by Rudolf Hackl, who reexamined the Tiryns floors in 1909.[9] In the large megaron at Tiryns, Hackl found evidence of a painted grid filled with marine motifs (octopuses and paired dolphins), and tricurved arches (Fig. 5.4). The arch pattern, Hackl contended, while originally designed to imitate rocky terrain with flowers, had become an explicitly textile motif by the LH III period (beginning in the early 14th century BC), as indicated by its appearance in painted depictions of clothing.[10] Marine motifs, Hackl further argued, also possessed a textile quality on account of their appearance in the so-called Curtain Fresco from Mycenae.[11] This painting, identified by Maria Shaw as containing depictions of ikria (or, stern screens constructed from fabric secured around wooden frames), featured a panel with bands of argonauts alongside more traditional textile patterns including concentric rhomboids (see Fig. 6.7).[12] Building upon Hackl's early work at Tiryns, Gerhart Rodenwaldt published an article in 1919 in which he outlined the carpet-like qualities of the painted floors of the megaron at Mycenae.[13]

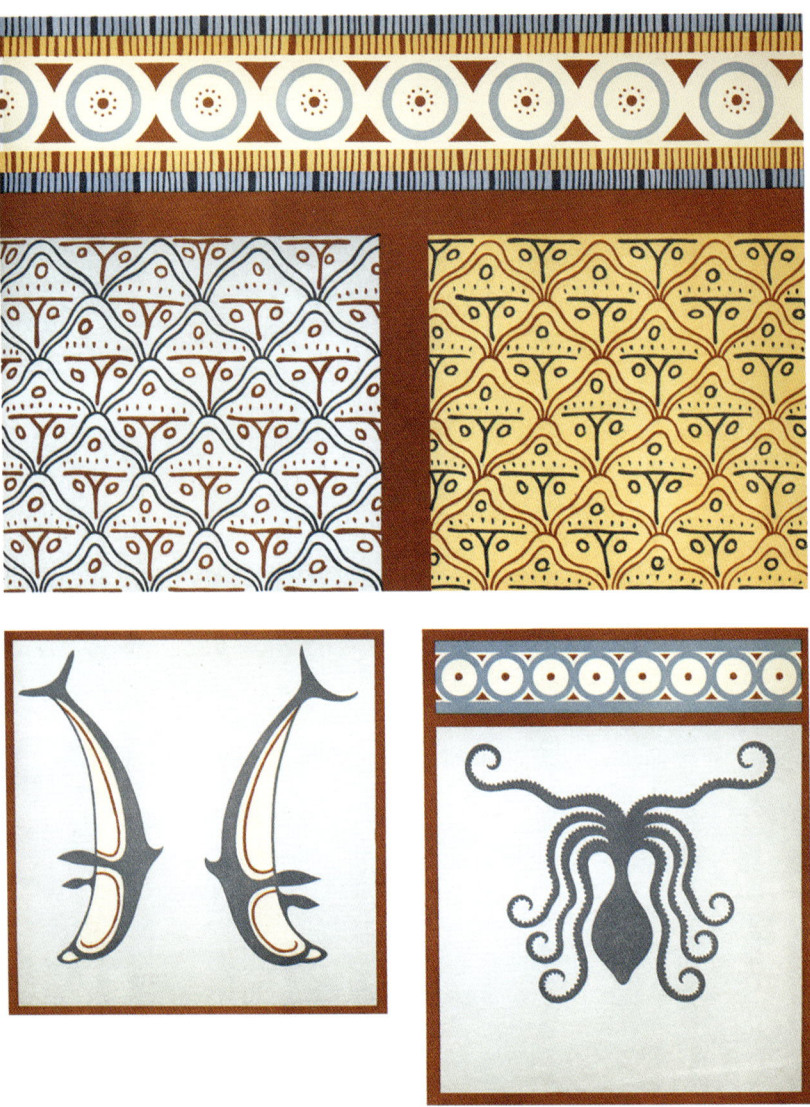

Fig. 5.4. Painted restoration of details of the floor decoration from the Tiryns megaron (Tiryns II, pl. xxi, courtesy the Deutsches Archäologisches Institut, Berlin)

5. Textile and Stone Patterns in the Painted Floors of the Mycenaean Palaces

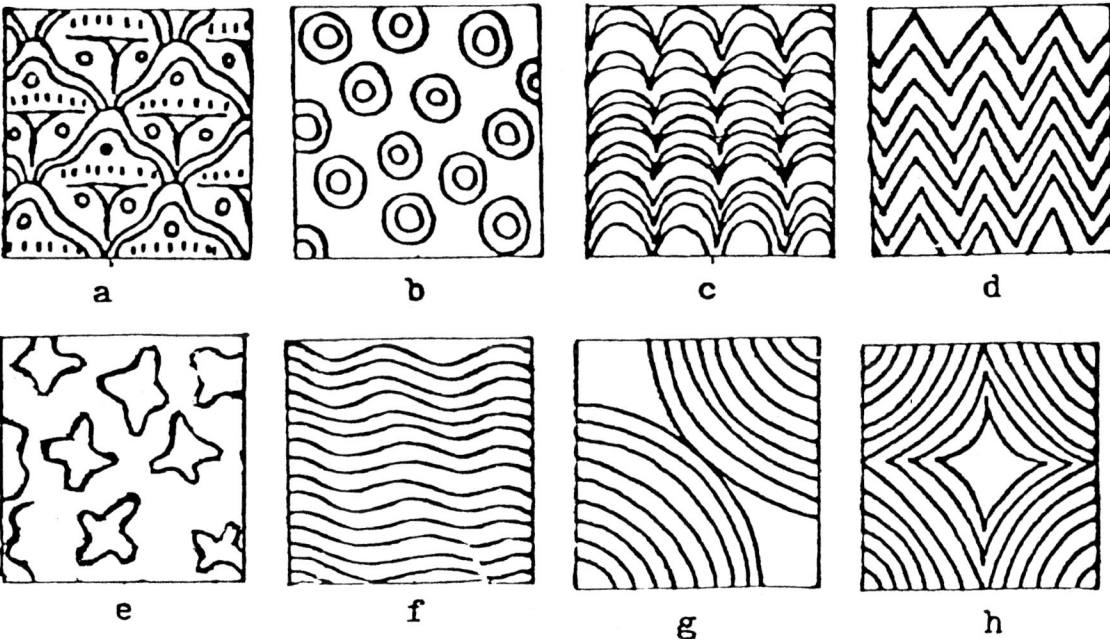

Fig. 5.5. Painted floor patterns found in Mycenaean palaces (Hirsch 1980, 458, fig. 3; drawing E. S. Hirsch, courtesy the Archaeological Institute of America and American Journal of Archaeology)

Half a century later, Hirsch employed two pieces of evidence in her challenge to this prevailing view. First were the visual similarities between the red grid lines in the floor paintings at Pylos (as well as at Mycenae and Tiryns), and the red-painted grout surrounding the stone floor slabs in Late Minoan palaces and/or elite residences on Crete at Phaistos, Knossos, Ayia Triada, Archanes, and Zakros.[14] Second were the close parallels between the floors' linear patterns (summarized by Hirsch in Fig. 5.5) and representations of different types of cut stone in Aegean Bronze Age wall and vase painting. The tricurved arch, Hirsch argued, represented a stylized version of a rocky landscape, or perhaps of conglomerate stone, as evidenced by similar patterns in one of the Pylos "Variegated Dadoes."[15] The circles with their central white dots, she suggested, could be identified as an imitation of liparite (a stone replicated in paint on Middle Minoan ceramics) or perhaps they depicted a composite design of small pebbles, also seen in the Variegated Dadoes. Hirsch associated the ridges with representations of rocky landscapes and/or "mineral imitations," and the parallel zigzag pattern, she argued, was used to represent water, wood graining, or most likely, veined alabaster or marble (thereby mimicking the use of this pattern at Mycenae). Scattered irregular shapes simulated "granulated rock-work" on Middle Minoan vessels. Wavy parallel lines, Hirsch observed, were similar to the banding used to depict stone vases in wall paintings from both Knossos and Pylos (Fig. 5.6), while concentric arc patterns were echoed in "Arc Dadoes" (see Fig. 5.3) from the same sites.[16]

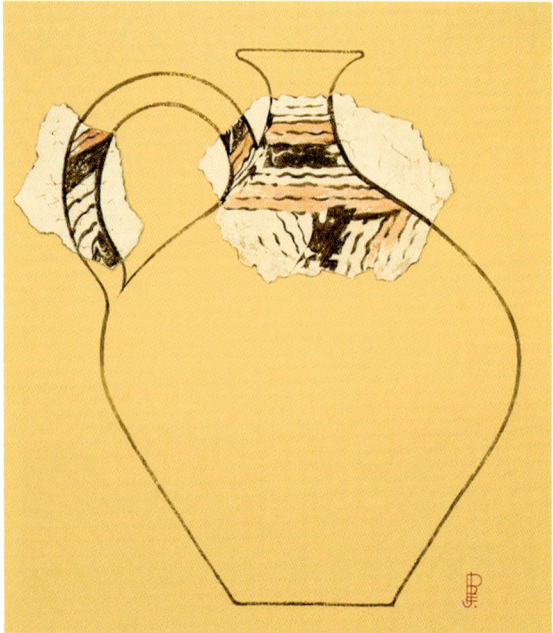 

Fig. 5.6. Reconstruction of a fresco representing a stone vase (2 M 6) from the Palace of Nestor (P. de Jong; color version of Pylos II, pl. 141, courtesy the Department of Classics, University of Cincinnati and the American School of Classical Studies at Athens; photo A. and J. Stephens)

Fig. 5.7. Reconstruction of the Throne Room floor from the Palace of Nestor (P. de Jong; Papadopoulos 2007, fig. 2c, courtesy the Trustees of the American School of Classical Studies at Athens; photo C. Mauzy)

The floors at Pylos reconsidered: evidence for hybrid construction

When considering the Pylos megaron's floors, Hirsch's first observation is convincing. The red grid lines in the megaron are indeed visually very similar to the strips of red-painted grout between stone floor slabs in several Cretan buildings. The connection Hirsch drew between floor patterns and painted representations of cut stone, however, is less compelling at Pylos, where the floor motifs are particularly complex. While it is undeniable that the circles, zigzags, wavy lines, and concentric arcs on the megaron's floors bear strong resemblance to the mottled and veined patterns of painted stone, the floors' other patterns – namely the tricurved arches, scales, interlocked quatrefoils, and cross-hatching – do not.

Part of the problem is that the linear motifs catalogued by Hirsch do not directly correlate with those found at Pylos. In Hirsch's summary (see Fig. 5.5), one motif, the scales (figure "C" in the drawing), is overly stylized, while the other three (the tricurved arches, interlocked quatrefoils, and cross-hatching) are intentionally omitted.[17] The reason for these omissions, as stated by Hirsch, was her lack of confidence in de Jong's watercolor of the floor of the Pylos megaron's Throne Room[18] (Fig. 5.7), the only context in which these three motifs appeared. The reasons for her lack of confidence were twofold. First, Hirsch argued that de Jong's rendering was inaccurate because of a discrepancy between his painted version of the decoration in one of the squares (E7, immediately north of the hearth) and the appearance of the same decoration in published photographs.[19] Second, she claimed that there was no "textual support" for the Throne Room's floor decoration on account of the fact that Blegen and Rawson chose to describe the motifs *en masse* in a list rather than enumerating them square-by-square as they had done for the painted floors in the megaron's Vestibule and Portico.[20]

In contrast to Hirsch's assessment, however, unpublished excavation documents from Pylos suggest that much of de Jong's painting of the Throne Room's floor is correct – including his representations of the three motifs omitted from Hirsch's list and his nonschematized version of the scales. The strongest evidence for de Jong's accuracy comes from accounts of his exceedingly careful examination of the floor (which he spent over a week studying *in situ* under the shade of a portable canopy)[21] and from a preliminary plan of the floor drawn by site architect Demetrios Theocharis in 1952. This plan, produced the year the floor was excavated, includes many of the decorative elements observed by de Jong a year later.[22]

Both of Hirsch's specific concerns can also be refuted. First, Hirsch's point that a list carries less weight (and suggests less certainty) than a square-by-square description, is unconvincing. It is equally likely that Blegen and Rawson refrained from individually describing every square with preserved decoration on the floor of the Pylos Throne Room simply because there were 75 of them and it would have been cumbersome to do so. In the cases of the Vestibule and the Portico, the floor squares with preserved decoration number only eight and 15, respectively. Small numbers of painted squares are also attested in other floors in the palace, and likely factored into Blegen and Rawson's decision to describe them individually. For instance, in the so-called Queen's Quarters, Corridor 49 retains evidence of painted designs in only ten squares, and in Room 50 decoration is extant in only 16 squares.[23] It would seem, therefore, that the painted squares of the Throne Room were presented as a list simply in an effort to convey the information in a more succinct, but no less significant, format.

Second, regarding Hirsch's objections to the motif in square E7, close inspection reveals that published photographs of this design[24] – which illustrate upper curves of two adjacent (but not touching) solid arcs, each with an underlying arc of dots – can be reconciled with the repeating scale pattern reconstructed by de Jong. While it is true that the spacing of the original and painted renditions is different, the actual form of the motif (curved arcs with underlying arcs of dots) is unchanged. De Jong's "inaccuracy" seems minimal.

By looking closely at the Pylos tricurved arches, scales, interlocked quatrefoils, and cross-hatching, a connection with textiles is immediately apparent. This link is evident both in the syntactical structures of each of these designs, all of which are examples of "expanding network" or rapport patterns,[25] and in the squares' motifs, which are echoed in Late Bronze Age painted representations of cloth. By far the strongest parallels come from Crete. The tricurved arch as a textile pattern appears at Knossos on the bodices worn by the LM I "Ladies in Blue," as well as on part of a kilt worn by a youth assigned by Mark Cameron to the LM II "Cupbearer class."[26] The scale pattern is present in LM I wall painting fragments associated with the Ladies in Blue as well as on the dress of the kneeling "adorant" from the LM I fresco in Room 14 of the Royal Villa at Ayia Triada (Fig. 3.11).[27] A fragmentary female figure from the Tiryns LH III Grand Procession of Women also features scales on her pleated skirt (Fig. 3.45), while a second figure's skirt preserves cross-hatching (albeit without filler squares).[28] A somewhat closer parallel for the cross-hatching (incorporating filler squares into roughly half of its design) appears in painted renditions of

Aegean textiles in Egypt, including an example found on the ceiling of the Theban tomb chapel of the Theban tomb chapel of Senmut, high steward during the reign of the pharaoh Hatshepsut.[29] Finally, interlocked quatrefoils are closely (if not identically) matched by patterns on the dress worn by the "dancing goddess" from Ayia Triada Room 14 (Fig. 3.9),[30] as well as on a kilt worn by one of the male offering bearers in the LH II/IIIA Knossian Procession Fresco (Fig. 3.18).[31] A second kilt in this fresco features a scale pattern nearly identical to that on the Pylos floors.

Within the Palace of Nestor itself, the floor squares with tricurved arches find a parallel in a fragment of the site's "papyrus net" wall painting (Figs 7.12, 7.13).[32] As Shaw has demonstrated, this wall composition, which consists of connected white tricurved arches with yellow papyrus fillers on a blue ground, bears the iconographic and syntactical hallmarks of a large painted textile. This identification is bolstered, Shaw argues, by the painting's use of an artist's grid – a technique employed frequently on Crete to facilitate the rendering of complex cloth patterns in Neopalatial and Final Palatial wall paintings.[33] Intriguingly, this same drafting device, which features prominently at Knossos, as well as at Pseira and Ayia Triada, is also evident in some of the Pylos floor squares painted with tricurved arches, scales, and quatrefoils. As I have argued previously, small orthogonal string-impressed lines, identified by Blegen and Rawson in nine of the Pylos Throne Room's grid squares, were not used to designate places for court officials to stand during "occasions of state," as the excavators originally suggested, but were instead artists' grids, employed to break down and evenly space the complex repeating designs that occupied certain squares.[34] While the use of this rather straightforward floor painting strategy may have been intuited, it is far more likely – given what we are now learning about the mobility of craftspeople and the exchange of technological and artistic ideas during the Bronze Age – that the technique was borrowed from wall painting.[35] The appearance of artists' grids on the Pylos Throne Room's floor, therefore, reinforces the link between its rapport patterns and depictions of painted textiles in murals.[36]

Collectively, this evidence suggests that not one, but *two* materials are represented by the linear designs in the Pylos megaron's painted floor grids. What makes this all the more notable is the realization that this same "hybrid construction" is echoed elsewhere in the Palace of Nestor. In Room 10, which was identified by Blegen and Rawson as a formal waiting area, the northeastern face of a plaster bench (Fig. 5.8) was painted to imitate veined stone, while its southeastern face was painted to resemble carved wooden legs and struts.[37] In Room 44 and in the Northwest Slope plaster dump there were also parts found of two unusual wall paintings featuring panels depicting both stone patterns (circles, wavy lines, and the so-called Easter Eggs) and rapport designs (tricurved arches and cross-hatching) separated by wavy lines. While Lang identified these paintings as highly stylized Variegated Dadoes[38] (Fig. 5.9), their motifs suggest that they, like the megaron's floor squares, were meant to represent hybrid mixtures of cloth and stone.

Fig. 5.8. East corner of the plaster bench (4 M 10) in Room 10 of the Palace of Nestor (Color version of Pylos II, pl. 109, courtesy the Department of Classics, University of Cincinnati)

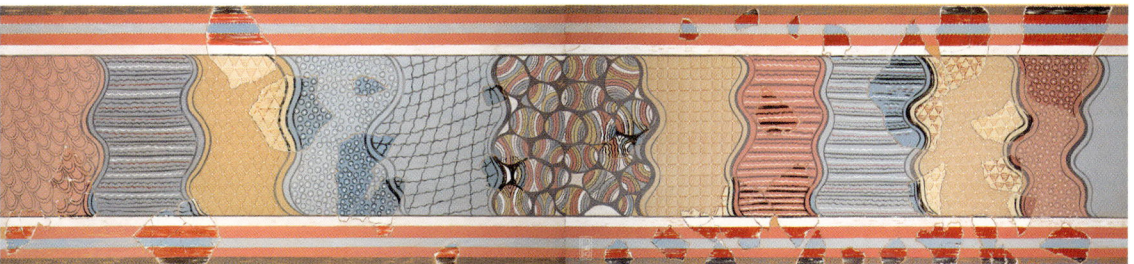

Fig. 5.9. Reconstruction of the "Variegated Dado" (14 D nws) from the Palace of Nestor (P. de Jong; Pylos II, pl. Q, courtesy the Department of Classics, University of Cincinnati)

In each of these cases, I would contend that the blending of materials was not a whimsical artistic expression but a meaningful choice intended to impart visual messages of the palace's opulence and innovation to the ancient viewer. Rather than slavishly replicating a realistic surface treatment constructed from (or covered with) a single material, the Pylian painters utilized the adaptable plaster medium to produce surfaces and structures that were physically impossible and, by extension, visually exciting. Indeed, in the case of the megaron's floors, it is this fundamental malleability of plaster that has been overlooked by scholars, who have routinely viewed the decoration as necessarily derivative – a strict imitation of a "real world" prototype.[39] Even Hirsch, who credited the "flexible" plaster medium for the innovative inclusion of emblematic sea creatures in the floor grids at Tiryns and Pylos, fell victim to this narrow viewpoint in her effort to make sense of the floors as a homogeneous type of canvas.[40] Furthermore, she attempted to bolster her argument that the floors were meant to replicate stone by citing the grid squares' "wall-to-wall" layout (seemingly "precision cut" to accommodate built features), as well as the logic that it would not have been expedient to place a carpet near a fire-burning hearth.[41] As works in plaster, however, painted floors are neither stone nor carpet, and their interpretation need not be restricted by expectations about what would have been appropriate or feasible for either of these materials.

Different effects at Mycenae and Tiryns

If the painted floors of the megaron at Pylos were designed as hybrids, this raises the question of the character of the floors in the other two palatial megara at Mycenae and Tiryns. It is worth noting that neither the floor at Mycenae nor the one at Tiryns has overtly hybrid characteristics. Instead, at Mycenae, the painted squares of the gridded floor seem to have imitated only stone, while those at Tiryns appear to have simulated textiles exclusively. At Mycenae, the floors' patterns, as identified by Rodenwaldt and Lamb in the Throne Room and the Vestibule, as well as in the adjacent courtyard, included parallel zigzags as well as scattered circles, parallel ridges, wavy lines and concentric arcs (Figs 5.10, 5.11).[42] Notably, no rapport patterns are represented.[43] Compounding this stony effect are the actual gypsum slabs set around the edges of both the Throne Room and the Vestibule's floors that seem to be a direct reference to the plaster squares' "parent material."[44]

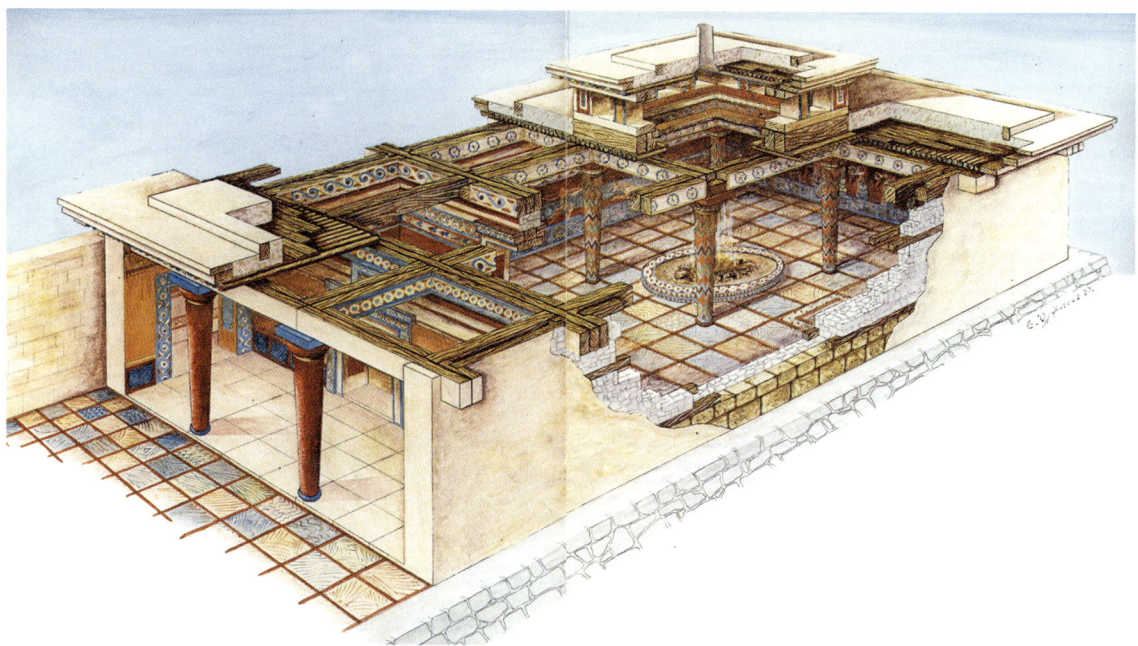

Fig. 5.10. Reconstruction of the megaron at Mycenae showing stone-like floor decorations (Mylonas 1983, 102–103, fig. 80)

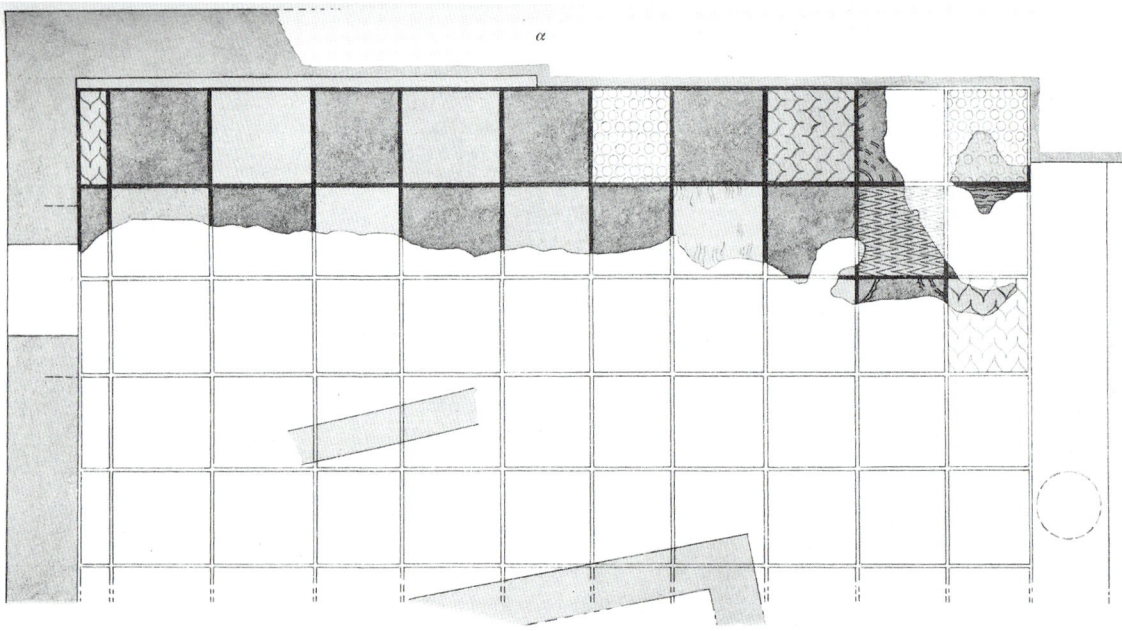

Fig. 5.11. The decorated floor of the courtyard of the megaron at Mycenae (Rodenwaldt 1919, pl. 7, courtesy the Deutsches Archäologisches Institut, Berlin)

5. Textile and Stone Patterns in the Painted Floors of the Mycenaean Palaces

In the Tiryns megaron, hybrids are also absent. Here, the floor patterns consist of only three designs: tricurved arches, paired dolphins, and octopuses, contained within a grid formed by wide bands of connected rosettes sandwiched between thinner bands of tooth ornament (Fig. 5.12 and see Fig. 5.4, above). The Tiryns tricurved arch and marine elements, as originally noted by Hackl, have strong parallels in painted depictions of textiles. This is also true of the tooth motif, composed of stacked barred bands of black on blue and red on yellow. Identified by Elizabeth Barber as a common textile border produced using a belt-weave, this pattern is frequently found in Aegean wall paintings depicted on the hems of women's garments.[45] At Tiryns, the elaborated version of this pattern found on the megaron's floors is matched identically on the bodice worn by a woman in the Procession Fresco (Fig. 3.45). Furthermore, the textile effect of the Tiryns floor is amplified by the strict arrangement of the painted designs in a repeating checkerboard pattern with alternating marine motifs (on blue backgrounds) and tricurved arches (on red backgrounds).[46] This regular disposition of patterns creates clear diagonal lines within the overall floor composition, transforming it into a monumental "marine style"[47] rapport design.

Intentionality behind the Pylian hybrids

The uniqueness of the Pylian floor hybrids prompts questions about the aims of the artist(s) who produced them. One explanation for their curious composition is that the blended floors were intended to be symbolic. In Aegean wall painting, the use of hybrids to create artificial, symbolic environments has been discussed by Anne Chapin, who contends that the inclusion of floral hybrids like the waz-lily[48] in Minoan Neopalatial painted landscapes (for example, those from the House of the Frescoes and the Minoan Unexplored Mansion at Knossos) imparted a supernatural quality

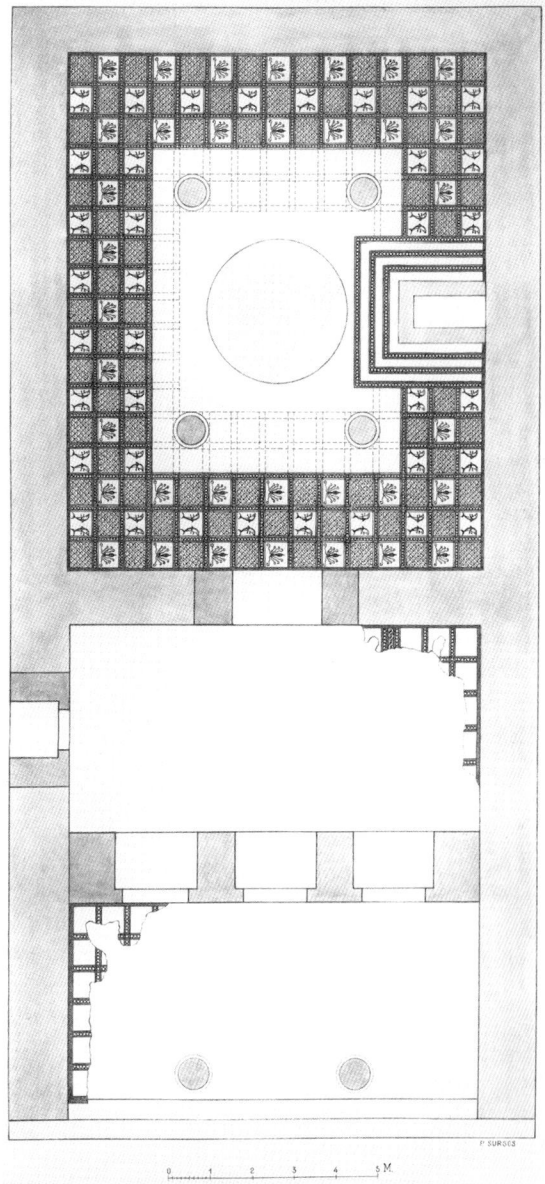

Fig. 5.12. Reconstruction of the painted floors of the megaron at Tiryns (Tiryns II, pl. xix, courtesy the Deutsches Archäologisches Institut, Berlin)

to these constructed environments.[49] Like animal hybrids such as griffins, the floral hybrids conveyed a special connection with the divine by betraying "an underlying belief in the power of divinity to act upon the natural world and to transform its appearance."[50] At Pylos, the impossible combination of two materials may have been equally meaningful, being representative of the ability of man (perhaps imbued with divine power) to supersede the limits of reality and impose his will on his surroundings.[51]

Additionally (and/or alternatively), the plurality of materials represented in the painted floor squares at Pylos may have been intended as a focusing device, operating in concert with the designs' erratic placement throughout the grids. Although Blegen and Rawson noted that the same patterns were often positioned one square over and two squares up from one another, corresponding to the "knight's move" in chess, this occurs only eight times in the Throne Room and twice in the Portico, and therefore can hardly be called a regular scheme.[52] Instead, the floor designs were arranged haphazardly, with the same patterns positioned in different parts of the floor and facing different directions. While the multiple orientations of patterns could be attributed to the production process (that is, the direction of a motif was determined by where the artist was standing when he painted it), it is equally likely that they were part of the floor's intended design and were meant, like the decorative patterns themselves, to attract and stimulate the viewer's eye. As argued by art historian Ernst Gombrich, surface patterns with straightforward repeating designs are by nature simpler to perceive and mentally process. Such visual monotony, however, also results in a certain mental "boredom." Using the example of a flagstone pavement, Gombrich writes, "We look at the grid and take it in at a glance as soon as we have grasped the underlying rule that all the flagstones are identical. ... When the expected happens in our field of vision we cease to attend and the arrangement sinks below the threshold of our awareness."[53] On the other hand, however, designs that are overly complex like a "crazy pavement" are visually overwhelming: "If monotony makes it difficult to attend, a surfeit of novelty will overload the system and cause us to give up."[54] By playing with pattern and orientation within the strict confines of a grid, therefore, the Pylian painters made the megaron's floors sufficiently complex so as to force the ancient viewer to engage with the designs, but not so complex that he or she might become optically beleaguered. Much the opposite, the viewer would have been captivated by the floors' intricate, fantastic appearance, which would have left him teetering excitedly (if precariously) on the thin line between comprehension and confusion.[55]

In practical terms, this heightened engagement would have translated into additional time spent viewing the floors. Enticed by the dynamic yet structured patterns, the viewer would have been encouraged both to absorb, and to respond to, the unique details of the floors' design. These details include both the different materials represented, and in the case of the Throne Room, the diagonal orientation of the southeastern part of the grid (Fig. 5.8). As I have argued previously, it is likely that this diagonal was *not* accidental (the result of a rushed job, as Blegen and Rawson, and many others have inferred), but intentional.[56] Evidence for this claim comes from the floor's artists' grids, which preserve corrected lines.[57] If care was taken to correct such small mistakes, it can hardly be imagined that a major "error," such as misaligning nearly half of the floor's painted decoration, would have been allowed.

The purpose of this diagonal may have been to encourage the viewer's movement through the space, specifically toward the throne. Similar effects have been identified in Roman mosaic

5. *Textile and Stone Patterns in the Painted Floors of the Mycenaean Palaces* 143

pavements, recently by Ruth Westgate, Ellen Swift, and Rebecca Molholt, and earlier by John Clarke, who coined the term, "kinesthetic address," to describe this type of interaction between viewer and surface decoration.[58] As defined by Clarke, kinesthetic address is the:

> power of the image to confront, affect [and] direct the viewer ... it is concerned with aspects of human perception: the actual physiology of seeing, the identification of the subject represented, and the psychology of following pattern (that is, human reaction to design directions).[59]

Using bath complexes at Ostia, Clarke demonstrated how black and white figural mosaics were instrumental to a visitor's experience of different rooms: in this instance, the changing orientations of the figures underfoot prompted the viewer to move "continuously around" in order to emulate and align himself with the figures' varying motions and positions.[60] In the case of the Pylos floor, the sharp diagonal orientation of the grid would have served much the same purpose, inviting the viewer to align himself with the floor and to move along a clear path toward the throne.

Positioned close to the doorway, the diagonal of this floor grid would have been visible *before* the throne itself came into view (as illustrated by the reconstruction in Figure 5.13), and would have helped the visitor to *anticipate* how to find a way through this important space. The fact that the sharp diagonal of the floor decoration would have been one of the first visible features of the Throne Room is significant when considering how the room was used. Indeed, this detail suggests that the Throne Room hosted events attended by visitors who may or may not have been familiar with the space; visual cues in the room's design and decoration would help orient all attendees to the royal throne.

Fig. 5.13. Reconstructed view into the Pylos Throne Room showing the visual effect of the diagonal grid lines (E. A. Markin and E. C. Egan)

Conclusion

The evidence presented here strongly suggests that the painted grids on the floors of the megaron at the Palace of Nestor were nothing short of an artistic marvel. Ingeniously conceived and executed, the floors surpassed the more realistic designs of comparable floors at Mycenae and Tiryns to create impossible, hybrid surfaces that would have attracted the eye of the viewer and helped him to navigate this most central part of the palace. These conclusions, derived from a close study of motifs and supported by theories of visual perception, underscore the immense importance of prehistoric floor paintings, which in the past have been not only under-interpreted but underestimated. More than simple decorations, the paintings were a dynamic expression of artistic ingenuity – actively dictating and shaping the ancient viewing experience from the ground up.

Notes

1. I wish to thank Anne Chapin and Maria Shaw for the kind invitation to contribute to this volume. This chapter is based on the paper, "Even better than the real thing: hybrids and painted floors at the Palace of Nestor," which I delivered at the 115th annual meeting of the Archaeological Institute of America in Chicago, IL, January 2-5, 2014. Special thanks go to Sharon Stocker and Jack Davis for permission to study the Pylos material, and for their tireless support and guidance, also offered by Hariclia Brecoulaki and Eleni Hatzaki. The focus herein is on the painted floors of the central megara of the Mycenaean palaces in which textile motifs are most evident. For discussions of other painted floors in Mycenaean palatial contexts, see de Ridder 1894, 291 (Room D at Gla); Hackl 1912, 222–234 (Corridor XII, Rooms XIII, XVIII, Corridor XX, Room XXI, and Corridor XXIII at Tiryns); Wace *et al.* 1921-1923, 186–188 (Room 52 at Mycenae); and *Pylos* I, 193, 211–215 (Stoa 44, Corridor 49, and Room 50 at Pylos).
2. The term "megara" (singular: "megaron") is the standard term for the canonical suite found at the core of a Mycenaean palace. The suite consists of three rooms: two anterooms known as the "Portico" (or "Porch") and the "Vestibule," and a main hall known as the "Throne Room," "Hearth Room," or simply the "Megaron" or "Megaron proper" (both with a capital "M"). This latter room is commonly marked by the presence of four columns surrounding a monumental central hearth.
3. *Pylos* I, 68–70, 74, 82–85.
4. *Pylos* II, 164–167.
5. Lamb 1921-1923b, 194.
6. Hirsch 1977a.
7. The term "pieced" refers to the construction of irregularly shaped carpets (in this case, designed to go around the megara's columns and hearths) by means of stitching together separately woven segments.
8. Dörpfeld 1885, 276. Notably, there are two megara at Tiryns: a large one (the "main megaron") positioned at the center of the palace, and a smaller secondary megaron, also with painted floor decoration, to the east (Schliemann 1885, 239–242; Müller 1930, 157–166). A similar secondary megaron, the so-called Queen's Hall (Hall 46) is found at Pylos. The floor of this latter room, however, is not painted (*Pylos* I, 197–203).
9. Hackl 1912, 222–234. For an overview of the difficulties involved in interpreting the meaning of Dörpfeld's comments about the incised patterns on the floors of the Tiryns megaron see Hirsch 1980, 453–454.
10. Hackl 1912, 226–228.
11. Hackl 1912, 232, 234; Rodenwaldt 1919, 102.
12. Shaw 1980; Chapter 6. For an alternative interpretation of the "curtains" as four distinct carpets, see Reusch 1945, 103.
13. Rodenwaldt 1919. In addition to the floors of the megaron, Rodenwaldt also examined those patterns extant on the adjacent plaster courtyard, discussed below in greater detail.
14. Hirsch 1977a, 46; 1980, 456–457.

15 See *Pylos* II, 173, 13 D 44d, and pl. 98.
16 Hirsch 1977a; 1980, 457–459.
17 Notably, "figure A" in Hirsch's summary is meant to depict the tricurved arch pattern at Tiryns only.
18 This is the term preferred by Blegen and Rawson (*Pylos* I, 76–92).
19 Hirsch 1977a, 32, citing *Pylos* I, figs 72, 73.
20 Hirsch 1977a, 32.
21 Documentation of de Jong's study of the painted floors of the Throne Room can be found in Blegen's 1953 field notebook (96–111) and in contemporary excavation photographs. De Jong's attention to detail helps to negate Blegen and Rawson's comment (*Pylos* I, 84, n. 32), noted by Hirsch, that many of the patterns on the Throne Room's floor were a matter of "individual interpretation and conjecture." While some degree of "interpretation" is to be expected given that the floor was heavily burnt, it is clear that de Jong made every effort to render its patterns accurately.
22 This preliminary plan of the floor of the Throne Room was found tucked into the back pocket of George Mylonas's 1952 excavation notebook. The plan is unsigned, but its author is identified as Theocharis in the notebook's entry for July 9. My understanding is that this drawing represents an early rendering of the Throne Room's painted floor patterns made at the time of their excavation in 1952, and was subsequently used as a model by de Jong (as indicated by his annotations to the original drawing) when he arrived at Pylos in 1953.
23 *Pylos* I, 212, 214.
24 *Pylos* I, fig. 72.
25 As defined by Shaw (2000, 53), who follows Barber (*PT*, 317), "rapport" refers to "a composition constructed along diagonal lines, which are drawn or implied by the orientation of the component motifs systematically arranged at regular distances." As employed in studies of the Aegean Bronze Age, this term originated in discussions of ceramic decoration. See in particular the work of Arne Furumark (1941, 114) and Gisela Walberg (1976, 83–84), who defines rapport as the result of criss-crossing torsional decoration arrayed in a net-like pattern.
26 *PM* I, 545–547, fig. 397; Cameron 1975, 327–328 and pl. 9b; see also Chapters 3 and 7.
27 *PM* II, 680–681, fig. 430c; Militello 1998, 68–77, 99–132, 250–282, pls 2, and C. Also see the scale pattern on an unassigned wall painting fragment of a textile from Room 14 (Militello 1998, pl. 1b).
28 Rodenwaldt 1912, 70–76, figs 27–28, 31, pl. viii.
29 Metropolitan Museum of Art 23.3.463. On the depiction of Minoan textiles on the painted ceilings of Egyptian tombs, see Shaw 1970; *PT*, 340–351; and Chapter 8.
30 Militello 1998, 104–107, fig. 27, pl. 3a.
31 See Chapter 7, and Shaw 2000, 54–57, Patterns B and E.
32 *Pylos* II, 186, pls 113, R (18 M ne).
33 Shaw 2010; Chapter 7.
34 Egan 2015. For Blegen and Rawson's argument that these grid lines might have been used to mark places of importance within the Pylos Throne Room, see *PN* I, 84–85. This argument has since been reiterated in a recent study of the Pylos megaron by Ulrich Thaler (2012, 199).
35 See, for example, studies of the relationship between wall painters and vase painters, e.g., Bloedow 1997; Boulotis 2000; and more recently, Egan 2008, 2012.
36 Artists' grids are also extant in two squares in the Pylos Vestibule that retain no painted decoration (*Pylos* I, 74). It is likely, however, that these too once held rapport (textile) designs. On the topic of patterns, it is important to note that the Cretan wall paintings featuring motifs similar to those on the Pylos floors predate the Pylian examples in some cases by more than a century (LH I–IIIA2 vs. LH IIIB). The decision of the Pylian artists to employ these particular motifs is significant and may represent deliberate archaicizing tendencies, and perhaps a conscious harkening back to the illustrious Minoan past. Pylian artists employed many distinctive architectural and ceramic forms derived from Crete well after newer styles had developed on the Greek mainland (see especially Hiller 1996). Reliance on explicitly Cretan textile designs may also explain the adoption of the artist's grid, which was perhaps viewed as a Cretan technique that came "part and parcel" with certain decorative patterns.

37 *Pylos* I, 103–105; *Pylos* II, 179, pl. 109.
38 *Pylos* II, 33–34, 146, 166–167.
39 The fact of plaster's "malleability" raises the important question of why the patterns on the floors must derive from stone and/or textile, and not another medium altogether – for example, painted ceramics. In response to this interesting (and very relevant) question, I would suggest that although the floor painters seemingly had endless options, they would have been inclined to reproduce materials that ultimately "made sense" as floor coverings. The ultimate goal of the artists, as I see it, was to push the limits of reality but not to break with it entirely by introducing a material that would have seemed incongruous in its physical setting. That said, it is certainly possible not only that ceramics were the medium by which motifs were transmitted into the floor painters' repertoire, but also that individual Mycenaean viewers perceived some of these patterns as "ceramic," a reaction I am continuing to investigate.
40 Hirsch 1980, 459.
41 Hirsch 1980, 456.
42 Rodenwaldt 1919, 89–91, 95–97; Lamb 1921–1923b, 193–195. Also see the discussions in Hirsch 1977a, 27–36; and 1977b. Notably, zigzags were the only pattern detected in the squares of the floors in the Mycenae Throne Room and Vestibule. Both of these floors, however, were badly damaged and it is likely that they originally included squares with scattered circles, ridges, wavy lines, and concentric arcs, all of which are extant on the floor of the courtyard.
43 This strict replication of stone patterns is also evident on a gridded plaster floor from the Middle Bronze Age settlement at Trianda (Ialysos) on Rhodes. As described by the excavator, Toula Marketou (2013), the squares of this floor, also divided by a red grid, were painted to imitate veined polychrome marble flagstones.
44 See the illustration in Rodenwaldt 1919, 90, fig. 3. The suggestion that the combination of actual cut gypsum and painted floor grids in the Vestibule and Throne Room of Mycenae is indicative of a relationship between the two materials was also made by Hirsch (1980, 456).
45 *PT*, 325–328. On the use of this motif (*Zahnornament*) at Tiryns, see Rodenwaldt 1919, 30–31; at Pylos, see *Pylos* II, 34, 160–162.
46 Hackl 1912, 223–224. Intriguingly, this same "checkerboard" arrangement and strict combination of two types of motifs is also evident in the floor decoration of Hall 611 of the late 17th century BC palace at Tel Kabri in Israel. This floor, analyzed by Wolf-Dietrich Niemeier (1996), is composed of a red-painted grid enclosing squares measuring *c*. 0.40 m on a side and painted alternately with yellow and white backgrounds. On account of the filler motifs, which include flowers (irises and crocuses) and marbled stone patterns, Niemeier (1996, 1252-1253) identified this floor as the work of an itinerant Minoan craftsman, and suggested that this style of plaster floor decoration (with a painted grid) may have originated on Crete despite the absence of excavated parallels on the island.
47 That the Tiryns gridded floor represented a "marine style" carpet is suggested by the homogeneity of its motifs. The octopuses and dolphins are clear allusions to the sea, as are the tricurved arches, which, as Janice Crowley (1991, 221–223) has observed, were often used to depict water. See in particular the use of this motif as part of the landscape on the silver Siege Rhyton from Shaft Grave IV at Mycenae.
48 For the identification of this motif, which combines formal characteristics of the lily and the papyrus flower, see *PM* IV, 322–329.
49 Chapin 2004. For the primary identification of floral hybrids in Neopalatial landscape scenes, see *PM* II, 466, figs 266a, d, 275c; Cameron 1967; 1984.
50 Chapin 2004, 56.
51 That the ruler at Pylos (designated as the *wanax* on the site's Linear B tablets) may have had a connection with divinities and/or that he himself was considered divine was first proposed by Emmett Bennett (1958, 26) and has been discussed more recently by Cynthia Shelmerdine (1985, 78), Thomas Palaima (1995), Joseph Maran and Eftychia Stavrianopoulou (2007), and Susan Lupack (2014).
52 *Pylos* I, 84. Examples of the "knight's move" appear in Throne Room squares A10 and B8, D4 and F3, D4 and E6, D7 and F9, E7 and F9, G3 and I2, J2 and L1, K7 and L9, and in Portico squares A6 and B8, and A8 and B6. The pitfalls of relying on the knight's move as a "key" for understanding patterning in Mycenaean floor

decoration are discussed by Hirsch (1977b), who critiques Lamb's revision of Rodenwaldt's reconstruction of the painted floor of the courtyard of the megaron at Mycenae.
53 Gombrich 1979, 8–9.
54 Gombrich 1979, 9.
55 That such visual "confusion" may have had the power to impact the experience of the ancient viewer is suggested by the work of Alfred Gell (1989, 73–83), who has observed the use of complex patterns in modern art as a means by which an artist can "trap" his audience. Recently, this idea has been explored by Ellen Swift (2009, 101–103), who has ingeniously inferred how complex grid patterns in 4th-century AD Roman mosaics could visually overstimulate the viewer, thereby subordinating him to the artist, and by extension, the patron. Such a connection could easily be drawn at Pylos, where the "patron" to whom viewers became socially indebted was the *wanax*.
56 Egan 2015. An unusual feature *not* present in the floor grids in the megara at Mycenae and Tiryns, this "curious irregularity" was interpreted by Blegen and Rawson (*Pylos* I, 83) as a mistake – a "miscalculation" made by floor painters whose quality of work had "fallen off appreciably at the end of Mycenaean [LH] IIIB." More recently, this idea of a mistake has persisted in the work of Jeremy Rutter (2005, 31, 34), who has proposed that this "eyesore" in the Pylos megaron's Throne Room floor decoration was the result of a "rush job" during a hurried attempt to get the megaron up and running at the beginning of LH IIIB.
57 These corrected mistakes, previously unknown, were identified in 2012 in square L10 of the Throne Room by myself and conservator Alexandros Zokos when we cleaned the eastern corner of the room's floor under the auspices of the Hora Apotheke Reorganization Project (HARP).
58 Clarke 1975, 1979; Westgate 2007; Swift 2009; Molholt 2011.
59 Clarke 1979, 20.
60 Clarke 1979, 20–29. See in particular the figural floor decoration in the *Terme dei Sette Sapienti Rotunda*, the *Terme dei Cisiarii*, the *Terme di Buticosus*, and the *Terme di Nettuno*.

6

Sailing the Shining Sea: Maritime Textiles of the Bronze Age Aegean

Maria C. Shaw and Anne P. Chapin

Introduction[1*]

Textiles were used extensively on Aegean ships for ropes and rigging, sails, awnings, ikria (stern screens), and other equipment, but like the wooden ships themselves, these items hardly survive.[2] Information on Aegean vessels and their textiles is therefore drawn from a variety of sources, including the excavation of Bronze Age shipwrecks[3] and the study of Linear B texts that reference maritime activities.[4] Water craft from contemporary Egypt and the Near East offer useful comparative material, as do archaeological, historical, and documentary evidence for seafaring from the Classical, medieval, and early modern eras. Even Homer's poetic descriptions of ships and sailing preserve details of early maritime activity.[5] The most useful information for Aegean seafaring, however, is iconographic. Artistic images of ships and other vessels made in various artistic media throughout the Bronze Age, such as ship models and images of water craft on frescoes, seals, sealings, rings, pottery, and even "frying pans," all offer significant information on Aegean ship design and construction, including how textiles were used and adapted to seafaring in all periods of the Bronze Age.[6] In particular, the Flotilla Fresco from the south wall of Room 5 of the West House at Akrotiri, together with its companion painting, the Shipwreck scene of the room's north wall, stand out for the quality of their painting, their intricate detail, and their fine state of preservation.[7]

The Theran paintings of seagoing vessels are clear and specific in detail: the Flotilla Fresco features seven large, highly decorated ships characterized by curved hulls with crescentic shapes and bows and sterns that rise elegantly from the water (Fig. 6.1). The vessels are equipped with bowsprits, horizontal projections on their sterns, steering oars, ikria (stern screens),[8] and "awnings" to protect passengers from the natural elements. Two of the ships have their masts raised; four others have unstepped, lowered masts, and one vessel, perhaps a cargo ship, is under sail. All but the vessel under sail are propelled by paddlers – an old-fashioned and inefficient form of propulsion considering that oars were known in the Late Bronze Age – but perhaps paddling offered greater maneuverability as the ships approach the harbor of the "Arrival Town." An oared boat and a canoe-like vessel, also with curving hulls, accompany the fleet; an additional five small craft are moored in the harbor. The Shipwreck scene, which is more fragmentary, depicts undecorated vessels of similar crescentic design engaged in maritime conflict: a bowsprit is broken, and drowned men drift in the sea.

The appearance of the Theran fleet, particularly the distinctively curved profiles of the hulls, suggest significant connections with Egyptian ship design, which is better understood from the remains of surviving boats and from the many boat models and depictions of boats that survive from Egyptian archaeological contexts.[9] Sailing vessels are evidenced in Egyptian art as early as the Naqada II (Gerzean) period, c. 3500–3200 BC,[10] and actual ships dating to the 1st Dynasty (c. 3000 BC) were excavated from boat graves at Abydos, where 14 wooden boats of planked, mortise-and-tenon construction accompanied the burial of an early pharaoh, perhaps Hor Aha.[11] But royal Egyptian boat burials are rare. Most famous is the burial of two disassembled solar boats in pits beside Khufu's pyramid at Giza in the Old Kingdom, c. 2550 BC (one is now reassembled).[12] Boat models from elite tombs of Old, Middle, and New Kingdom date offer evidence for different types of river craft; noteworthy is a collection of 35 boat models buried in the Valley of the Kings (Thebes) with King Tutankhamun (ruled 1361–1352 BC). These models reproduce various types of vessels and state barges that would have been used by the pharaoh in real life, together with the solar boats that are symbolic of his passage to the afterlife.[13]

In the past, knowledge of Egyptian seagoing sailing vessels was limited to historical texts and rare artistic depictions of extended voyages, such as Queen Hatshepsut's (ruled 1504–1482 BC) trading expedition to Punt depicted on the walls of her mortuary temple at Deir el-Bahri.[14] This knowledge is now supplemented by the results of new excavations of ancient Egyptian shipyards at the Mersa/Wadi Gawasis seaport on the Red Sea.[15] Based on the results of various excavations and much study, an experimental replica seagoing craft, the Min of the Desert, was constructed and successfully sailed on the Red Sea; its speed, performance, and reliability in open-water testing exceeded expectations and show that modern scholars have significantly underestimated the seafaring capabilities of early mariners.[16] Similarly, the information gleaned from Aegean artistic and archaeological sources allowed modern-day ship builders to recreate a Minoan seagoing vessel, dubbed the Minoa. The replica ship, now housed in the Maritime

Fig. 6.1. Ship 1 from the Flotilla Fresco, West House, Akrotiri (drawing A. Chapin after Televantou 2000, pl. 3)

Museum of Crete in Chania, is 17 m long and 4 m wide, and is built on a keel 22 m long of bent cypress wood. The hull is made of split cypress logs lashed together ("sewn") with ropes. The hull was sealed externally with layers of cloth soaked in a fat-resin mixture, which gave the hull a white base color (though it is unknown whether ancient Minoan ship builders did the same). The hull was then painted to resemble the decorated hulls depicted in the Flotilla Fresco. The deck was partially planked, stem and stern, and furnished with an ikrion; benches for rowers were placed midship. The Minoa was equipped with oars; a central mast was installed and secured by fore- and backstays. Its white linen sail, like other sails depicted in Bronze Age art, is square in shape, reinforced by a network of lines (see below), and secured between upper and lower yards (booms). In celebration of the 2004 Olympic Games, the Minoa was sailed 210 nautical miles from Chania, Crete, to Piraeus, the port of Athens, with a volunteer crew of 24, who worked the oars in shifts and made the journey in 10 days. The average speed while rowing was 2.4 knots; that under sail, 3.2 knots. The experimental voyage confirmed the sea worthiness of the ancient Minoan design.[17]

These advances in our knowledge of Bronze Age seafaring underscore the need to consider how textiles were used on Aegean vessels. This brief review outlines the use of lines and rigging made of string, cord, and rope, together with evidence for equipment made from woven textiles – specifically, sails, tarpaulins, wind screens and/or awnings, and ikria. There are also depictions in Egyptian art of maritime textiles that may reflect imports from the Aegean. The Aegean evidence will be addressed here; Aegean exports to Egypt are discussed by Elizabeth Barber (Chapter 8).

Cordage

All ships have rigging and lines, and in the Bronze Age as today, these were made with strings, cords, and ropes of various weights and materials. The ship under sail in the Flotilla Fresco from Thera, together with accompanying depictions of masts and rigging on other ships in the naval procession, permits a detailed restoration of the mast (Fig. 6.2), the masthead with its rows of sheaves, the upper and lower yards (the boom) used to support the square sail, and the rigging, with two pairs of lifts (rope lines) for the upper yard (A) and the boom (B), halyards (C), braces to adjust the yards (D), and sheets to trim the sail (E).[18] Not depicted on the sailing ship, but probably an essential piece of rigging, are fore- and backstays which stabilize the mast longitudinally. It has been suggested that these were omitted from the painting either because the voyage was short, or for reasons of artistic convention.[19]

Evidence for ropes and rigging in antiquity comes from both archaeological remains and historical sources, and indicates that ropes around the Mediterranean were made from a variety of materials, including fibers from the date palm, the Doum (Dum) palm, esparto grass of southern Europe and north Africa, halfa grass of north Africa, papyrus of Egypt, and flax, grass, reed, and hemp.[20] Large ropes made of culms (hollow stems) of papyrus were recently discovered in Middle Kingdom (12th Dynasty) contexts in a cave at the ancient Egyptian harbor of Mersa/Wadi Gawasis, on the Red Sea. The papyrus (*Cyperus papyrus L.*) grows in the Nile Valley, so the ropes were brought to the coastal harbor, perhaps for an expedition to Punt.[21] Elsewhere, the shipwreck at Cape Gelidonya, Turkey, dated by radiocarbon to *c.* 1200 BC, preserves the remains of a Syrian vessel that was equipped with cordage made of Doum palm fibers, a plant

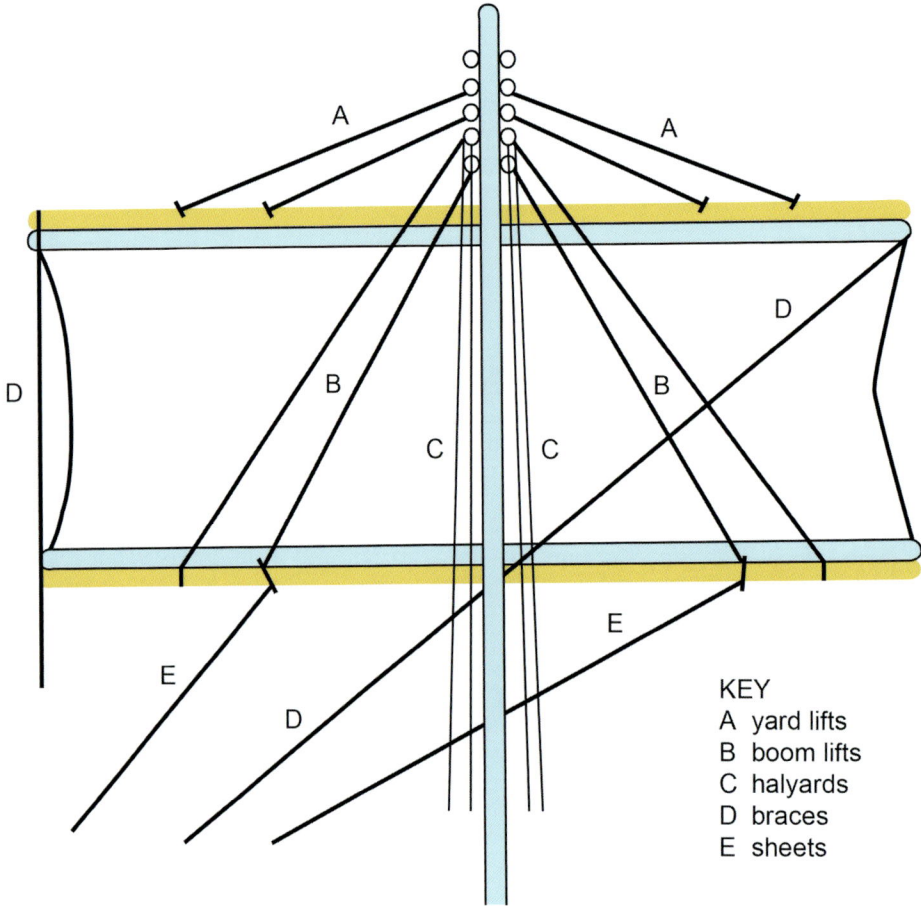

Fig. 6.2. The rigging of the sailing ship from the Flotilla Fresco, West House, Akrotiri, Thera (drawing A. Chapin and J. Silvia after Morgan 1988, fig. 71 and Wachsmann 2009, fig. 6.20)

native to Egypt and the Arabian peninsula.[22] As ships moved around the Mediterranean, they were probably regularly refitted in foreign ports, sometimes with ropes made of local materials, sometimes with imports.

Sails

Evidence for sails first appears in Aegean art in the EM III period, just before *c.* 2000 BC, when seals from Palaikastro and Adromyloi (near Praisos, Crete) depict ships with raised masts supported by fore- and backstays.[23] Later artistic depictions indicate that these sails would have been square or rectangular in shape and were stretched between a yard and a fixed boom, as is the case for the vessel depicted under sail in the Flotilla Fresco.[24] Ships such as these sail most effectively before the wind, but have limited ability to sail into the wind. For this reason, ships with fixed booms were often equipped with oars. After *c.* 1200 BC, this sailing technology was eliminated in favor of loose-footed, brailed sails, in which the boom was eliminated and replaced by lines attached to the foot of the sail. Brailed sails were easier to furl, and thus ships were more responsive to changing conditions of wind and sea.[25]

More information on early sails can be gleaned from artistic representations, textual references, and archaeological remains of sails that survive from northern European sites. The cargo ship in the Flotilla Fresco is depicted with a raised sail, plain white in color, which is presumably a reference to its manufacture from undyed linen. In the Classical era, sails were usually made of linen, but papyrus or rush sails were also known.[26] Naval inventories of the Classical era record that sails on Athenian triremes were of two kinds, heavy and light, perhaps made of flax and papyrus respectively.[27] Braided mats of goat hair made *c.* 2000 BC in Mesopotamia may also have served as sails for river boats,[28] and woolen sails are known from historical and archaeological sources in northern Europe as early as the 9th century AD.[29] Presumably these same materials were available for Bronze Age Aegean sail manufacture as well.

The size of prehistoric sails can be estimated from the vessel under sail in the Flotilla Fresco, which has a raised sail extending about half the length of the ship's hull.[30] Iris Tzachili estimates the length of the hull at 30 m, making its sail about 15 m in width; its height would have been about half its width.[31] It can be inferred from these estimates that sails were too large to weave on a single loom and had to be assembled from strips of sailcloth sewn together. This inference seems confirmed by the vertical seams indicated on the linen sails preserved on two boat models in King Tutankhamun's tomb.[32] And, since sails can tear in strong, gusting winds, it is possible that sails were reinforced, perhaps with leather or cordage, as evidenced by Minoan images of ships with cross-hatched sails found on some seals and sealings.[33] Cross-cultural comparison with reinforced woolen sails preserved in Scandanavian archaeological contexts suggests further that the cross-hatching on Aegean glyptic imagery may refer to woolen sails, which are more elastic than those made of linen.[34] This body of evidence reinforces the sense that Aegean mariners probably employed a variety of materials for sail manufacture.

Decorated sails

It is likely that most sails in the Bronze Age were undecorated, as in historical times, and served as functional, essential pieces of seafaring equipment. Lionel Casson, however, in his overview of ancient seafaring, observes that early sails were sometimes "painted" – that is, colored, whether purple in the case of a royal vessel, or black, as in the myth of Theseus in which the black sail declared the hero's demise as the ship returned home to Athens.[35] In Egypt, two model boats from King Tuthankhamun's tomb have linen sails dyed red with madder (*Rubia tinctorum*).[36] Pictorial evidence for patterned sails, however, is rare even in ancient Egypt. Notable is a wall relief carved on King Sahure's (Old Kingdom, 5th Dynasty) temple at Abusir that portrays the pharaoh's ship of state.[37] This depiction was among scenes of the king's trip to Syria, whence he returned home accompanied by "friendly" Asiatics aboard his ship.[38] The sail of the king's ship is divided into a regular pattern of small squares, each filled with a rosette-like motif which conceivably evokes a solar or stellar symbol. One might surmise that in this instance the elaborately decorated sail denotes the high status of the king. A second depiction of an Egyptian patterned sail comes from the late Old Kingdom tomb of Unas Ankh, the overseer of Upper Egypt (Theban Tomb 413); this sail is made of sewn strips of cloth, each woven with parallel zigzag designs in reddish brown.[39]

Aegean evidence for ornamental sails is even scarcer. In the Flotilla Fresco, a ship with a raised mast (Morgan's Ship 1[40]) could depict a furled sail decorated with black and red wavy bands (Fig. 6.1). Lifts running from the upper yard to the masthead look like they could raise a sail, and thus the rigging supports the furled sail identification. Alternately, however, the "sail" could also be viewed as a sunshade supported by a row of posts. The pictorial evidence is ambiguous, and neither reading is clearly more correct. If the wavy bands do belong to a sail, then it is possible they were made from other materials, perhaps leather to reinforce the upper and lower edges of the sail. When viewed against the highly decorative character of the ships in the naval procession, with each craft boasting an ornamental hull, bow, and stern, functional decoration on sails would not seem out of place. Additionally, the cross-hatched pattern of sails depicted on certain sealstones, mentioned above, here interpreted as reinforcement to sailcloth, could also have conceivably served a decorative function.

Awnings, wind screens, or deckhouses?

Details of the Flotilla Fresco indicate that the large processional ships and some of the smaller water craft were equipped with framed structures covered in textiles. It is generally assumed that these represent awnings to protect high-ranking passengers from the sun.[41] The awning of the smaller, oared ship painted just below the Departure Town is plain white, but the awnings of the large processional ships are decorated with colorful reverse-S stripes (in red and black on yellow ground) and wavy bands (red on white, and black on blue). According to this view, these patterns describe the edges of woven textiles that were stretched across wooden frameworks to create the sunshade, the full image of which cannot be seen because the ships were painted in profile.

An alternative view (presented here) posits that the thick upper edges of the "awnings" shown in the Flotilla Fresco represent bolts of cloth that could be unrolled during a sea voyage to protect the travelers from wind and the waves. Upon arriving at the harbor, such cloth would have been rolled up (as it appears in the fresco) and placed within what seem to be crutches at the tops of poles seen in the painting. This construction would have served, then, not just as an awning/sunshade, but also as a lightweight, portable deckhouse erected on wooden frames. A parallel can be found in contemporary Egypt, where large traveling ships used by high-ranking officials and royalty were often equipped with deckhouses built amidships using timber frames and covered with decorative tent-cloth embellished with lively patterns.[42] Checkerboard designs were popular, but spiral decoration suggesting Aegean influence is also known (Chapter 8). Later in the Classical era, side screens made of linen and hair were standard issue on Athenian triremes in the 4th century BC, as evidenced by their being listed in naval inventory inscriptions.[43]

The ceremonial context of these structures is also of interest. Fresco fragments of the miniature fresco from Ayia Irini, Keos (a composition which depicts preparations for feasting, among other vignettes), preserves a ship procession in which passengers are seated under a decorative "awning."[44] Another ship with a framed structure (an awning?) appears on a gold signet ring discovered near Tiryns that was decorated with ritual imagery.[45] The ritual and processional iconography collectively associated with this ship imagery indicates that "awnings" were probably detachable structures erected for special, ceremonial events that were taken apart afterwards to return the ship to its more functional uses.[46] Reinterpreting these structures as semi-open, tent-like deckhouses, rather than "awnings," simply transfers the ceremonial connections to deckhouses.

Ikria

Ikria, or stern screens, were important pieces of ceremonial deck furniture placed toward the sterns of large sailing ships. As pictured in art, and as analyzed by the two articles that follow, ikria were unroofed, lightweight, portable structures made from wooden frames, open at the front, and partially covered with oxhide or textiles that protected its occupant from wind and waves without obstructing the view forward (Fig. 6.12, below).[47] Ikria superficially resemble the aftercastles of large Egyptian ships, but differ in that they are lightweight and roofless whereas the Egyptian cabins are sturdy and roofed.[48] For this reason, the term "captain's cabin" is problematic, as a cabin is roofed and it is assumed, but not proven, that captains sat in them. The ikria's significance in the Neopalatial period as a symbol of naval power and elite social status is highlighted by a frieze of eight large-scale ikria painted in Room 4 of the West House at Akrotiri, adjacent to the Flotilla Fresco (Fig. 6.11).[49] The iconographic prominence of ikria in the West House suggested to the excavator that a sea captain owned and occupied the building.[50]

Ikria in Mycenaean painting passed largely unrecognized until the two studies by Maria C. Shaw presented here identified a frieze of four ikria among fresco fragments excavated by Christos Tsountas in 1886 from the palace at Mycenae; these fragments had previously been misidentified as a "Curtain Fresco". Shaw's detailed investigation of these fresco fragments, together with her carefully drawn inferences and conclusions, are now widely accepted by scholars of maritime history as indicative of the structure, materials, and function of ikria in both the Minoan and Mycenaean eras.[51] Indeed, these studies, together with another that investigates frescoed ship imagery at Pylos,[52] contributed to a greater appreciation of naval themes in Mycenaean art, the importance of which is now underscored by the recent identification (mentioned above) of a ship procession fresco in Hall 64 at Pylos (one of the ships preserves an ikrion) and newly excavated fresco fragments with ship imagery from Iklaina.

Painted "Ikria"[53] at Mycenae?

M. C. Shaw 1980, 167–179

In 1886, in the excavations which led to the discovery of the Late Bronze Age palace on the hilltop of the citadel of Mycenae (Fig. 6.4), Christos Tsountas found two important groups of painted stucco fragments, one in the Megaron, the other in a room directly to its north.[54] The first group of fragments was the basis for extensive discussion and partial restoration of the figurative composition which once adorned the walls of the megaron.[55] The wall painting of the other room, depicting a series of intriguing objects decorated with various patterns was, however, never restored or sufficiently analyzed, although a substantial part of it had been preserved (Fig. 6.3).[56] The present article proposes that the objects depicted in it might be interpreted on the basis of frescoes found in the LC I site of Akrotiri on Thera, and offers a pictorial restoration (Fig. 6.10) which may serve as an aid for further consideration of the motifs by others.[57]

Fig. 6.3. Fresco fragments from Room 33 of the palace at Mycenae (ArchEph 1887, pl. 12)

The "panels" and their restoration

A study of Figure 6.3, from a watercolor copy of the fragments as published by Tsountas (1887), makes it clear that we are dealing with four separate and similar objects, the features of which are rendered with a certain consistency, suggesting that the artist had a well-known model in mind. The drafting in the ancient painting is clear, but not always careful, to judge from some uneven lines and irregularities in the patterns. The palette is varied, consisting of black, red, yellow, blue, a grayish green, and one instance of brown, but the colors are applied flatly without any attempt to produce textures or shading.[58] Outlines and interior details are rendered in thick black lines and the general background is off-white.

6. *Sailing the Shining Sea: Maritime Textiles of the Bronze Age Aegean*   157

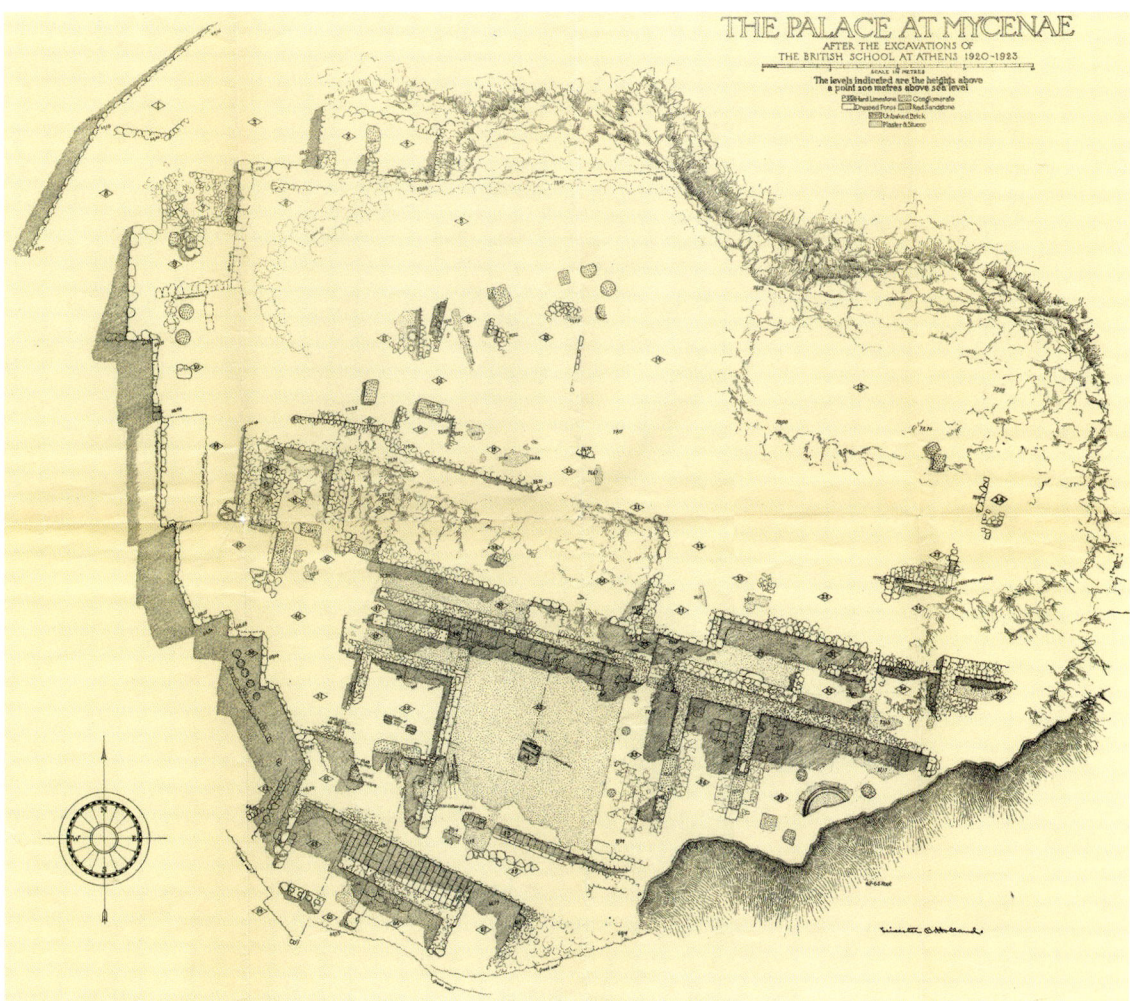

Fig. 6.4. Plan of palace area, Mycenae (after Wace 1921-1923, pl. 2)

Figures 6.6–6.9 offer restorations of the individual objects, Figure 6.10, a suggestion of a possible overall composition. The labels A–D refer to the individual objects which, for the sake of convenience, will be referred to as "panels," while the arabic numerals on the sides refer to the decorative patterns. Color is indicated only on the preserved fragments to help differentiate them from the restored parts (see Fig. 6.5, Key to Colors).

The most instructive of the painted fragments is that at the top right of Figure 6.3, in which we recognize a rectangular object, our Panel A, painted against an off-white background. The form can be briefly described as a wide rectangle, divided horizontally into four patterned friezes separated from each other by narrow yellow bands, and vertically by an upright, upwards-tapering, black "pole," marked by off-white Xs between off-white horizontal lines, forming hourglass patterns. The "pole" starts from the black band at the bottom of the panel and projects beyond a red band at the very top. It probably stops just below the first of a series of horizontal bands and stripes seen at the top of the fragment. Diagonal and perhaps also horizontal black "lines" starting at the top part of the pole help "attach" it to the sides of the panel. These are better preserved on the right side, where we see two slanting lines, one short, one long, reaching to the side. There seems to be a more complicated arrangement on the left, but, because of the poor preservation of paint and plaster at this point, the situation is impossible to define further. A tentative solution is suggested in Figure 6.6. Another odd detail is that the top narrow blue band of the panel is left unpainted where it extends on the left of the pole. The "scale" pattern (pattern no. 1) of the top frieze is repeated twice, in both cases on a blue ground; the "marbled" pattern of the second frieze (pattern no. 2) appears once, but is balanced by what may be a related design (pattern no. 3) in the bottom frieze, in both cases on a red ground. In general, the friezes are of somewhat unequal heights.

Panels B–D (Figs 6.3, 6.7–6.9) would seem to preserve bottom sections of the objects. From them we can derive further information: that the pole starts just at or within the bottom band, but does not extend below it; that this bottom band, which is clearly the base of the object, projects slightly beyond the sides of the panel with rounded ends and is painted sometimes in a solid color, sometimes with patterns; that narrow yellow bands between friezes are typical; that the patterns of each panel are rendered against two alternating colors, blue and red in Panel A, red and a grayish green in the other panels; and that there seem to be two basic patterns on each of these three panels. We should finally note that some of the friezes are made taller by means of what we might call "accessory" narrow bands in the same color that are sometimes plain, or sometimes with a different design of a simple nature, such as a wavy line or vertical bars. It seems quite likely that these accessory bands are a device to help the artist maintain the desired overall height of each object, when the character of the pattern does not allow a frieze to be tall enough.

With this information in mind, we can now restore Panel A with a degree of certainty, as in Figure 6.6. As restored, and without counting the projection of the pole at the top, this rectangle is c. 0.63 m high and 0.55 m wide.[59] Ambiguous points in its restoration are the details of the "attachment" of the "pole" to the left side of the panel, for reasons explained above, and the extent of the course of the horizontal black lines extending outwards from the top of the pole. These could extend indefinitely, linking the individual panels, or they could terminate at the sides of the panels, as in Figure 6.6. The completed Panel A, nevertheless, can now act as a model with the help of which one can easily complete the other, much more poorly preserved, panels, although a few points of doubt still remain.

Panel B (Fig. 6.7) is a most interesting one, because of the use, in two of its friezes, of nautili, the only figurative motif to appear in the entire painting. The nautili appear on a grayish green ground, which also serves as the color of the base; the other pattern, a network of lozenges, is preserved in two friezes on red. The higher nautilus frieze is further provided with an accessory band, marked by vertical bars, which brings it up to the combined height of the bottom nautilus frieze and the base of the panel. Because of the relatively small heights of the friezes here it was necessary to restore

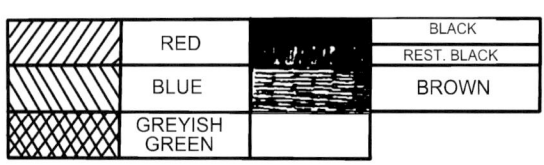

Fig. 6.5. Key to colors used in restorations

6. *Sailing the Shining Sea: Maritime Textiles of the Bronze Age Aegean* 159

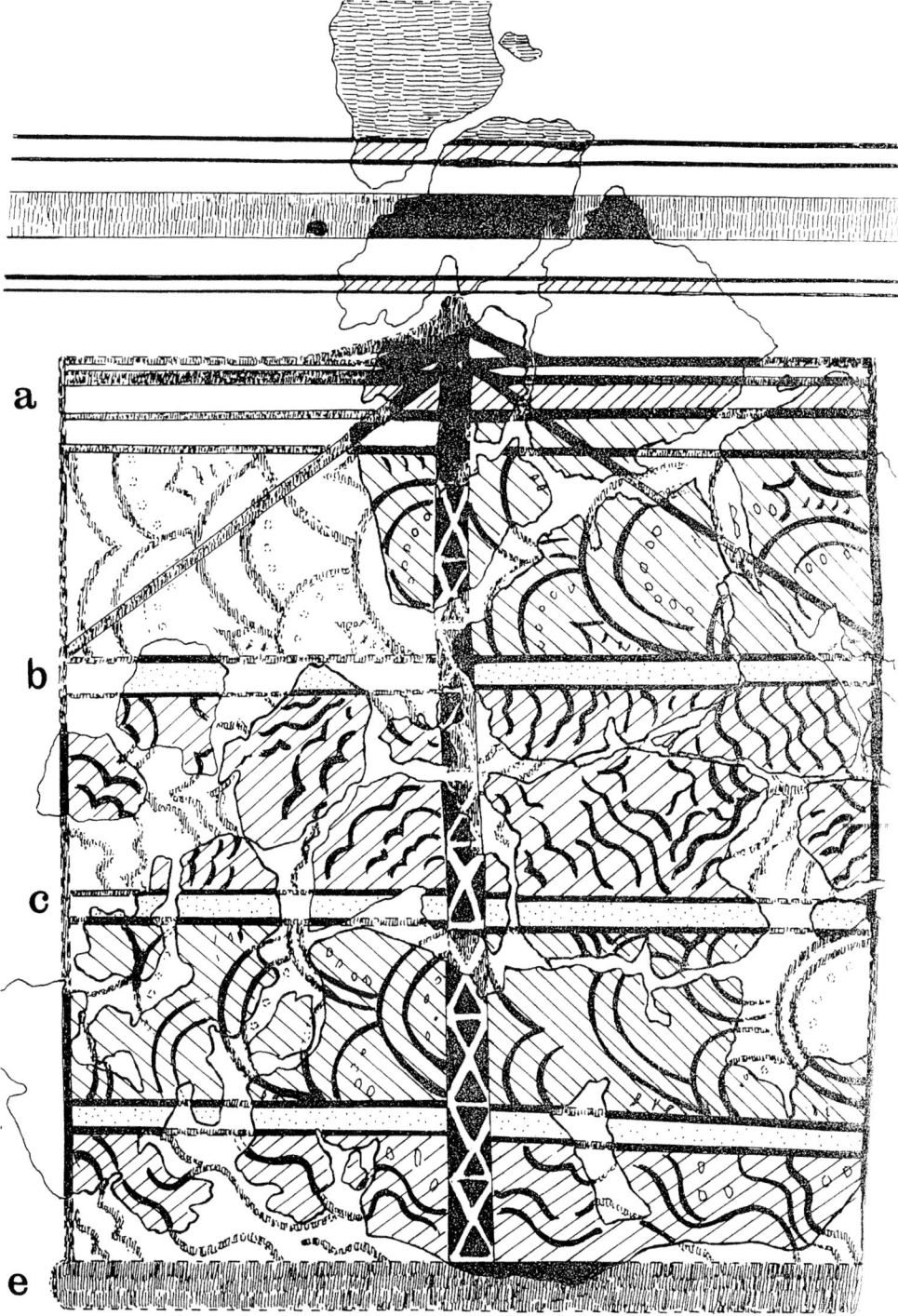

Fig. 6.6. Ikrion Panel A from Room 33 of the palace at Mycenae (restoration M. C. Shaw)

Fig. 6.7. Ikrion Panel B from Room 33 of the palace at Mycenae (restoration M. C. Shaw)

five of them, in contrast to the four taller ones of Panel A. On the basis of the bisection of each panel in equal halves by the pole, we can estimate the width of this panel, which proves to be slightly narrower than that of Panel A.

Panel C (Fig. 6.8) preserves the complete height of two friezes and part of a third. These are of unequal heights, the bottom one being approximately half of the one with the wavy lines, including the latter's barred borders. The narrow frieze on the bottom is in red, which is also the color of the patterned base; the taller one is grayish green. On the principle of an alternation of colors, patterns and, here, heights of friezes, it has been necessary to restore a total of four friezes to bring this panel up to the height of Panel A. The width of the rectangle cannot be estimated, but would probably be roughly that of A.

Panel D (Fig. 6.9), which is very poorly preserved, utilizes the same pattern of wavy lines that we encountered in Panel C, here also on a grayish green background. The accessory band is marked by a wavy line, instead of being barred as in C. The motif of its second frieze is uncertain, but it might depict stylized leaves, as suggested in the reconstruction. Four friezes have been restored for this panel, of which the top one is half the presumed height of the one with the leaf patterns. That halving the height of a frieze does not go against the principles of construction is attested in our panels, and can be seen in Panel A, where the bottom frieze is essentially half the height of the third one from the bottom, which has the marbled design.

For the overall composition we must rely on certain pictorial clues provided in a number of the fragments, and also on certain presumed principles of pictorial arrangement. The section of the fresco with Panel A (Figs 6.3, 6.6) preserves at the top a series of painted bands which undoubtedly form the upper framework of the painting. Starting at the top, there are: a broad brown band, edged by a narrower red one below it; a mediumsized band in black, separated from the two bands above and from the next narrow red band below it by strips of unpainted plaster. Unpainted plaster also occurs at the top and the sides of Panel A and at the bottom and left side of Panel B (Fig. 6.7). The sections of the fresco with Panels C and D (Figs 6.8–6.9) moreover, instruct us that below an unpainted area, ranging from 0.04 m to 0.07 m in height, at the bottom of each panel, there is a fairly broad black band, of which the full height is not preserved. This band, which presumably forms part of the lower framework of the composition, could have been part of a series of bands, like those at the top, or, more likely, since bands at the bottom of a composition are unusual in Aegean painting, it could have been part of a painted dado. Fragments within Panels C and D provide the crucial information that the objects depicted were aligned, although their distance from each other cannot now be estimated. The repetition of identical objects and their alignment suggest a frieze (Fig. 6.10), of which the maximum preserved height is about 1.0 m with the upper and lower bands included, or *c.* 0.70 m without them. With the maximum preserved margin between panels being *c.* 0.08 m and the width of each panel being *c.* 0.55 m, the maximum preserved length of the frieze is about 2.60 m.

Where exactly on the walls of a room such a frieze might have appeared is now a matter of conjecture. If it had been the only wall painting in the room, as one might be tempted to conclude from the lack of other fresco fragments, it might have occupied a prominent position, perhaps at or below the level of the lintels of any windows and doors. There it could have run as an independent frieze, or it may have surmounted a painted dado. Unfortunately, little is known about the heights of doors in Aegean buildings. In Minoan architecture, in

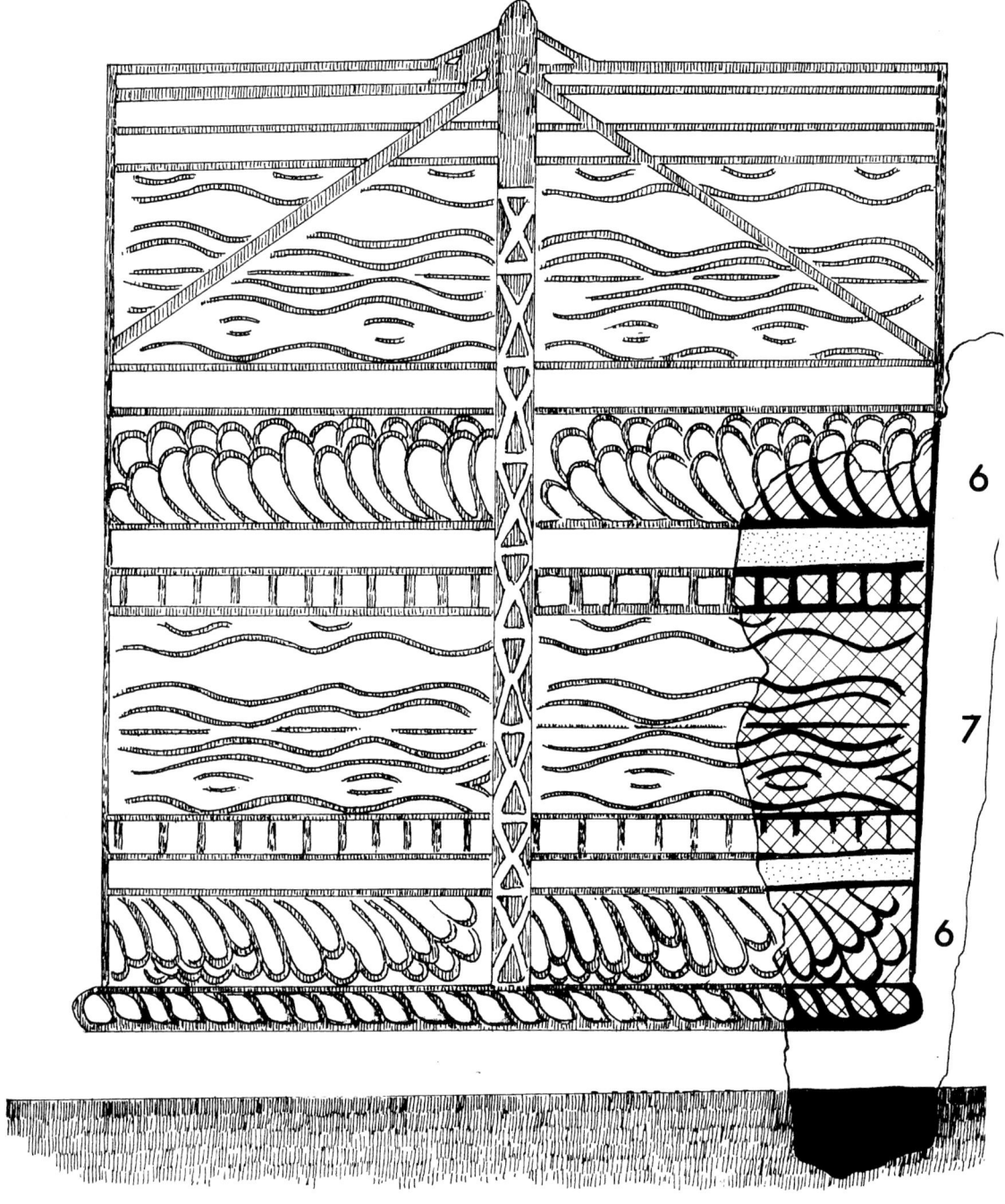

Fig. 6.8. Ikrion Panel C from Room 33 of the palace at Mycenae (restoration M. C. Shaw)

6. Sailing the Shining Sea: Maritime Textiles of the Bronze Age Aegean

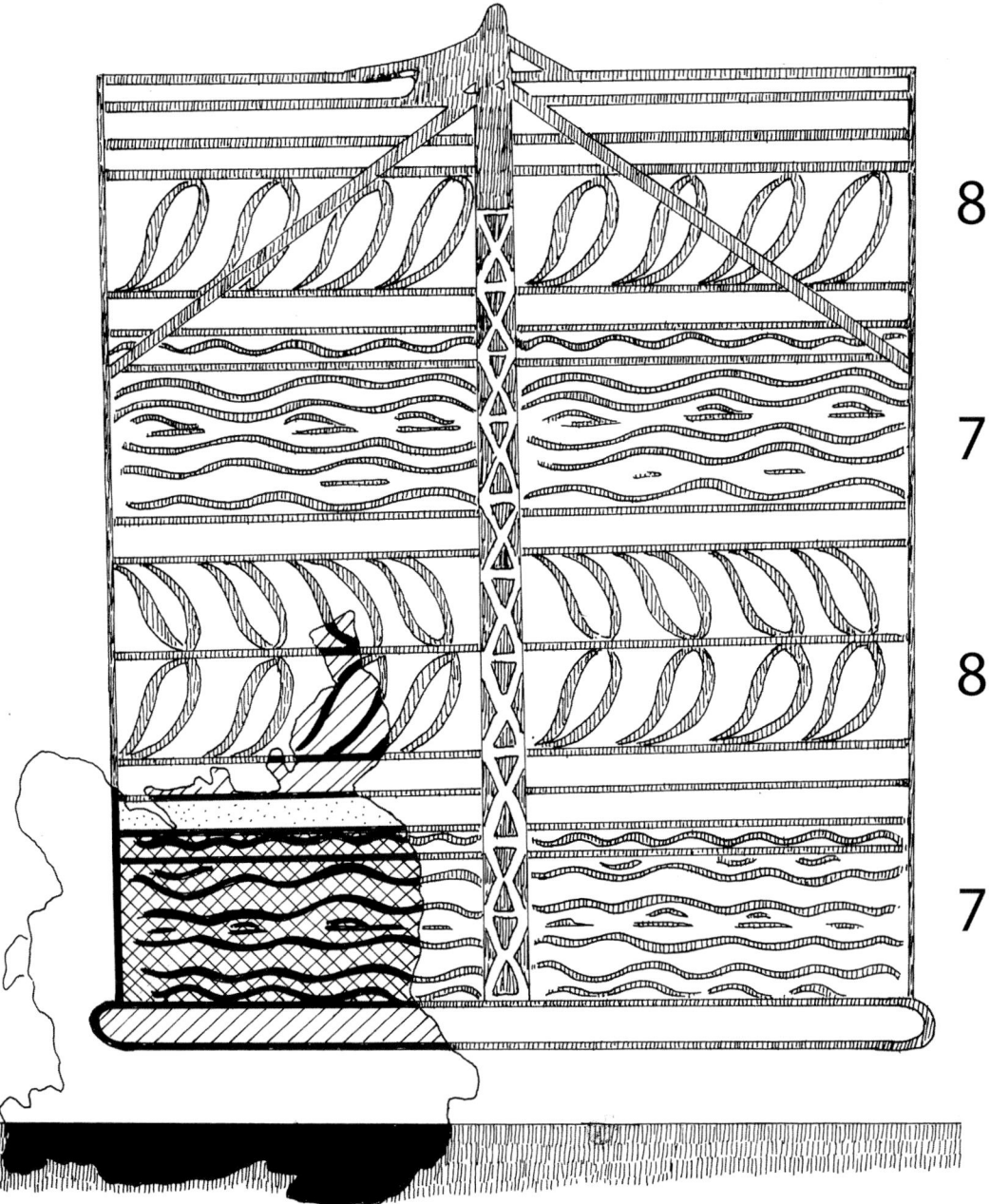

Fig. 6.9. Ikrion Panel D from Room 33 of the palace at Mycenae (restoration M. C. Shaw)

Fig. 6.10. Ikrion Fresco from Room 33 of the palace at Mycenae as a frieze (restoration M. C. Shaw)

a few examples where the height could be estimated, it ranged from 1.60 m to 2.00 m.[60] If this range also applies to Mycenaean architecture, a dado here might cover an area of some 0.60–1.00 m in height. As a frieze, the painting could well have appeared on more than one wall.

The specific architectural context of the painting is important in any consideration of its meaning and purpose. Here we are confronted with a problem. Although Tsountas, in his *Praktika* report, was quite specific and clear about the room in which the stucco fragments were found, namely in Room 30 (Fig. 6.4), subsequent authors (Winifred Lamb, Alan Wace, and others) attributed the painting to a more elaborate nearby room with plastered benches (Fig. 6.4, Room 33).[61] Since the discrepancy in attribution was not argued or explained, I assume that it was a matter of misunderstanding, perhaps from overreliance on the brief *Ephemeris* report, where the reference to the location of the painting is indeed vague.[62]

Room 30 (Fig. 6.4), which, trusting the word of the excavator, I believe to be the correct context of the painting, is about 4.80 m square (an estimate derived from the plan) and somewhat irregular in shape. The walls are too badly preserved for us to know much about its original appearance; we are better informed about rooms in its immediate vicinity. Directly to its west were three adjacent units: the large room with benches, already mentioned above and described by Wace as the "Anteroom of the Domestic Quarters" (Room 33), and two contiguous narrow spaces (Rooms 31–32), generally thought to be the location of the two flights of a staircase which led to the apartments on the higher terrace north of the Megaron, and possibly also to the top of the Megaron.[63] A *sottoscala* (Room 32), acting as a storeroom, under the southern flight could apparently be entered from the little Room 30, where there is a break in the west wall. The main entrance to the latter is unknown. Wace suggested that it was entered from the east, where the wall is now destroyed, while Tsountas proposed that one could descend into it from the rooms to the north.[64] Its location, nevertheless, indicates clearly that it was part of the domestic quarters; Tsountas offered the suggestion that it may have served as a bedroom.[65] The south wall of the room is the easternmost end of the north wall of the Megaron, while its north wall is closer in orientation to the north walls of the rooms to the west, rather than to the Megaron, but set a little further south than they. To its east there is a narrow room of unknown function with very thick walls.

From this survey one derives the impression that the room was created between pre-existing structures, and this in turn suggests that its construction may belong to an architectural phase late in the history of the palace. Wace, in fact, attributes the room to the latest remodelings of the palace, which would make it later than the construction of the Megaron, the Antechamber to the Domestic Quarters and the associated staircase, as well as the blocking of the east end of the South Corridor, which had once continued further east.[66] Assuming that the fresco was still on the walls of the room in the final days of the palace, we can reasonably infer a date as late as the LH IIIB period. Lamb notes that the stucco fragments were somewhat affected by fire (like those of the frieze in the Megaron), and this may well have been the fire generally associated with the final destruction of the palace at the end of the LH IIIB period.[67]

The frequent occurrence of the patterns adorning our panels in LH IIIB contexts at other Mycenaean sites further supports the proposed date, although there is at present no systematic study of such patterns which can provide definite dating criteria. Eight major patterns occur, mostly well defined, some less so because of poor preservation. The patterns have been examined in some detail by Helga Reusch in her doctoral dissertation,[68] but a few comments can be added here and the information brought up to date.

Pattern no. 1 on a blue ground occurs twice in Panel A (Fig. 6.6); it consists of overlapping arcs, each outlined twice and with a series of little circles along the interior outline. The junctions of the arcs are occasionally marked by a filling ornament consisting of a chevron and small parallel lines. The pattern corresponds most closely to Arne Furumark's "concentric arcs," but it also relates to the generally more regularly rendered scale pattern, familiar from representations of textiles, but known also now from a painted floor in the LH IIIB palace at Pylos.[69]

Pattern no. 2, occurring once in Panel A on a red ground, simulates veined stone, such as alabaster or some types of marble (Fig. 6.6). It consists of four groups of wavy and/or scalloped parallel lines, each group occupying a corner of the panels created by the bisection of each patterned frieze by the pole. This arrangement leaves a roughly lozenge-shaped gap at the center. The closest parallel is on painted floors of Mycenaean palaces, but also, to use one of Furumark's expressions, in a version which represents an "excerpt" of the more complete pattern, it is encountered frequently on painted dadoes as well, in Minoan, Theran, and Mycenaean buildings.[70]

Pattern no. 3, appearing in the bottom frieze of Panel A, also seems like an abbreviated version of pattern no. 2, or else one closely related (Fig. 6.6). Successive wavy lines cross the frieze diagonally, possibly converging in groups towards the bottom. In the spaces in between are, here and there, a few small circles. The closest example in painting, but much more polychrome than our example and without the small circles, is the pattern of the painted dado in the Throne Room at Knossos.[71] The pattern also occurs in figurative scenes as a subsidiary landscape feature, as for instance in one of the Vapheio cups, in which it clearly indicates a rocky terrain.[72] A derivative of this pattern appears in LH III pottery, examples of which are to be found under Furumark's "Rock Pattern III."[73] Although suitable for dadoes, to my knowledge the motif has not appeared on painted floors or as a textile design. It lends itself best to a narrow, frieze-like arrangement, and was probably selected for the same reason here, in lieu of the related broader rock pattern higher up in the same panel.

Pattern no. 4, a frieze of nautili, appropriately appearing on a greenish ground, perhaps evoking the sea, occurs twice in Panel B (Fig. 6.7). The motif is one of the most favored in Aegean art and its use in various media hardly needs illustration. Suffice it here to state that in the present form, as a frieze, it occurs in several wall paintings in the Palace at Pylos, where it frequently is a subsidiary band to larger scenes.[74] Nautili are also known to have appeared along with octopuses and types of fish on painted panels of Mycenaean floors, as pointed out by Reusch.[75] To my knowledge it is not known as a pattern, woven or embroidered, on clothing, although such use is not inconceivable since birds, flying fish, butterflies, and various fictitious animals occur among such patterns.[76]

Pattern no. 5, which appears twice on a red ground in Panel B, is an ennobled version of Furumark's Diaper Net pattern (Fig. 6.7).[77] It is formed of small independent lozenges set within a network of larger ones, clearly an expansible surface pattern, curtailed here to fit the narrow frieze. A stem with leaves appears in each lozenge as a filling ornament. Although networks with rhomboidal units occur on painted Mycenaean floors, there is no exact parallel for the rectilinear version with floral motif.[78] On the other hand, rectilinear variants of the motif, with or without a floral ornament, occur frequently in Aegean representations of dress; one such motif has also been used on a painted dado at Pylos, and on what looks like the facade of a building in a miniature fresco recently found at Thebes.[79]

Pattern no. 6, which occurs once, possibly twice, against a red background in the existing part of Panel C is difficult to identify, as only a small part of it is preserved (Fig. 6.8). In the bottom frieze we note some leaf- or petal-like patterns in two overlapping layers, somewhat reminiscent of half-rosette friezes in carved stone or painting, where such an overlapping of petals occurs.[80] Whether the "petals" were arranged in one direction, as shown in our restoration, or met antithetically at the median point at the pole is now impossible to determine.

Pattern no. 7, occurring in both Panels C and D on a grayish green ground, has been identified by Reusch as a simulation of wood graining, which is quite plausible (Fig. 6.8).[81] More realistic versions than the ones on our panels occur in broad bands in wall painting where their meaning as wood is clear.[82] Whatever its original derivation, the pattern also occurs in a close variant on belts and bands of Aegean dress.[83]

Pattern no 8, appearing on a red ground in Panel D, calls for little comment, since its restoration is largely conjectural (Fig. 6.9). It could be a ribbed pattern, as Reusch suggested, or a foliate band of which examples abound in vase painting from LM IA on.[84] It is known both in a simple version, as restored in the top frieze of Panel D, or as leaves arranged on either side of a stem, as in the broader, lower frieze. In its simple form it seems to relate to a pattern used in bands in Aegean dress.[85]

In addition to the above patterns there are some subsidiary motifs, such as the barred border above the nautili of Panel B, a wavy line above the wood graining of Panel D and what looks like a rope pattern at the base of Panel C. The simplicity of design of the first two makes it unnecessary to discuss them in detail. Suffice it here to note that they both occur in narrow bands and borders in representations of female dress and that the barred border, in this simple or in more elaborate versions, is a common motif for bands framing various painted representations.[86] The rope pattern occurs occasionally as a border pattern in carved metal and ivory objects, but is also known to have been used as a decorative band in Theran representations of dress.[87]

There remains the hourglass pattern on the black poles, preserved in Panels A and B (Figs 6.6, 6.7). Reusch suggested that the pole was tied with light-colored strings and the closest parallel in Mycenaean painting I can discover is the strapping of the yoke in chariots, as in the Hunting Scene fresco from the Palace of Tiryns.[88]

Interpretation

With these general conclusions about the appearance, context, and possible date of the composition in mind, we can now turn to the problem of interpretation. One of the difficulties at the time of discovery, as Rodenwaldt pointed out, was that this type of wall decoration was unparalleled, and the main clue had to be provided by the patterns associated with textile and painted floor decoration. On the basis of the latter type, Rodenwaldt offered the best interpretation possible at that time, namely that the painting depicts a series of wall hangings. This explanation supported his earlier suggestion that Mycenaean painted floors simulated patterned carpets, especially as the same odd combination of marine and abstract designs was found both on floors and in the painting under consideration.[89] This interpretation was generally accepted, first by Reusch, who concentrated on an analysis of the motifs, later by Lamb, and also by Wace who referred to the objects depicted as "curtains," a convenient nickname which was adopted in later literature.[90]

There is no doubt that Rodenwaldt's interpretation points in the right direction, but, as he himself admitted, there are difficulties associated with it. The theory that these are curtains, whether seriously meant or not, is not plausible, for the heavy framework of the objects would be superfluous in that case. So far there is no evidence for depictions of curtains in Aegean painting, except perhaps for one case in a small bedroom in the Old Palace of Phaistos, where Doro Levi interpreted traces of a simple wall painting behind a bench-bed as possibly simulating curtains.[91] The character of the framework in our objects, on the other hand, raises doubts also about Rodenwaldt's interpretation of suspended rugs. If we bear in mind that the attachment of the central pole to the sides implies further upright supports there, presumably concealed behind the fabric, and that the base, the top band, and the yellow bands between friezes may have been in wood, rather than in cloth or other flexible material, we are confronted with an unnecessarily heavy frame. A tapestry or rug could easily be kept taut if suspended from the two upper corners and, if excessively bulky and heavy, could be supported further by a couple of wooden cross-bars at top and bottom. The pole was clearly not used for carrying the object around, in the mode of a standard, for then it should have extended below the base. Such a sturdy framework suggests to me that we have a three-dimensional object of which we only see one side.

On the above reasoning, I would like to propose that the Mycenae Fresco represents ikria (formerly defined by myself as boat cabins, but better recognized today as stern screens) and are Mainland versions of the Late Bronze (LB) I type familiar now from two wall paintings from Thera: one a miniature fresco depicting, among other scenes, a fleet, the other a frieze in which ikria in one-to-one scale are the exclusive motif.[92] Two samples of Theran ikria are reproduced here in Figures 6.11 and 6.12, one from each painting.

A glance at the Theran examples and our reconstructed motifs from Mycenae shows that there are admittedly some radical divergencies but also crucial similarities between the two sets. These can be summarized as follows: 1. the overall proportions, those of a wide rectangle, are typical in both cases; 2. the "walls" in each case are made of a flexible material, clearly leather at Thera, possibly woven fabric at Mycenae; 3. narrow horizontal bands divide the surfaces, in the case of Thera supporting a continuous piece of hide, at Mycenae acting both as a frame and as a point of junction of separate segments of fabric; 4. instead of the three poles we see in the Theran examples, only a central one occurs at Mycenae, although, as pointed out above, there may have been further upright supports at the sides.

From the above observations, it is clear that at Mycenae we may have a humbler version of a stern screen, the main objection being that structurally there is a basic difference in the termination of the objects at the top. It is now difficult to determine whether the walls

Fig. 6.11. Ikria Fresco from Room 4, West House, Akrotiri (reconstruction A. Kassios, courtesy C. Palyvou)

6. *Sailing the Shining Sea: Maritime Textiles of the Bronze Age Aegean* 169

Fig. 6.12. Detail of a boat from the Flotilla Fresco of Room 5, West House, Akrotiri (M. C. Shaw after Gray 1974, pl. xiii)

of the Mycenaean ikrion were shorter than the height of a man seated inside, and open at the top, like the Theran ones, or whether the later model, if this is what we have, may have been roofed. A stone vase with relief decoration from Epidaurus depicts a boat with an unroofed "cabin," while the well-known signet ring from Tiryns featuring a ship shows a roofed structure set amidships with two figures seated inside (Figs 6.13 and 6.14).[93] This could be a cabin, or a schematized rendering of the awning which appears in several of the ships in the Theran miniature fresco. In view of the ambiguity one wonders whether our Mycenaean ikria also combine various elements in one hybrid symbol, and whether the central pole with the strappings evokes the idea of a mast overlapping a cabin amidships or a sail.

The matter of composition could not be without significance and we might note in this connection that in both cases the ikria are arranged in a frieze-like manner, marked by a series of stripes at the top and with a dark band at the bottom – the beginning of a dado at Thera, but of uncertain restoration at Mycenae. As pointed out above, the decorative nature of the Mycenaean composition requires that the objects be placed at a limited distance from each other, much in the fashion of the ikria in the Theran frieze.

One final note of interest is that both frescoes may have decorated rooms of comparable function. The Theran frieze ran on the four walls of a unit interpreted by the excavator as a bedroom, and the same function was suggested for the little Mycenaean room by Tsountas.[94] Whether a bedroom or not, the latter was certainly located within the domestic quarters of the palace at Mycenae and was not a state or formal room.

Up to the time of the discovery of the West House frescoes at Thera, the predominant symbols of Aegean military power known from wall painting had been those connected with

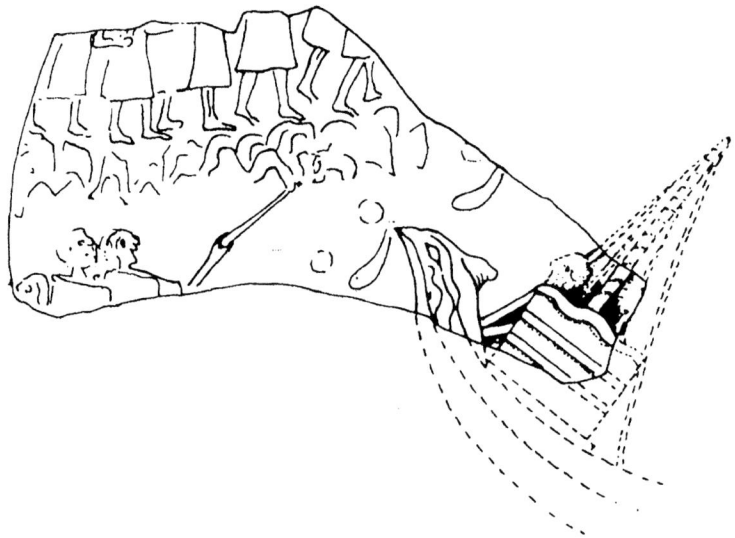

Fig. 6.13. Stone vase with relief decoration depicting a boat with an ikrion, from Epidauros (drawing M. C. Shaw after RA 1971, fig. 3, with restorations added in dotted lines)

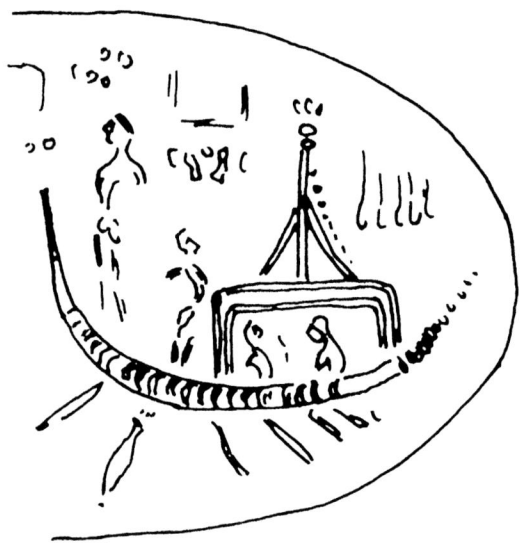

Fig. 6.14. Signet ring with ship, from Tiryns. (after PM II, 245, fig. 142)

armor, the outstanding example of which is the figure-eight shield, familiar from both narrative scenes and decorative friezes, where it was used in a manner comparable to that of the ikria in the Theran frieze.[95] Considering that the Aegean peoples had been primarily seafarers and that maritime supremacy, echoed in historical times in the writings of Thucydides, was clearly a crucial matter, it was surprising that ostentation of naval power did not play a more significant role in painted scenes. The existence of such representations has, however, long been suspected through possible reflections in the minor arts, foremost among which is the scene on the silver Siege Rhyton from the shaft graves at Mycenae, and to a limited extent through depictions of ships on rings and seals.[96] Thera now, in part, fills this gap and raises the hope that more of such scenes may be found in future excavations, and also that some unrecognized remnants of these themes will be understood with the help of this new information.

I would also like to propose two more identifications of fresco scenes, one certain, the other rather speculative but worth mentioning. In light of the well-executed ships in the Theran Fleet

Fresco, we can now recognize the painting of a ship on two stucco fragments from the Palace of Pylos, preserving part of the mast with rigging, an upper framework in checkerboard pattern, and part of a sail or central awning (Fig. 6.15).[97] Tentatively, two fresco fragments found with the Processional Fresco in the Palace at Thebes, which Reusch had already noted as possibly relating to the "wall hangings" at Mycenae (Fig. 6.16),[98] may be identified as parts of ikria or sails. In Figure 6.16b one sees what may be the edge of a woman's skirt next to an object with a straight vertical side marked by a thick black line. This defines an area to the left divided horizontally by narrow and broad bands marked by decorative patterns, some easily attributable to the type of ornament that is characteristic of patterned fabrics. Fragment *a* seems to be part of the same or a similar object, and it further includes the interesting detail of a narrow yellow band, next to a barred border above and a band with a rock-pattern below. If indeed the fragments preserve a depiction of objects similar to those at Mycenae, and if both are screens or sails, we would have here a convenient chronological intermediary between the Theran example and the LH IIIB version at Mycenae.[99]

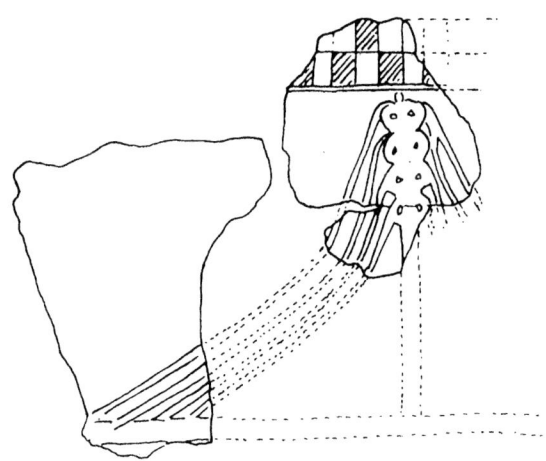

Fig. 6.15. Fresco with ship from the palace at Pylos (restoration M. C. Shaw after fragments depicted in Pylos II, pl. 113)

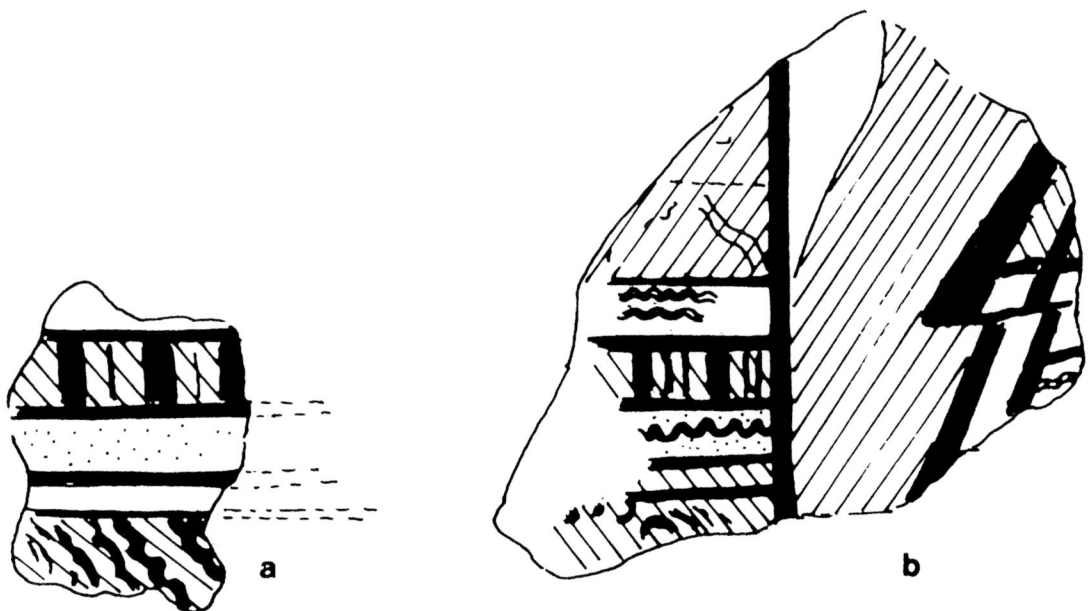

Fig. 6.16. Two fresco fragments from the palace at Thebes (after Reusch 1956, pl. 13, figs 38, 39)

It is not surprising that ikria should stand as a symbol of naval power and authority. In a recent article discussing the character of the Theran boats, Lyvia Morgan Brown argues for an analogous use of deck cabins as symbols of authority in some Egyptian painted scenes, some of her examples demonstrating even certain structural similarities with the Theran ones.[100] To her Egyptian parallels I would like to add a rather striking example, particularly relevant as it appears on one of the painted relief scenes in the Temple of Hatshepsut at Deir el-Bahri, dating to the beginning of the 15th century, and therefore not far removed in time from the Theran settlement. The scene depicts the transport by boats of Hatshepsut's colossal statue. One of the "cabins," or perhaps more properly a baldachin, shown on one of the ships, and reproduced here in Figure 6.17, is striking indeed.[101] Like the Theran ikria on the ships of the miniature frieze, it has three posts on the visible side and a parapet wall, low enough to reveal the seated officer's head, shoulders, and raised arm holding a staff. As at Thera, the baldachin rests on a raised platform and is movable: it can be used on and off the ship.[102] As at Thera, a guard or attendant is seated outside the ikrion, knees bent, his back resting against its wall. One also wonders whether the presence of an animal, here a sphinx symbolic of the Queen's power, next to the baldachin, may not perform a function similar to that of the animals appearing by the ikria at the stern of the ships in the Theran miniature fresco.

In a wider context, the Fleet Fresco of Thera and, by association, the possible painted ikria at Mycenae, may be seen as a pictorial display of naval power analogous to the ostentatious enumeration of contingents in the Homeric Catalogue of Ships. This is not to imply that there is a direct and strict thematic connection, but simply that both painted fleets and ikria and the recited list of ships deal with subject matter deemed crucial and appealing to painter and poet alike and to an Aegean "public" that enjoyed and patronized these arts. On the analogy of the use of the theme of a besieged city, it would not be surprising if more of such motifs were shared. In the light of this thought, we might even see the tropical scene of the miniature fresco at Thera, whatever its historical locale and occasion, as an artist's vision of seamen's tales about far off and exotic lands, a vision romantically exaggerated like the enchanted places encountered by Homer's Odysseus.[103] Thus Thera can be seen as part of that world which is in the background of the epics. This world, as the excavations now reveal, was one of intense internationalism and cultural ferment, in which Thera, and possibly more of the Cyclades, may have played the role of a center in which multiple cultural streams converged.[104]

Fig. 6.17. Detail of a ship, Deir el-Bahri (after Naville 1906, pl. 12)

Ship Cabins of the Bronze Age Aegean

M. C. Shaw 1982, 53–58

The impressive plethora of articles and studies relating to the LC I frescoes (*c.* 1550–1500 BC) discovered since 1967 on the island of Thera, is eloquent testimony to the treasure house of information on that era, now made available to us with these representations. Not least among the beneficiaries is the field of marine archaeology, for which two frescoes have been particularly crucial. One, the Flotilla Fresco rendered in miniature style in Room 5 of the West House at Akrotiri, depicts a fleet of ships and boats moving to the right against a backdrop of coastal towns and a hilly horizon.[105] The other is a decorative frieze of ikria from Room 4 of the same building, in which ship "cabins" (better identified as stern screens) are arrayed one next to the other and repeated on identical scale, rather in the fashion of ornamental wallpaper (Fig. 6.11).[106] It is of interest that both frescoes were found in the same house, which, as already suggested, must belong to a person deeply involved with the sea and seafaring.

It was the scrutiny of the details of these two representations which helped me lately to identify a fragmentary fresco from the LH III (*c.* 1300–1200 BC) palace at Mycenae, discovered in 1886,[107] as possibly depicting a frieze of ikria (stern screens) comparable in composition to the ikria panels of the West House at Thera. Figures 6.10 and 6.11 reproduce examples of ikria from the friezes at Mycenae and Thera. In both cases the artist shows only one side, that is, he has not rendered it as a three-dimensional object. Of these, the Mycenaean depiction, incompletely preserved in the original painting, has been restored both by analogy to the complete Theran examples and by additional details preserved in three more ikria in the Mycenaean fresco.[108]

A model of the "cabin" – that is, the stern screen (Fig. 6.18) – shows more clearly how this, and possibly other Bronze Age ikria, may have been constructed and used. Each side of the Mycenaean structure, for instance, was made up of joined rectangular pieces of patterned material, probably heavy cloth, attached to a wooden framework. From fresco fragments restored as Panels A–D (Figs 6.6–6.9), it is evident that this framework consisted of three horizontal slats (all rendered in yellow), placed over the joints of the textile bands, with a fourth one at the very top of each side wall. A fifth horizontal band at the bottom projects slightly beyond the vertical edges of the ikrion wall in rounded ends. This could be the edge of a wooden floor, or, for reasons to be discussed below, it may be just a wooden base, part of the framework. The only visible vertical support is a central pole, or rod, tapering slightly upwards and projecting just above the top of the ikrion wall. The pole does not project below the bottom of the structure, thus excluding the possibility

Fig. 6.18. Model of a Mycenaean ship cabin (ikrion) (M. C. Shaw)

that the objects portrayed are banners or standards that would be carried by a long pole.[109] The pole, which is painted black, is marked by superimposed X patterns divided by horizontal lines, all rendered in white, that suggest they were bound with string or leather straps.[110] Next we note that there is a series of straight, black diagonal lines at the top, which, in my view, also depict straps, or strings, used for holding the central pole in place. The first set of lines at the top joins the vertical pole to the black horizontal line, or string; further down, the other set reaches down to the topmost of the yellow slats. The straps/strings are rendered in white against the black pole, but in black beyond it, presumably for pictorial clarity.

In drawing an analogy with the ikria in the Theran Ikria Fresco (Fig. 6.11), one can note the overall similarity in shape, the presence of horizontal slats, and the slat or beam at the base of both sets of structures. The fact that the upper corners of the ikria are nowhere preserved in the extant Mycenaean fresco fragments leaves it unclear whether there were side poles here also, as at Thera. It is possible that such poles existed and that they were partially covered, at least in the lower part of the screen walls, by the textile bands, which would have been placed over them. The tops of these poles may have been visible in the original complete painting. The actual height of the Mycenaean ikria could have been the same as that of the Theran one, the latter with walls short enough to allow at least the head of the captain seated inside to be visible from the outside, as we learn from the miniature fresco from the same site, where ikria are actually depicted on ships (Fig. 6.12). Unfortunately, the Mycenaean fresco is devoid of human figures and there is no relevant supplementary information from other Mycenaean frescoes.

One definite difference between the two examples, however, still remains: that the walls of the Theran ikria are curved at the top, while those at Mycenae are straight. Also, in the latter site the vertical pole was placed on the exterior of the ikrion wall and was entirely visible.

A possible explanation for the former difference is the use of different materials for making an ikrion in each case. In the Mycenaean example I suggested, above, that bands of cloth joined horizontally and were covered by wooden slats. In the Theran frieze it is obvious that each side was made of a single piece of ox hide, since the dappling continues uninterruptedly above and below the slats of the framework. The arched upper rim may have, to some extent, been determined by the use of the skin of a single animal in each case, especially if the trimmed foreparts marked the top. In this connection one should recall the analogous curving termination of one of the two types of Bronze Age body-shield, the best example being depicted on the inlaid lion-hunt dagger from the Shaft Grave 4 at Mycenae.[111] Shields of this size are described in Homer as made of ox hide in noun-epithet formulas considered by philologists to be of great antiquity, and we must remember that the related and contemporary figure-eight shield is generally depicted in art with dappling patterns.[112]

The second difference still remains: that at Mycenae the central pole was placed on the exterior and that the side poles (if any) may not have been visible. There is, however, a detail which, if correctly interpreted, may tone down these discrepancies, or even explain them away. Particularly crucial are the long diagonal lines, which, I assumed above, tied the central pole to the vertical sides of the ikrion wall at the level of the top yellow slat. In a discussion with students of why the pole was "tied" there, rather than at the very top slat, the idea emerged that this choice may have been significant and that this would leave the top cloth band free to be folded or rolled down, as shown in the model (Fig. 6.18).[113] The detachment of the central pole, from at least the upper part of the ikrion, and its attachment by means of strings/straps, rather than nails or pegs, would make the pole more resilient and flexible, to allow the cloth to be rolled down as easily by one sitting inside the structure, as by one outside. The top textile band (here painted blue with scale patterns outlined in black) could be kept up in rough weather, hooked somehow to the upper parts of the side poles, which we must now assume in this scheme, or rolled down manually, probably along with the top slat to which it was attached. In such an arrangement the diagonal strings would obviously have nothing to do with hoisting. In the model, the bottom end of the central pole was set into a mortise in the base and the pole stands slightly away from the slats. The upper parts of all three poles are now visible, as in the Theran ikria, and the box-like appearance of the ikrion, as it appears in the Mycenaean painting, is toned down. Some differences still exist, but should be expected, as ikria must have changed over the years and when locality differed.

A few more points deserve some consideration. Figure 6.18 suggests a rather light, flexible structure which could be easily assembled, or, when not in use, conveniently collapsed and stored away, like a tent. The depiction of ikria on and off ships suggests that they were mobile furniture. The only permanent connected fixture may have been a wooden platform, for which there is clear evidence in the Theran miniature. How the ikrion was secured to the platform, or the platform to the boat, is not clear from the paintings. A few conclusions as to other features of LB I ikria and platforms can, nevertheless, be inferred from the Theran evidence. The placement of figures in and near ikria in the miniature frieze provides some clues.

First, it is clear that the floor of the "cabin" was elevated, as one might expect, above the deck boards, as is confirmed by the fact that the heads of the standing steersman and the presumably seated captain are usually at the same level. As pointed out above, the bottom band in the Mycenaean example may have just been a base, part of the ikrion wall framework, or the edge of a platform serving as a floor. At least at Thera there is clear evidence on this matter, for the bottom beams of the ikria usually project in front of the "guard" who sits with his back to the structure. If this had been the edge of a projecting floor the seated man would have been further to the right, with his back away from the front side of the ikrion. The posture and height of this guard provide a further clue for an unseen part of the platform. A comparison with another guard, one seated across and with his back against a small box-like structure, reveals that the former sits in a more erect manner and that his long robe falls more vertically than that of the latter, who is squatter, with legs apparently more flexed and head at a lower level than that of the other man. The only conclusion I can draw here is that the man next to the ikrion is sitting on a step. This step would have served as a transition between the platform top, serving as a floor, and the deck.[114]

Next, overlapping the guard seated on the left is a structure which I like to interpret as a railing. In the best-preserved example, the flagship of the Flotilla Fresco (Fig. 6.12), we note a horizontal member supported by two concave curving forms, the whole structure painted yellow.[115] I consider this railing, presumably of wood, to have been separate from the platform and, in fact, in the illustrated example a blue area is seen continuing above and below the horizontal member and in between the curving legs, suggesting that this structure stood in front of the platform, which in this example is painted blue with a hatching of fine black lines.[116] It is reasonable to assume that a corresponding railing existed on the opposite side, providing protection for the ikrion and creating a fenced corridor for safe circulation on either side at that lofty and precarious spot. In Figure 6.19 I took the liberty of incorporating the railing and

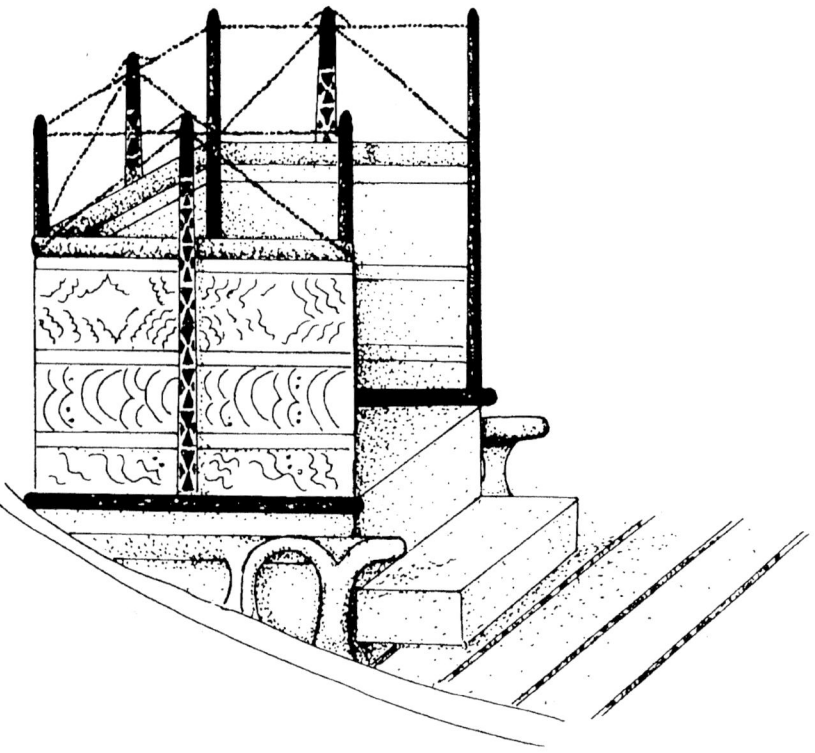

Fig. 6.19. Tentative reconstruction of the Mycenaean cabin as set on a ship (drawing M. C. Shaw)

the platform with a step in the reconstruction of the Mycenaean ikrion, for, although the same exact features were not necessarily present there, related structures and methods may have been used.

If any of the few conclusions offered above are correct, it would be only because the Theran artists seem to have rendered certain practical details with surprising consistency. This, along with the differentiations in human physiognomies, a touch of realism, and the care shown for minute detail, are qualities which increase all the more one's faith in the dependability of the Theran miniature fresco as an accurate record of at least a few aspects of life in the Aegean Bronze Age, despite any artistic conventions, or the spicing of scenes with fabulous elements.

Notes

1   In addition to offering an introduction to the use of textiles on Aegean ships, this chapter presents reprints of two articles by Shaw (1980; 1982) on a Mycenaean fresco from the palace at Mycenae depicting ikria (stern screens) made of elaborately patterned textiles. The articles are here adapted for republication by Anne P. Chapin. To the greatest extent possible, the text preserves its original published form. Notes and illustrations are renumbered according to the new arrangement; in-text citations are reformatted as endnotes and edited for consistency; and spelling is adjusted as needed. Many thanks are given to Prof. William Murray, who kindly offered his thoughts on an earlier version of the introduction. All remaining errors are those of the authors. Texts reproduced by permission.

2   The scholarship on prehistoric ships and seamanship is large and growing. Important works include Wedde 2000; Wachsman 2009, 2013; Tartaron 2013.
3   Shipwrecks dating to the Minoan era (LM/LH I and earlier) include the MM IIB Pseira wreck (Hadjidaki and Betancourt 2005–2006) and the Koulenti wreck on the Laconian coast (Spondylis 2012). Aegean wrecks dating to LM/LH III (the Mycenaean period) include the Ulu Burun shipwreck (Pulak 1998), the Cape Gelidonya wreck (Bass 1961; 1967), and the Point Iria wreck (Phelps, Lolos and Vichos 1999).
4   Palaima 1991.
5   Wright 2008; Tartaron 2013, 35–37.
6   For a full review of the pictorial evidence, see Wedde 2000.
7   *Thera* VI, 38–60, pls 91–94, 96–108, 110, col. pls 7–9; Morgan 1988; Doumas 1992, pls 26–48; Televantou 1994. Ship frescoes are also preserved at LC I Ayia Irini, on Keos (Morgan 1998), at Iklaina (Messenia) in LH IIB/IIIA1 contexts (Cosmopoulos 2011) and in Hall 64 in the Mycenaean palace at Pylos (Brecoulaki *et al.* 2015). These frescoes are less well preserved but confirm the symbolic significance of naval imagery in the prehistoric Aegean in both the Minoan and Mycenaean eras.
8   See below, p. 155, and n. 53 on the modern use of the Homeric term.
9   On Egyptian ships and boats, see Landström 1970; Casson 1995; Ward 2000.
10 London BM 35324; Landström 1970, 13, fig. 15; Tartaron 2013, 52, fig. 3.5.
11 O'Connor 2009, 183–194.
12 Lipke 1984.
13 Reeves 2007, 142–145.
14 M. C. Shaw 2000.
15 Ward and Zazzaro 2010; Ward 2012.
16 Ward 2012.
17 http://www.mar-mus-crete.gr/index.php/en/the-minoa-ship/reconstruction, accessed 22 August 2014.
18 Morgan 1988, 124, fig. 71; Wachsmann 2009, 96–97, fig. 6.20.
19 Morgan 1988, 126; Wachsmann 2009, 96.
20 Wachsmann 2009, 254, with further references.
21 Borojevic and Mountain 2011.
22 Wachsmann 2009, 367, n. 43.
23 *CMS* II.2, 261b, 276b; Wedde 2000, 80, 331, nos 701, 702.
24 Doumas 1992, pl. 37; Wedde 2000, 80–85; Tartaron 2013, 53. Note that the sail is turned 90°, according to early artistic convention, so that it can be seen and recognized by the viewer in the image.
25 Wedde 2000, 85–87; Wachsmann 2009, 251–254; Tartaron 2013, 53.
26 Morgan 1988, 124.
27 Morrison, Coates and Rankov 2000, 190.
28 Waetzoldt 2010, 117–118.
29 Möller-Wiering 2010, 122–123.
30 Note the artistic convention which shows the sail facing the viewer, when in fact it would have been perpendicular to the viewer in order to catch a following wind.
31 Tzachili 1999. If Waetzoldt is correct that the goat hair objects mentioned in Mesopotamian texts are sails, then they are significantly smaller (at 9 m in length) than Tzachili's estimate of 15 m for the sail depicted in the Flotilla Fresco. But if the Theran ship was closer in size to the 17 m long modern replica craft, the *Minoa*, then the measurements are remarkably close.
32 Landström 1970, 103; Reeves 2007, 144. On sails with vertical seams, see Möller-Wiering 2010.
33 The recreated *Minoa* uses a rope net to protect its linen sail from tearing in gusting winds.
34 Möller-Wiering 2010, 123–125, clarifying Tzachili 1999.
35 Casson 1971, 234–235.
36 Landström 1970, 103; Reeves 2007, 144. On madder, see Chapter 2 and *PT*, 236–238.
37 Borchardt 1913, pl. 9; Smith 1965, 150–151, fig. 188.
38 Smith 1965, 150, fig. 6.
39 Landström 1970, 49, fig. l. 133; Saleh 1977.

40 Morgan 1988, 121.
41 Doumas 1992, pls 35–38. On ship awnings, see Morgan 1988, 137–141; Wedde 2000, 130–132. Awnings may now also be identified on some of the ships in the naval frescoes from Iklaina and Pylos (Tartaron 2013, 60).
42 Landström 1970, 99.
43 Morrison, Coates and Rankov 2000, 150. The linen side screens probably shielded men from the weather whereas the side screens of hair offered protection from attack.
44 Morgan 1998; Wedde 2000, nos 675, 676; Tartaron 2013, 54.
45 *CMS* I, 180; Morgan 1988, 122–123, fig. 8; Wedde 2000, 132, no. 911; Tartaron 2013, 54, fig. 3.7.
46 Morgan 1988, 137–141; Tartaron 2013, 54–55.
47 Morgan 1988, 137–142; Wedde 2000, 132–134; Tartaron 2013, 55–56.
48 Wedde 2000, 134.
49 Doumas 1992, pls 49–58.
50 *Thera* VI, 34–60, where Spyridon Marinatos identifies Room 4 as the "admiral's bedroom."
51 Wedde 2000, 133; Wachsmann 2009, 94; Tartaron 2013, 55–56.
52 M. C. Shaw 2001.
53 The Homeric term ikria, used by Spyridon Marinatos (*Thera* V, 35) in reference to the ship "cabins" in frescoes, I generally adopted as a convenient term. The term ikria in Homer seems to describe deck beams or planks, forming partial decks at the fore and aft of a ship (Cunliffe 1924). Since the decks could and did serve as a platform on which to stand, walk, and sleep, the word ikria came to mean in post-Homeric times "scaffolding" or "benches/seats" on which to accommodate theater spectators (Chantraine 1970, *s.v.*). It should be made clear that nowhere in Homer does the term clearly imply "cabins," and indeed, since the Bronze Age Aegean ikria appear to have been unroofed, the term "cabin," however familiar to the reader, seems inadequate.
54 These and other frescoes, found elsewhere at that time on the Mycenaean acropolis, were noted by Tsountas (1886, 59–79; 1887, 160–169, pls 10–12).
55 Besides Tsountas, the main studies on the fresco are made by Gerhart Rodenwaldt (1911, 231–250; 1921; 1941). With the resumption of excavations in the area of the palace by British archaeologists, further fragments were found in the Megaron (Lamb 1921–1923, 164–171, 249–255; reviewed by Rodenwalt 1926, 241–247).
56 Tsountas 1887, pl. 12. Panel C is here shown upside down. Besides Tsountas's mentions of the fresco, there are brief discussions by Rodenwaldt (*Tiryns* II, 232, 234; 1919, 102); Reusch (1945, 103–105); Lamb (1921–1923, 257–259).
57 My article was much improved thanks to useful comments by Joseph W. Shaw and Professor J. Walter Graham.
58 The exact hue and quality of the brushwork of the painting and details of technique cannot be discussed here since the present study had to rely exclusively on the modern color copy which is, nevertheless, adequate enough for an iconographic study. The only "panel" with a clear blue color is A, in the friezes of the "arc" or "scale" patterns (Fig. 6.6, pattern no. 1).
59 All measurements given in the present study are deduced from Tsountas's color copy (Fig. 6.3), which must have been 1/3 the original scale of the actual fragments, rather than 2/3, as Tsountas states (1887, 168). This apparent discrepancy is inferred from the specific measurements of the sections illustrated as quoted by Lamb (1921–1923, 258):

    1. Nat. Mus. Athens, No. 2786: 0.85 × 0.62 m: here Panel A.
    2. Nat. Mus. Athens, No. 2787: 0.52 × 0.39 m: here Panel B.
    3. Nat. Mus. Athens, No. 2788: 0.55 × 0.17 m: here Panel C.
    4. Nat. Mus. Athens, No. 2789: 0.39 × 0.31 m: here Panel D.

60 Graham 1962, 172.
61 Tsountas 1886, 70–71, 73, pl. 4. The plan of Figure 6.4 was made by L. Holland in Wace 1921–1923, pl. 2. The first difference in attribution occurred in Lamb 1921–1923, 258, then in 1964, 79. Subsequently, George Mylonas (1966, 68) referred to the room with the plastered benches (room 33 in the plan) as the "Gallery of the Curtains."

62 Tsountas 1887, 168–169. There is also a chance that the fragments may have fallen here from a nearby room. Professor Graham has suggested the possibility that they may have fallen from the Megaron itself.
63 Tsountas (1886, 70) had previously suggested a single staircase here. For Wace's discussion of this area of the palace and the staircase, see Wace 1921–1923, 257–263.
64 Tsountas 1886, 70; Wace 1964, 79.
65 Tsountas 1886, 71. Patches of the clayish floor of the room are preserved at a level about 1.50 m higher than that of the floor of the Megaron, and similar to that of the floor of the Anteroom to the West. A low, badly preserved rubble structure, of which the function is now unclear, appears against the base of the north and east walls; benches or additional wall foundations have been suggested (Wace 1921–1923, 261–262).
66 Wace 1921–1923, 261–262.
67 Lamb 1921–1923, 258. For the date of the destruction of the palace see Mylonas 1966, 77.
68 Reusch 1945, 103–105. The lack of illustrations in this dissertation greatly hampers its usefulness.
69 Furumark 1972, 345, fig. 58, 348–350 (pattern no. 44). There is also a scale pattern marked by little circles on the kilt of one of the processional male figures in a painting at Knossos (*PM* II, 729, fig. 456b). For further examples of the motif on Aegean dress, see Sapouna-Sakellaraki 1971, 168–169 and M. C. Shaw 1978, 30–31. For the floor pattern from Pylos, see Blegen and Rawson 1966, fig. 73 and Chapter 5.
70 For its use as a floor pattern, see Blegen and Rawson 1966, fig. 73. Examples on dadoes abound: *Thera* VI, pls 3–5; *PM* I, 356, fig. 255; *Pylos* II.
71 *PM* IV, 921, fig. 895.
72 Marinatos and Hirmer 1960, pl. 182.
73 Furumark 1972, 327, fig. 55 (pattern no. 34), where Furumark also illustrates a parallel used in a scene of swimmers on an inlaid dagger from Pylos.
74 *Pylos* II, pl. R, top.
75 Reusch 1945, 104; *Tiryns* II, 225. No illustration has been provided for the latter.
76 For birds and fictitious animals on Aegean dress see *PM* III, 37–42; for fish and birds see M. C. Shaw 1978, 32; for butterflies see Sapouna-Sakellaraki 1971, 163, fig. 68.
77 Furumark 1972, 383, fig. 67 (pattern no. 54).
78 For a comparable network pattern on a floor see Blegen and Rawson 1966, pl. I.
79 For general parallels in clothing patterns see *PM* II, 729, fig. 456c and 731, fig. 457a; *Thera* II, 54, fig. 44 and VII, col. pls B–E. For the use of the motif in architectural ornament see *Pylos* II, pl. Q, bottom, where it appears on a dado and Spyropoulos 1971, pl. 25, illustrating fragments of the miniature scene.
80 Lang (*Pylos* II, 146–147, pl. 139) summarizes the uses of the half-rosette pattern in dadoes and/or friezes.
81 Reusch 1945, 104.
82 *Tiryns* II, pl. 8; *Pylos* II, pl. 137, bottom, and color pl. L, 4 bB 1 and 6 B 32.
83 The pattern can be seen on the belt and on an ornamental band on a woman's dress in a painting from Thera (*Thera* VII, pl. I, fig. 64). Evidence that it may well derive from wood graining is provided by another painting from Thera where it appears on the trunk of a tree (*Thera* II, 53, fig. 43).
84 Reusch 1945, 103; Furumark 1972, 396–400 and p. 397, fig. 69; Popham 1967, 339, 338, fig. **1**. For a carved example on a gold double-axe from the LM I period see Marinatos 1960, fig. 110, top.
85 Sapouna-Sakellaraki 1971, 160, 163, figs 66, 68.
86 For its use in female dress see *Pylos* II, pls N, **O**; as a picture frame see Tsountas 1887, pl. 10.2.
87 Marinatos (1960, pl. 222, top) shows the use of the motif on a carved ivory handle of a Mycenaean mirror from Rutsi, near Pylos. *Thera* plate vol. VI, pl. 5, right and *Thera* VII, pl. 64 illustrate its use as a dress trim.
88 Reusch 1945, 103. For the chariot yokes see *Tiryns* II, 105, fig. 42, pl. 12.
89 *Tiryns* II, 232, 234; Rodenwaldt 1919, 102.
90 Lamb 1921–1923; Wace 1921–1923, 79; Mylonas 1966; *Pylos* II, 235.
91 Levi 1976, 86.
92 See now Doumas 1992, pls. 35, 39–40, 49–56. The ikria in this frieze are 1.83 m high and 1.01 m wide (*Thera* V, 41). Figure 6.11 the ikrion from the large frieze within their architectural context (see also Doumas 1992, pl. 54); Figure 6.12 gives part of a drawing of one of the best-preserved ships from the miniature frieze (see now, Doumas 1992, pl. 36).

93 The relief on the stone vase fragment from Epidaurus is of uncertain date, but is probably not far removed in time from the Theran examples of boats. It is of interest that no lateral poles seem to be indicated. I believe that the dolphin here is not meant as a live one, but is rather a figurehead attached to the ship (Sakellariou 1971). For the signet ring see *CMS* I, 180 (inv. no. 6209).
94 *Thera* VI, 24; Tsountas 1886, 71.
95 Painted friezes with shields in a variety of sizes occur at Knossos, Thebes, Tiryns, and Mycenae. For references to these, see Kritseli-Providi 1973.
96 Vermeule (1964, pl. 14) offers a line drawing of the scene on the silver rhyton. For representations of ships on seals and elsewhere see S. Marinatos 1933 and Morgan 1988, 121–142. To these examples must now be added a representation of a ship with the prow in the shape of a bird and with a steersman, engraved on an agate seal, found in the MM III shrine at Anemospilia at Archanes (Sakellarakis and Sapouna-Sakellaraki 1997, 692–694, fig. 793).
97 *Pylos* II, 186 (19 M ne), pl. 113. The details of the mast can be compared both with the ship on the Tiryns signet ring (Fig. 6.14) and with a painting of a boat on a LB IIIB Cypriot vase (Vermeule 1964, pl. 32A).
98 Reusch 1956, 29–30, pl. 13, figs 38–39. Colors are indicated as in Figure 6.5.
99 For a summary of views on the date of the Theban frescoes see Vermeule 1964, 341, n. 1. That these are representations of items other than clothing, such as robes, is also supported by Christos Boulotis, who has had access to the original fresco fragments illustrated by Reusch and to other related pieces (pers. comm. August 1979).
100 Morgan Brown 1978, 639–641. I had earlier prepared a similar study with similar general conclusions, as part of this article.
101 Naville 1906, pl. 124: see the cabin on the boat at top right.
102 So also Morgan Brown 1978, 639. The presence of half decks at stern and prow and possibly amidships facilitated the easy placement and removal of ikria.
103 See now Doumas 1992, pls 30–34.
104 Mark A. S. Cameron (1978) argues that the Cyclades played an intermediary role in the transmission of the art of wall painting from Minoans to Mycenaeans, but this should certainly not deny direct artistic contacts between Mycenaeans and Minoans. Sara Immerwahr (1977) and Iakovides (1979) see the presence of Mycenaeans in the Theran population.
105 Marinatos 1974, pl. 9; Doumas 1992, pls 35–48.
106 Marinatos 1974, pl. 4; Doumas 1992, pls 49–62.
107 Tsountas 1886; 1887, 160–169, pl. 12.
108 Shaw 1980.
109 That the pole does not extend below the base is inferred from the painting of another ikrion at Mycenae (Figs 6.3, 6.7). The accuracy of this detail in the watercolor copy was kindly confirmed for me by Dr Sara Immerwahr, who examined the actual fresco in the storeroom of the National Archaeological Museum of Athens (pers. comm. July 1980).
110 Compare *PM* II, 2, figs 502d, 503.
111 Marinatos and Hirmer 1960, pl. 36; *PM* III, 95, fig. 53.
112 For discussions of the date and materials of body-shields, see Lorimer 1950, 132–192; Page 1963, 232–235; and Buchholz *et al.* 1977, 1–4. The use of ox hide for Theran shields is attested in the miniature paintings (S. Marinatos 1973, 494–497, col. pl. iii). The use of wood, probably in combination with leather, is suggested by a bed reconstructed from a plaster mold in one of the Late Bronze Age I houses at Thera (*Thera* IV, pls 102, 103).
113 I am particularly indebted to Mrs. Sarah Sharp, an undergraduate student at the University of Toronto, for this idea. Here I would also like to thank my husband, Prof. Joseph W. Shaw and Prof. Lionel Casson for helpful advice.
114 It is useful to compare the platform and step combination (suggested here) with analogous Egyptian examples. See an example in the model of a ship from Tutankhamun's tomb in Landstrom 1970, 102–103, figs 323, 325).

115 The drawing is a copy of this detail from the "Flag Ship," as published in Gray 1974, pl. xiii, but with the large left fragment of the ikrion shifted slightly, according to the later correction by the restorers, as seen in Marinatos 1974, pl. 9.
116 The curving legs are interesting. Refinements in carpentry are also attested by a stool with curving legs, reconstructed from a plaster mold in a house at Thera (*Thera* V, pls 102, 103). For a painted representation of comparable furniture, see the table on which lies the sacrificed bull on the Ayia Triadha sarcophagus (Marinatos and Hirmer, 1960, pl. xxviii).

7

String Lines, the Artist's Grid, and the Representation of Textiles in Fresco

Excerpts from Maria C. Shaw 1998; 2000a; 2003; 2010 adapted by Anne P. Chapin[1]

Introduction

Excerpt from M. C. Shaw 2003, 179–185

Even in our era of digital technology with its capacity for amazing visual imaging, there is still a sense of embarrassment when "mechanical" drafting devices are mentioned in connection with Ancient Art – as if its purity were marred by the notion of artifice.[2] It must have been such a desire for purity that in the earlier part of the 20th century led Geerto A. S. Snijder to interpret Minoan art as the product of pure intuition, made by "eidetic" artists.[3] The term, deriving from the Greek word είδος (form/essence/species), was first introduced in the 19th century in clinical psychology and in theories of phenomenology and mnemonic ability. Such a trait, it was believed, enabled "concrete visualization" by the endowed individual, even in the absence of the subject thus recalled.[4] It was Snijder's view that it was the eidetic ability and a complete reliance on intuition that explain the life-like quality of the art produced by both the Minoan artists and the Paleolithic cave painters.[5]

Yet, and while intuition and inborn talent do partially explain naturalism, the notion of being "eidetic" as part of an artist's condition should be used with caution so as not to discourage the search for this art's other dimensions. Luckily, research in the last several decades in the area of Aegean art, and especially in wall painting, has flourished by doing the latter. The often drawn conclusion is that, far from being merely spontaneous, this art was shaped by much preliminary planning, including the careful selection of iconography and theme, since such elements also served as vehicles of ideological message for the Aegean elite.[6]

Here, I turn to another aspect of wall painting that demanded much attention and preliminary planning related to the actual execution of the mural. This was a long process that started with the stuccoing of the wall and ended with the final painting of the pictorial subject. These stages are usually discussed under the rubric of "Technique," the latter covering a wider range of aspects that lie beyond the scope of this paper.[7]

My article focuses instead on a limited aspect of the overall planning, and specifically on two types of drafting devices (what I refer to below as *guidelines* and *grids*) used at an early stage, basically before paint was placed on the wall. These devices, which make use of lines that help the painter render the subject, have left traces, some incised, others impressed onto the plaster surface while it was still damp and fairly soft but already smoothed and polished (the polishing was sometimes repeated after the colors were added). No paint was used to draw or emphasize these lines that were meant to become eventually virtually invisible. Their basic functions were to mark out the position of the composition on the wall, to provide indicators for the allocation of selected pictorial elements and, especially in the case of the grid, to help produce designs in a professionally accurate way.

Some of the lines were "drawn" free-hand, especially when used for partial sketching – or even as trials or practice on the part of the artist. Others required the assistance of drafting tools. One such tool was the compass; another was a length of string stretched taut and snapped across the plastered surface so as to leave its imprint. The imprints are usually referred to as "string-impressed lines" or, briefly, "string lines". String was also used to construct what is often referred to as an *artist's grid* (or simply *grid*), its horizontal and vertical lines crossing at regular intervals at right angles to form "squares." In the Aegean, the grid was usually limited to specific parts of a painting rather than being used for the entire composition. The incised lines were "drawn" with some kind of pointed tool, either freehand or with the assistance of devices such as a straight edge, or apparently stencils and French curves.[8] In all these cases, and as Cameron incisively remarked in his dissertation, the lines had the advantage of still being visible even after a thin slip of fine plaster was added to the surface just before starting the actual painting.[9]

Since the equivalent of the Aegean guidelines and especially the artist's grid were also used in Egypt, these devices provide a common ground on which to base cross-cultural comparisons with the purpose of exploring possible artistic influences and exchanges. The geographical area on which this paper concentrates is the Aegean, and the period is that of the Second Palaces of Crete, when the art of wall painting was at its prime. There is less emphasis on the Mycenaean mainland,[10] both because it is later and because the devices considered here were used more sporadically – in some cases, as that of the grid, extremely rarely. Pride of place as far as innovation is concerned, goes to Crete, with most processes documented there coming from the Palace of Knossos, because of its extensive program of figural and other representational wall decoration.[11] Paintings, however, existed in the remaining palaces, villas, and houses too. The degree to which the evidence from both areas is available archaeologically, or accessible to the scholar, naturally varies.

Most of these sites were excavated around the start and earlier part of the 20th century, the greatest influx of new information since then being that from the magnificent and well-preserved frescoes found chiefly in the 1960s by Spyridon Marinatos in his excavations of the thriving LC IA settlement at Akrotiri.[12] His and subsequent publications, accompanied by superb illustrations of the paintings,[13] have much facilitated my research, even for details of draftsmanship. Fewer, but still of value for information on technical matters, are frescoes from other Aegean islands, such as Kea (at Ayia Irini) and Melos (at Phylakopi).[14]

The presentation below is divided into three parts. The first deals with guidelines, according to categories; the second with the artist's grid; and the third with comparisons between the Aegean and Egypt, particularly with respect to the use of the grid.

7. String Lines, the Artist's Grid, and the Representation of Textiles in Fresco

Guidelines

String-impressed lines used to mark out the perimeters of a composition

The two main formats of Aegean mural composition were a narrow frieze that was usually placed directly below the ceiling, and a painting that usually started at a certain distance above the floor and ended at the top some distance below the ceiling. More rarely, the composition covers the entire wall; such examples are so far known from Thera only.[15] Marking the upper and lower limits of a composition must have been one of the first tasks undertaken by the artist. In the earlier Neopalatial period, the upper border usually consisted of consecutive multicolored bands or stripes of varying widths, for which I provide an illustrated example from the LM I Palace at Kommos (Fig. 7.1).[16] These bands were almost invariably separated from each other by a string line, a practice encountered in frescoes on other Aegean islands as well.[17] A horizontal line – sometimes impressed with a string, sometimes merely painted – also marked the base of the composition. Below it, was often a painted dado. Whether impressed vertical lines ever defined the lateral sides of the composition remains uncertain. Some information can be found in the better-preserved paintings of Thera, where the tendency seems to be to terminate the composition at the very end of the wall or at a vertical architectural element such as the wooden post of a door.[18] More often than not, however, the Theran paintings spread uninterruptedly across adjoining walls, as those in Crete must have done.

Fig. 7.1. Fresco fragments with string-impressed lines and multicolor bands, from Kommos (courtesy Kommos Excavations)

Imprinting the lines that separate the bands of the horizontal borders, especially in cases where a composition spread over more than one wall, must have been a task that required some skill, but it is still a method that we can only speculate about. Clearly, at least two, or, more likely, three individuals must have been involved for each wall. Two would have held the ends of the string, keeping it taut, while the third may have been placed centrally to help level it every time the string was moved to produce the next horizontal line. The paintings of Thera show an amazing consistency in the width of each successive band, but such accuracy was not universal.[19]

String-impressed lines used to delineate elements of architecture in the pictorial area
Here the most instructive examples are the miniature building facades in paintings from the Palace of Knossos. String-impressed lines mark both vertical and horizontal features, such as courses in the masonry or edges of buildings in the Tripartite Shrine Fresco.[20] Cameron[21] mentions impressed and incised lines as having been used for delineating pathways, such as those depicted in the Sacred Grove and Dance Fresco,[22] and as having been used for alignment purposes – for instance, to mark the vertical axis of column shafts. An example he quotes is a miniature columnar façade decorated with double axes, the fragments of which were found in the Thirteenth Magazine of the West Wing of the Palace of Knossos, which means that it is possibly one of the earliest such representations.[23]

Architectural representations occur also in the paintings of Thera and Kea, and later in those of the Mycenaean mainland.[24] Impressed or incised lines drawn to help render architecture are used rather erratically in the Mycenaean period, and by all appearances they are virtually absent in the paintings of Thera – at least in the one main such representation, the frieze in the West House depicting coastal towns.[25] Therefore, it makes it all the more surprising and interesting that the method is frequently used in architectural fresco depictions in Kea, though from what I could judge from published photographs, the lines mark the vertical edges only.[26]

String-impressed and incised lines used to delineate other elements in the pictorial area
My first example here is a decorative band on the hem of the long robe worn by a female figure in the Procession Fresco from the Palace of Knossos (Fig. 7.2). Her importance is clear from the fact that a number of processional figures carrying gifts converge on her.[27] The band consists of several stripes separated from each other by string lines and decorated alternately with architectural elements, such as a row of beam ends, or religious objects, namely pairs of incurved altars rendered abstractly. A similar type of decoration was imitated later in the Mycenaean Palace at Pylos, where it was used again on the hem of a female figure, labeled by Mabel Lang as "The White Goddess."[28] String lines also render the cords of a sash that hangs in front of the kilts worn by men in the Procession Fresco from Knossos.[29] The impressed lines here were later painted black.

Impressed lines were used to define rectilinear elements in two floral compositions in frescoes found at Amnissos.[30] In one case they render the horizontal lines of planters depicted in the shape of incurved altars whose sides were incised;[31] in the other they outline a stepped structure enclosing a clump of white Madonna lilies shown against a red background.[32]

There is another floral fresco that made use of impressed lines, this in the West House at Thera. Two vases with flowers, shown as if standing on a windowsill made of a veined type of

7. *String Lines, the Artist's Grid, and the Representation of Textiles in Fresco* 187

Fig. 7.2. Decorative band on skirt of female figure in the Procession Fresco, from the palace at Knossos (photo J. W. Shaw. Archaeological Museum of Heraklion, Hellenic Ministry of Culture and Sports – Archaeological Receipts Fund)

stone, were depicted on the plastered jambs of a real window. Impressed lines here define both the horizontal edges of the simulated sill and the vertical ones of the jambs, which are rendered in yellow, implying wood.[33]

Incised, straight, curvilinear, and circular lines used for sketching and visual accent
Sketching in Minoan and other Aegean painting is of two kinds and carried out at different stages. In the later stage the subject is roughly sketched mostly in red or, more rarely, yellow – the so-called cartoons.[34] Here I shall concentrate on the earlier stage of sketching, which uses no color and renders only a small area of the subject. Perhaps, as Cameron thought, the few contours provided in such cases (such as part of the feather of a bird, or a bull's leg, or even the shin of a processional figure) were intended as markers of where important motifs should be eventually fully painted within the composition.[35]

To my knowledge the most complete example of an incised sketch in Minoan painting is a detail on the painted stucco relief of the so-called Priest-King from the Palace at Knossos. Here, Evans detected "slightly incised sinuous lines" on the torso, above and below the figure's clenched hand. He interpreted them as the contours of a lock of hair running down the center of the torso whose original color had entirely disappeared since antiquity.[36]

Sketching with a pointed instrument is quite extensive in Thera, but, again, it does not ever seem to delineate the entire subject.[37] An example is the painting of the so-called Priestess from the West House, shown carrying a brazier. As can be seen in a partial drawing (Fig. 7.3),[38] the artist seems to have had problems with rendering the arms and hands, perhaps because of the fairly complicated gesture and the fact that the object had to be rendered in some detail. Some of the curved lines (shown dotted in the drawing) coincide with the final outlines. These were eventually retraced in dark paint; other lines represent failed attempts or trials. Perhaps this was not an experienced artist. In the same figure, the forelock is the only part of the hair to have been preliminarily outlined by an incised line. Such is also the case for the forelock in the case of the seated goddess in the famous painting from Xeste 3, perhaps another instance of marking ahead of time where the figure should be located on the wall – though there is some partial outlining of the goddess's remaining hair – the twisted locks running down her back.[39]

Partial and rather erratic is the use of incised sketching in the case of the frieze of ikria, or stern screens, also from the West House. Though the ikria differ in decorative details, their main features remain the same. Among the latter are the lateral posts, part of the cabin's frame, each of which ended at the top with a finial in the form of a *waz* lily. While examining the illustrations of several ikria, I was able to detect incision only in one case,[40] and, if my impression was correct, this might mean that the choice to use such lines was made by one particular artist, who felt insecure and chose the safest way to render the particular details, particularly the curving volutes of the lily.

Fig. 7.3. Preliminary sketch (indicated in dotted lines) in the "Priestess" Fresco from the West House, Akrotiri (drawing G. Bianco after Doumas 1992, pl. 25)

In the examples that follow, there are the rare cases where the sketching is not as partial and sometimes renders a complete outline. In a fresco from Xeste 3 in Thera,[41] at least the outer coils of the volute of a running spiral were regularly incised; a stencil or a French curve were likely used – devices that were apparently employed even for the rendering of selected contours of the human figure in Thera.[42] Another mechanical aid may have been a straight edge. This may have been employed when the outline of a long sword wielded by a monkey shown in a fresco from Xeste 3 was incised.[43]

The fact that a simple compass was used in Aegean painting was mentioned above. One example of its use is found in the frieze of rosettes in the architectural columnar façade already mentioned in connection with impressed lines that help with the alignment of certain pictorial details. The use of a compass in the frieze can be detected from a tiny round depression visible at the center of each rosette, clearly representing where the pivot of the compass was set. Elsewhere, the evidence is an incised circumference, this drawn with a pointed tool attached to a pivot with a length of string.[44]

In some cases I was unable to discern whether the artist used incision or impressed lines made with a very fine string. This is the case with the beaded net decorated with blue stars, likely a wall hanging, shown at the top of a ritual scene with women in a painting from the House of the Ladies on Thera.[45] Strung beads are also known from Sinclair Hood's excavations of the Royal Road at Knossos, but there the spiraling beaded lines were clearly incised freehand.[46] Ambiguity also surrounds the exact function and when the incised lines were made in another network pattern, this from Xeste 3, also representing some kind of wall hanging.[47] Here, an incised line was used to separate two white bands, which described the undulating contours of the large lozenges in which the net was divided. One possibility is that the incised line was made before the two bands were rendered in stucco relief (as a guideline), and it remained visible intentionally for the sake of definition and emphasis in the design.

I close with a sub-category of the types examined in this part of the paper, namely the use of incision purely for visual accent. As opposed to guidelines, incised lines of this kind were made at a later stage in the production of the painting. One example is the decoration one can see on the plumes in relief of the crown of the Priest King from Knossos – wavy, parallel lines arrayed diagonally.[48] Another is the definition of individual curls *after* the color of the hair of the individual portrayed was painted and dried. The technique occurs in the case of two of the Crocus Gatherers in Xeste 3,[49] and, later, at Mycenae in the painting of the so-called Mykenaia in the Cult Center, found by George Mylonas,[50] where it anticipates the use of incision to pick out details used centuries later in black-figure vase painting in Archaic Greece.

Artist's grid

This is the most elaborate and sophisticated among the drafting devices considered here.[51] It is encountered predominantly in paintings in Crete, where it essentially helped render textile patterns seen on clothing, male and female. The selective examples that follow are intended to illustrate how the grid worked and to demonstrate the chronological range of its use.

The textile designs on the painted plaster reliefs from Building AC, Pseira
Excerpt from M. C. Shaw 1998, 64–65
The reliefs from Pseira are among the masterpieces of Aegean art both technically and artistically (Figs 3.5, 3.6). The plaster is very fine; the surface is polished, and, as in the best of Minoan frescoes, it is covered with a fine slip (*c.* 2 mm thick) upon which the pigments were applied. Some of the colors on the displayed panels still retain their brilliance, such as the sky blue, or the bright yellow ocher that in places assumes warm red tones. Black and white complete the palette, which, despite its limited range, achieves surprisingly polychromatic effects through a rich interplay and varied juxtaposition between the few colors. Fine black lines are frequently used for delineation, outline, and emphasis in patterns, while dots and disks in different colors (red, black, white) and in light-on-dark and dark-on-light schemes produce luxurious decorative effects. There is meticulous attention to detail and draftsmanship even on the smallest scale.

Most interesting is the use of a grid in Panel A to ensure the correct replication of the individual motifs (rosettes) that make up the textile pattern on the sleeve (Fig. 7.4). The grid was impressed on the curving surface of the arm by use of a taut string that was allowed merely to touch the still-damp plaster. Its lines did not follow the curvature of the body, as they might have, had they been incised by hand. They are vertical and horizontal, and they almost disappear where the relief recedes. This grid is one of the finest preserved in Minoan painting, made up of tiny (7 mm wide) squares. Grids used to render textile patterns in other Minoan frescoes are occasionally set obliquely, either because they reflect the movement of the part of the human body that the

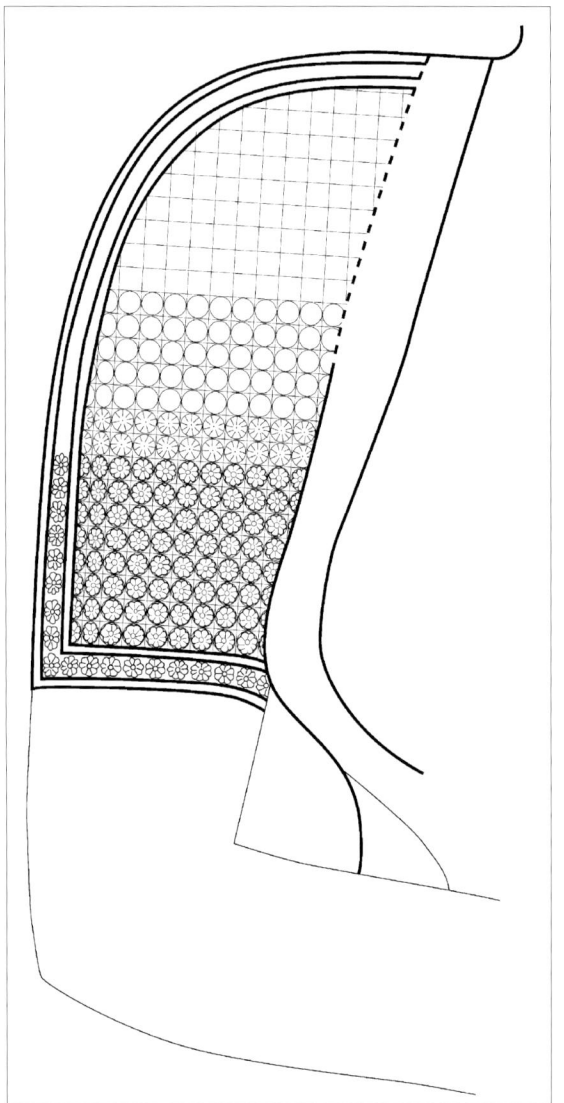
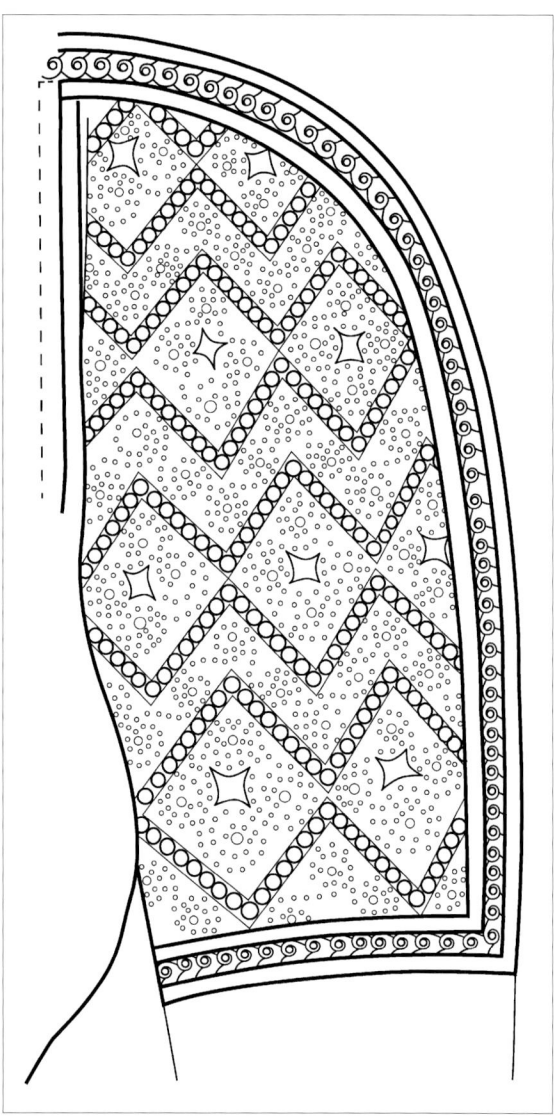

Fig. 7.4. Pattern and grid on bodice of Lady A from Building AC at Pseira (drawing M. C. Shaw and G. Bianco)

Fig. 7.5. The ungridded textile design of Lady B from Building AC at Pseira (drawing M. C. Shaw and G. Bianco) cloth covers, or because of the particular configuration of the surface pattern, usually one that combines different repeated motifs. Examples of the obliquely placed pattern can be seen in the dress of the "goddess" in the fresco at Ayia Triada (see Fig. 7.6, below); the Lady in Red (Fig. 3.2) and the Ladies in Blue at Knossos (Fig. 9.1); the kilts of the men in the Procession Fresco at Knossos (discussed below); and a fresco at Chania.[52]

The rosettes in the Pseiran relief are set one per grid square and are surrounded by a black circle. The circles were probably drawn with the help of a stencil. Where the paint is worn there is evidence underneath of delicate red lines, part of the preliminary sketch. Fig. 7.4 illustrates the stages involved in producing the textile pattern, and one wonders at the need for such a painstaking process, especially as several dozens of rosettes were needed to fill the bodice. Whether the skirt had the same or a similar pattern will never be known, because the original surface slip carrying the color has entirely disappeared.

No grid is apparent on the other figure at Pseira, a puzzling matter in view of the elaborate patterns on both sleeve and skirt (Figs 3.5, 3.6, 7.5). One reason for such an omission may be that the chevrons on the sleeve intrinsically dictate their own ordering and eliminate the need for a grid. In the case of the network of undulating lozenges on the skirt, it is not clear how this may have been worked out. One possibility is that a grid was used, but was drawn rather than impressed, and is now covered by the painting. An impressed grid may have been avoided to prevent it from marring the final design. To my knowledge no grid, impressed or drawn, has been detected in the very similar pattern with lozenges enclosing rosettes from Thera (Fig. 7.3). It is to be noted that the latter design also uses black for the outline and the details of the patterns. Black outlining, then, need not be taken as a sign of a later date at Pseira.

Grids used for the textile designs in the frescoes from Ayia Triada

Excerpt from M. C. Shaw 2003, 185

The next example takes us to the Villa at Ayia Triada – also destroyed in LM IB – and to the magnificent painting in Room 14 (Figs 3.8–3.11).[53] Here the grid is used to render the very complex pattern on the skirt of the so-called goddess. The grid's units are nearly square, ranging from c. 1.7 cm to 2.1 cm on a side. A diagrammatic drawing here helps illustrate the role played by the grid (Fig. 7.6). The grid's "squares" provide fixed spaces which accommodate the basic elements of the quatrefoil motif: its core and the petals. Spaces between the petals of each quatrefoil in turn accommodate those of adjacent quatrefoils, the repetition of the motif and the controlled alignment producing what Elizabeth Barber calls an "interlock" pattern.[54] There is also

Fig. 7.6. Pattern and grid on the costume of the "Goddess" from the Royal Villa at Ayia Triada (drawing M. C. Shaw and G. Bianco)

a colorist trick used by the artist, in that two different colors were used alternately to create diagonal lines. The result is one of variable visual illusions that were surely intended to dazzle the viewer, and, perhaps, inspire awe of the goddess. A principle to note here and in other examples is that the lines of the Minoan grid do not coincide with or reiterate those of the design itself. The grid is a tool whose usefulness ends the moment the design has been rendered with its help. No attempt is ever made to erase the impressed lines, either. These are eventually covered over by the colors of the background and the motifs painted on it. In the example presently under consideration, the interlocking pattern leaves little room for a proper background. Even the interspaces between the quatrefoils, the little lozenges with incurved sides, constitute still another decorative motif.

The Procession Fresco from the Palace at Knossos
Excerpts from M. C. Shaw 2000a, 52, 53-58
Pertinent here are five patterns (A–E) from the Procession Fresco, the complexity of which required the use of grids. Five drawings (Figs 7.7–7.11) illustrate the stages of the composition of each pattern: the laying out of the grid (at the top of each drawing), the outlining of the basic motifs (middle), and the addition of final details (bottom). Figures 3.15, 3.18 and 3.19 show the patterns with the figures, and without the grids,to allow for an uncluttered appreciation of them. There is naturally a degree of schematization inherent in the computer-generated illustrations, but the primary aim here was to clarify the character of the patterns, rather than to produce a record of irregularities normal in any execution carried out by the human hand. Absolute accuracy in the reproduction of color cannot be claimed either, given the unevenness in the degree of dilution of pigments and their poor preservation in the original. As usual, the Minoan palette is simple, consisting of four basic colors besides black and white: red, blue, ochre and orange. The last was used on the kilt of the Cupbearer and was apparently produced by applying diluted red strokes over an ochre background.[55]

Pattern A: skirt of a woman (no. 7) in the Procession Fresco, Knossos
This pattern appears on the lower part of the skirt of a woman moving right followed by six male figures in long robes, at the start of the procession (Figs 3.15, 7.7).[56]

PATTERN AND GRID
Because of poor preservation, there are uncertainties. This motif, repeated in rows, approximates the shape of a tricurved arch, but the fact that the open ends curl inwards, as if to form spirals, also suggests an ivy motif.[57] It is painted in white impasto added onto what appears to have originally been a blue background, later turned grey and black by a fire. It is not clear whether another color was added over the white, as may have been the case in Pattern C discussed below.

Exceptionally in Pattern A, the artist's grid is made up of rectangular rather than square units. The size of each varies: 1.4–1.8 cm for the horizontal side, and 1.9–2.2 cm for the vertical, depending on the location in the area of the dress thus decorated. In Figure 7.7, I have adopted the more standard size, 1.5 × 2.1 cm. Each arch/ivy spans two units width-wise, the dividing line between them marking its center. Its height is equal to one unit, except for the curling ends, which drop slightly below the bottom line of the row of grid units. It is clear that the grid here helps maintain the bilateral symmetry of each motif and its correct alignment in horizontal and vertical rows. The resulting overall pattern is rather simple, compared, for instance, with the network effect of the scale pattern seen in Figure 7.8, which is created by alternating the vertical alignment of the motifs every other line.

COMPARANDA

Given the ambiguity of the shape of the repeated motif, it is difficult to find comparanda for it. If intended as a tricurved arch, examples of the motif range in Minoan pottery from LM IB to the later LM III periods. The same dates apply to cases where the motif is detached and shown either singly or in rows, and mostly pendant, rather than with the points up as in the fresco.[58]

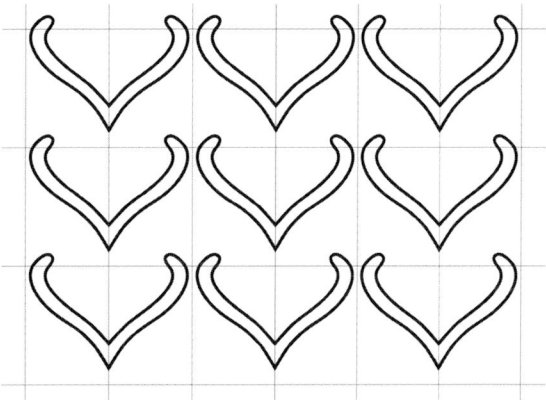

Fig. 7.7. Creation of Pattern A on a female figure (no. 7) in the Procession Fresco, from the palace at Knossos (drawing M. C. Shaw and M. Nelson)

Pattern B: kilt of male figure (No. 20) in the Procession Fresco, Knossos

The pattern appears on the kilt of one of three men walking to the right (Figs 3.18, 7.8).[59] The curving hems of the kilts are decorated with the barred pattern, at the end of which seems to be attached a net that hangs down the front of the kilt, which is weighted down by beaded floral ornaments attached to the ends of the strings. The lines of the net were incised with the help of a straight edge and painted black. Little horizontal lines and dots fill the corners of each lozenge. Incision, rather than string lines, was used, for the aim was to render the way strings can bunch up in a real net, while the organization of the woven pattern is fixed.

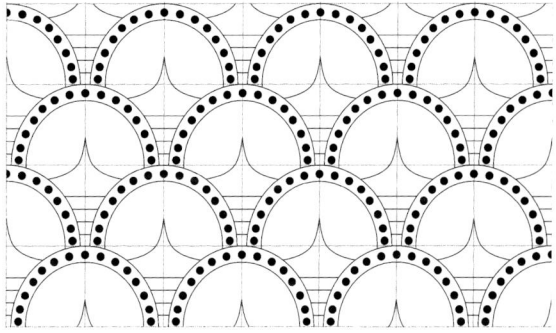

Fig. 7.8. Creation of Pattern B on male figure (no. 20) in the Procession Fresco, from the palace at Knossos (drawing M. C. Shaw and M. Nelson)

PATTERN AND GRID

The repeated element is an arch outlined twice, the resulting band filled with round red dots. A red painted schematized derivative of the foliate plant motif is the filler, its basal leaves curved, the rest drawn as parallel lines. The scale pattern is created by avoiding a vertical alignment between arches in the successive rows.[60] The background is ochre. The grid consists of square units that vary from 1.5 cm to 1.7 cm, the more standard size, 1.6 cm, being the one adopted in Figure 7.8. Each semicircle occupies two units of the grid both width and height-wise, again with the dividing line between the grid units defining the center of the motif.

COMPARANDA

The scale pattern is among the most commonly used in Minoan and other Aegean art. In Minoan pottery it goes back to Kamares ware and continues into the Postpalatial period.[61] In fresco, the pattern occurs on fragments found under the floor of the Corridor of the Processions (Fig. 3.13); Evans suggested that those fragments probably had the same theme as the later fresco.[62] It also occurs on the dress of the kneeling woman in the fresco from the Villa at Ayia Triada (Fig. 3.11), there rendered in dots, and on the plaster fragments without clear contexts from scattered locations in the Palace of Knossos.[63] On the Mycenaean mainland, one of the earliest examples may be the fresco fragments depicting patterned dresses under the Ramp House on the Acropolis of Mycenae.[64] In a form that closely resembles the Knossian example, the scale pattern is also found on the painted floor of the megaron of the palace at Pylos.[65]

Pattern C: kilt of male figure (No. 21) in the Procession Fresco, Knossos
This pattern appears on the kilt of a man just ahead of the one noted under Pattern B.[66]

PATTERN AND GRID

This is a diagonal or rapport pattern, each of its lozenges containing a pendant ivy, the spiral ends of which are linked by two curving bands, bordered below by dots (Figs 3.18, 7.9). A vertical blade-like element, perhaps a bud, marks the center of each ivy, the upper tip of which is marked by a dot. All dots in this pattern, including those at the junctions of the lines of the diagonal grid, were painted in impasto against a blue background. The cores of the spirals of the ivies are ochre, and traces of yellow ochre appear here and there on other parts of the motif, raising the possibility that it was originally painted entirely in ochre. If so, one wonders if a golden ornament, a pendant, is implied.

The whole design was executed with the help of a squared grid. The units were nearly square (just *c.* 1 mm larger height-wise in each case), and they varied in size, depending on their location on the kilt, from 2.1 cm to 2.7 cm. Their more standard size, height 2.4 cm and width 2.3 cm, is adopted in Figure 7.9. As can be seen in that illustration, the diagonal pattern was obtained by drawing the diagonals of each of the grid's square units. Ivies and dots were added next. Interestingly, the diagonals of the pattern were also impressed like those of the grid. Whether a string dipped in black paint was used for the diagonal lines, or the impressions of an undipped string were later painted over, is not clear without microscopic inspection.

It is possible that the pattern represents an actual net. One thinks of beaded nets worn over dresses by women in Egypt.[67] If it is a net, we could also raise the related question of whether the ivies represent gold ornaments attached to the fabric of a kilt. The possibility of using attachments as decorations in Aegean dress has been discussed by others.[68]

Fig. 7.9. Creation of Pattern C on male figure (no. 21) in the Procession Fresco, from the palace at Knossos (drawing M. C. Shaw and M. Nelson)

7. String Lines, the Artist's Grid, and the Representation of Textiles in Fresco

COMPARANDA

I am unaware of parallels for the particular rapport arrangement of the ivies in other frescoes. In pottery they usually appear in bands (but set sideways rather than pendant), the beaded variety being rather prominent in LM I.[69] The arrangement in bands is also encountered in textile depictions in other frescoes.[70]

The pattern of a diagonal grid finds a parallel in the Lady in Red (Fig. 3.2).[71] In this, the red background was painted first, then the lines were impressed and painted black, their points of junction covered by a dot in white impasto. Horizontal parallel red lines fill half of each lozenge. A similar pattern occurs in what may be a predecessor of the Procession Fresco (Fig. 3.13).[72] The lines were again incised, as they are in some Mycenaean frescoes (see Fig. 7.12, below).

Pattern D: kilt of male figure (No. 22) in the Procession Fresco, Knossos
This figure is directly ahead of the one discussed under Pattern C. Only a tiny part of the kilt is preserved.[73]

PATTERN AND GRID

This is an elaborate interlocking pattern (Figs 3.18, 7.10). The repeated motif is a curvilinear quatrefoil set within a cross, itself a rectilinear version of the quatrefoil. The units of the artist's grid vary from 1.5 cm to 2.0 cm, the more standard size, 1.6 cm, being adopted in Fig. 7.10. The crosses, painted in black, essentially follow the lines of the grid, each arm of the cross occupying one grid unit. They are arranged in a staggered or diagonal fashion and their points of junction (four in each case) are each marked by a solid black lozenge with incurved sides. The curvilinear quatrefoils within the crosses are also outlined in black, their centers occupied by a lozenge with incurved sides, which is also outlined rather than solid.

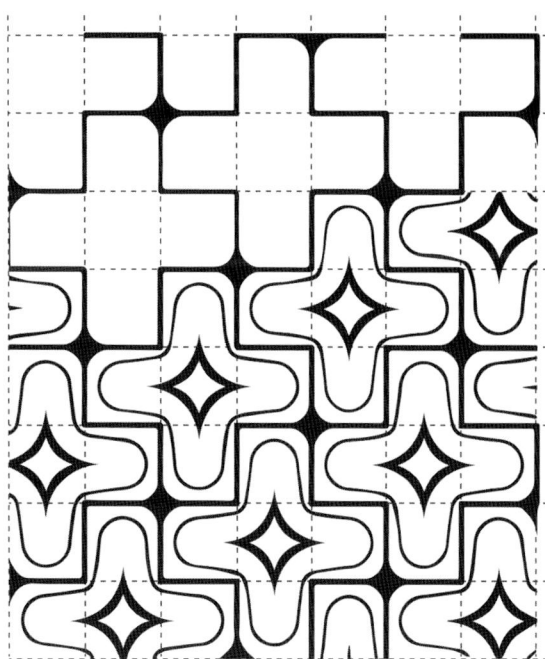

COMPARANDA

I am unaware of the particular combination of curvilinear and rectilinear quatrefoils in a rapport arrangement in pottery. To avoid repetition, I shall here limit myself to comparanda for the cruciform motif, the curvilinear one to be considered under Pattern E, for I do not know of another example of the combination of the two in frescoes. Examples of interlocked crosses occur in loose plaster fragments from a variety of locations in the palace of Knossos, though from undateable contexts. Sometimes alternating rows are painted in different colors, blue and ochre being one such combination.

Fig. 7.10. Creation of Pattern D on male figure (no. 22) in the Procession Fresco, from the palace at Knossos (drawing M. C. Shaw and M. Nelson)

Fig. 7.11. Creation of Pattern E on the Cupbearer in the Procession Fresco, from the palace at Knossos (drawing M. C. Shaw and M. Nelson)

Filling ornaments also vary.[74] On a fragment of a Mycenaean fresco depicting a woman's dress, there is a pattern of crosses enclosed in crosses (see Fig. 3.40).[75]

Pattern E: kilt of the Cupbearer in the Procession Fresco, Knossos

This pattern appears on the kilt of the famous Cupbearer, the only processional figure for which the head and most of the body are preserved (Figs 3.19, 7.11).[76] Fragments of this figure were found in the South Propylon of the Palace, but the similarity in scale and in other respects with the figures from the Procession Corridor makes Evans' suggestion that it belonged to a continuation of the same fresco quite convincing.[77]

PATTERN AND GRID

The pattern uses interlocked curvilinear quatrefoils, the spaces between them being painted solid black, creating a meandering band that surrounds them. The grid units vary in size from 1.7 cm to 2.0 cm, depending on their location on the kilt. The more average size of 1.8 cm is adopted in Figure 7.11. Once the grid was impressed, the quatrefoils were outlined, each arm being the size of a grid unit. The spaces between them were then painted black. A circle surrounded by dots in red was placed at the center, the motif was then repeated within the ends of the quatrefoil arms in blue. The background was painted ochre over which one can see diagonal brushstrokes in red.

COMPARANDA

The use of curvilinear quatrefoil in an interlocking pattern is not, to my knowledge, encountered in pottery, but it occurs in Minoan frescoes,[78] and again in a very similar version on the painted floor of the megaron of the palace at Pylos.[79] An interesting and rich variation of Pattern E is the interlocking pattern on the dress of the so-called goddess in the fresco from the Villa at Ayia Triada, which was discussed above (Fig. 7.6).

The textile designs on the frescoes from Xeste 3, Akrotiri, Thera
Excerpt from M. C. Shaw 2003, 186

Outside of Crete, but still in the Aegean, the most interesting example of grids occurs on the clothing of two female figures who are part of a procession fresco still under restoration from Xeste 3 on the 2nd floor of Room 3B. I was most fortunate to have been invited to look at these examples (September 2002) and to be generously allowed by Andreas Vlachopoulos, who is publishing the frescoes of this building, to include the information that follows. The grid units in each case are a different size: the larger is 4 × 3.95 cm, the smaller 3.70 × 3.90 cm. At this point

7. String Lines, the Artist's Grid, and the Representation of Textiles in Fresco   197

in the process of cleaning and restoration, it is very difficult to discern and figure out how the grid helps render the particular textile design. All that can be said about the latter is that it is of a simple type in the more visible case of the dotted arches.

The grid in textile designs on Mycenaean frescoes: the palace at Pylos
Excerpt from M. C. Shaw 2010, 316, 318, 319–320

The design from Pylos (18 M ne) consists of a floral motif, which was repeated at regular intervals, resulting in a "surface pattern" that is rather reminiscent of modern wall paper. Two illustrations are provided here, both based on those published in the original publication of the piece by Mabel Lang. One is a drawing which illustrates the artist's grid that was used to render it (Fig. 7.12), the other is a digital restoration in color (Fig. 7.13). As shown in the former, the grid was divided into what we shall call "grid units." The size of each unit is fairly consistent: 3.1 cm width-wise, and 3.5 cm on the vertical side.

As we can see from our drawing of its pattern, the repeated ornament, the flower, is bisected by the vertical and horizontal lines of four grid units that ensure its bilateral symmetry, while the central, vertical line marks the point where the wavy tricurved bands join, allowing them to be repeated regularly in successive horizontal rows. Both the net bands, which are white, and papyrus flowers, which are yellow, were first outlined in red, and shown against a background of sky-blue color. As is evident from this description, this design was rendered by a meticulous artist for whom symmetry and fine detail mattered. To my knowledge, the device used to render it has not been encountered anywhere else in Mycenaean painting. Given the numerous Minoan examples, it seems likely that using a grid at Pylos was a concept that was conveyed from Crete to the Mycenaean mainland.

Fig. 7.12. Creation of the gridded textile pattern in fresco fragment 18 M ne from the palace at Pylos (drawing M. C. Shaw and M. Nelson)

Fig. 7.13. Color drawing of the textile pattern in fresco fragment 18 M ne, from the palace at Pylos (drawing M. C. Shaw and M. Nelson)

A comparison can be made to a pattern on a piece of plaster found at Mycenae, seen in the upper part of a skirt, below a belt, that was decorated with interlocked crosses in a network pattern. The design can be seen here in an illustration based on a tracing of the original fragment made by the present writer (Fig. 3.40).[80] The design is clearly complex enough that, had the painting been in Crete, it would surely have required the use of an artist's grid. However, and as was typical outside Crete, no grid was used in this case. The importance of this piece from Mycenae, I believe, lies in the very fact of this omission, for it represents a stage in the fascinating process of artistic transmission from Crete to the mainland, where attention to technical detail gradually fades away. One is tempted to see the painting of the textile at Pylos has having been made earlier than the one at Mycenae: the former used a grid, the latter did not. Ultimately, the Mycenaeans found it even easier to use textile patterns that were simple enough to be rendered entirely freehand.

With the evidence in mind of the manner of transmission suggested above, we need to return to one of my original questions: whether the textile pattern from Pylos came from human attire or some other kind of domestic cloth. My preference is that the examples quoted above, and the lack of evidence for elaborately patterned male attire in the Mycenaean mainland, make it more likely that the garment at Pylos belonged to a woman. My only hesitation as far as interpreting it as a textile pattern for dress (male or female) is because of the size of the grid units used to render the patterns. The average size of each grid unit at Pylos was 3.2 × 3.2 cm, while those of the Procession Fresco do not exceed 2 cm square. Some people may not find the difference sufficient to eliminate the option of its being part of a patterned garment, and I would have to concede to such an interpretation. As an alternative, one could offer that this fresco may represent cloth for other uses, for instance, ikria and such. But the fact is that such items, as we know from available evidence, were executed without the use of a grid.[81]

Guidelines and grids in Egypt: comparisons with the Aegean[82]

Excerpt from M. C. Shaw 2003, 186–188

This section does not aim to be a detailed exposé of the named techniques in Egypt, but rather to stress points of comparison and difference with related ones in the Aegean. One cause for some of the differences is because the Egyptians used the tempera technique, rather than the fresco used in the Aegean.[83] All preliminary guidelines in Egypt were painted, rather than incised or impressed. Like the eventual final painting, the paint for the guidelines was applied on a dry surface, the bonding effected through adhesive agents that were mixed with the pigments. Painted guidelines, including grid lines, were thus eventually covered over by the final painting, and we would not have known about them had it not been for two conditions. One is that the top painted layer has flaked off in a number of cases, due to adverse atmospheric conditions. The other, particularly useful for our knowledge of the grid, is that paintings were not always completed, particularly in the case of ambitious decorative programs attempted in private tombs.

Like the inhabitants of the Aegean, the Egyptians started by marking the position of the composition on the prepared wall surface. They first drew the upper border, the baseline, and the vertical sides, using red paint.[84] The top and lateral sides were further marked by a distinct decorative border, which emphasizes that they treated each wall as an independent entity, contra the Aegean continuity, especially of friezes, from wall to wall.[85] All the guidelines mentioned, including the ones used to divide the wall into registers, were made using a string dipped in red paint and snapped against the wall, marking it with a red line.

The main difference with the Minoans, aside from what was mentioned above, has to do with the particular use of the grid. The main aim of a grid in Egyptian painting was to help maintain a canon of proportions the Egyptians found ideal for both the human figures and the material world surrounding them. One of the results is that the Egyptian grid generally covers the entire composition, whereas the Minoan one is restricted to a specified area, basically that of patterned cloth. Also, it is likely that the Egyptians used a grid to help them paint their ceilings, when these were decorated. The decoration then consisted of abstract patterns simulating matting and woven textiles – the patterned types generally believed to have been imported from Crete, where wool, rather than the linen favored by the Egyptians, lent itself more easily to the use of color and thus patterned weaving.[86]

Unfortunately, exactly when the grid was first used in Egypt is difficult to determine, as Gay Robins explains in her recent investigation of the issue.[87] Her conclusion, based mainly on evidence from figural scenes painted on walls, is that it is first attested in the reign of Sesostris I (1971–1926 BC).[88] This is interesting from the Aegean point of view, for the Minoan/izing textile-like patterns on Egyptian ceilings also first appear in that period,[89] but the two events could have been independent. Whichever the case, and whether judged in terms of the now competing systems of High or Low Aegean Chronology, the appearance of the grid in Egypt seems to precede the one in the Aegean. Yet, and given the distinct characteristics in the function of the grid in the two areas, one should wonder if the Minoan grid is necessarily derivative.[90]

Besides the typically Egyptian grids, there is one that seems to be a foreign intruder. I refer to one that helped render a pattern in one of the now famous wall paintings found over 10 years ago by Manfred Bietak at Tell el-Dab'a, or Ancient Avaris, the capital of the Hyksos,[91] with their clear Aegean connections. The pattern served as a backdrop in a bull-leaping scene,[92] and it creates the impression of a maze – a nickname by which the painting is now often referred to

Fig. 7.14. The Bull and Maze Fresco, from Tell el-Dab'a, Egypt (restoration C. Palyvou)

Fig. 7.15. Creation of the gridded maze pattern in the Bull and Maze Fresco, from Tell el-Dab'a, Egypt (drawing M. C. Shaw and M. Nelson)

(Fig. 7.14). Like Minoan patterned textiles, the maze relies on repetition, here of a single motif that can be described as a capital letter I, which is painted in red (Fig. 7.15). What creates the impression of a maze is the meandering black line that weaves its way around and between the repeated I motif.

Close scrutiny of detailed photographs of the grid published by the excavator[93] allows one to see traces of technical details. What is puzzling, should one undertake a serious comparison with the Minoan grid, is that the lines of the Tell el-Dab'a grid (horizontal and vertical) do not seem to be continuous within the designated area. In Minoan painting, on the other hand, the impressed grid lines are never erased, wholly or in part. Erasing too would not have been necessary in a fresco technique, which is said to have been used at this site in Egypt, since impressed lines are colorless and thus essentially invisible in the finished painting. Erasing grid lines in the tight spaces would in fact have been tricky and risked accidental loss of parts of the design itself. Lastly, there is the difference that Minoan grid lines generally cut across the elements of the design to ensure bilateral symmetry, while in the Tell el-Dab'a painting they seem to sketch or parallel the elements of the final design, as guidelines do in Minoan painting.[94]

The detailed analysis above is not meant as an exercise in splitting hairs. Yet acknowledging differences, however subtle, is warranted if we are to explore the kinds of historical circumstances that may have led to this amazing case of the apparent transferring of predominantly Minoan/Aegean traits to such a remote place in Egypt at such an early date. One of the explanations I offered in an early article[95] is that the makers of the Tell el-Dab'a frescoes could have been itinerant artists (conceivably Minoan, or other people from the Aegean trained in the Minoan tradition). As itinerant artists, I argued, they may have spent some years away from Crete, trading their talents beyond the Aegean, while also picking up new techniques and representational ways in the process, before they reached Tell el-Dab'a. A sense of passage of time and distance from the Minoan models implied by this theory perhaps finds new support in the purported recent redating of the frescoes by Bietak to the era of Hatsepsut and Thutmose III, instead of to that of Ahmose as upheld previously.[96]

Interesting too, from an archaeological and art historical perspective in the scenario just proposed is that, in the paintings from Tell el-Dab'a we may be catching a glimpse of the very process that led to the interplay of various strands of technical and iconographic traditions in other areas of the Aegean and the Eastern Mediterranean. In principle, and in varying degrees and strengths, this is a process to which we can also attribute the formation of the various Aegean regional schools: that of Minoan Crete, the Cyclades, and later of the Mycenaean world.[97] Rather than regret the variations, I would exclaim: "Vive la difference!" for it provides further testimony to the versatility, invention, and ingenuity of the Bronze Age painters everywhere.

Notes

1. This chapter synthesizes selections from articles previously published by M. C. Shaw with the goal to create a concise overview of grids and other drafting devices used by Aegean artists to depict textiles. As much as possible, the texts preserve their original published form. Notes and illustrations are renumbered according to the new arrangement; notes are reformatted for consistency; and spelling is adjusted as needed. Texts reproduced by permission.
2. One such instance was the reaction on the part of some members of the audience to an oral presentation later published in Birtacha and Zacharioudakis 2000.
3. Snijder 1936, *passim* and 50–78.
4. The above definitions of "eidetic" rely on *The Encyclopedia Britannica* (1979) III, 813; 11, 890–891 (this is the derivation of my quote); *Webster's Third New International Dictionary* (1993) 727. For comments on Snijder's views see Cameron 1975 and Immerwahr 1990, 14, 160, 207 n. 3, 209 n. 5.
5. Snijder 1936, 67–68.
6. To cite but one instance – this one especially pointed because of its apparently innocuous theme – is the Minoan landscape, which is viewed today as having been used for ideological propaganda. The main proponent of this view is Anne Chapin (2004).
7. For reviews of the technique in its entire spectrum, particularly important is Cameron 1975 *passim* and 280.284, based on firsthand observation of numerous plasters from the Palace of Knossos. For support of his view that the technique was *buon* fresco see comments by R. E. Jones in Evely 1999, 148–153. In this technique, the pigments were applied on a relatively damp plaster surface with which they bonded in the process of drying. On the technique of Aegean frescoes, see Bysbaert 2008. A useful section is also provided in Evely 2000, 470–484. Here I must thank Joseph W. Shaw and Paul Rehak for the useful comments they made on the manuscript version of this paper.
8. For this possibility, see Birtacha and Zachariuodakis 2000.
9. Cameron 1975, 280.
10. For Mycenaean technique, see Lang 1969, especially 10–25; Kritseli-Providi 1982, *passim* and 93–99; Immerwahr 1990, 11–19.
11. Basic to my observations on the devices considered here – though less so those on the grid – are discussions in Cameron 1975, 281–284.
12. Naturally, pictures have their limitations, but my purpose here is not to trace each and every case of drafting, but rather representative examples. Of great help in my research were *Thera* I–VII; *WPT*; Immerwahr 1990; and Doumas 1992 with its brilliant color illustrations.
13. This is particularly true of Doumas 1992.
14. For Kea, see Coleman 1973; Abramovitz 1980. For Phylakopi, see Immerwahr 1990, 180; Morgan 2007.
15. Doumas 1992, 86–87, pls 50–51, and 124, pl. 90.
16. J. W. Shaw 1984, pl. 54f. More elaborate versions occur at Akrotiri, such as one incorporating a row of antithetical ivy leaves, and another a running spiral, in Doumas 1992, 117, pl. 83, and 121, pl. 86.
17. For Kea, see Coleman 1973, pl. 54b and Abramovitz 1980, pl. 10. The Flying Fish Fresco from Phylakopi has single bands, one at the top, and the other at the bottom, separated from the pictorial area by a string line. Immerwahr 1990, pl. 16.
18. Doumas 1992, 117, pl. 83. Elsewhere, the wooden doorjamb is simulated in painting, as at the real doorway between two rooms in the House of the Ladies. Some three vertical lines (impressed or incised) appear on the jamb, and are best visible in Warren 1975, 117.
19. An example of imperfect bands – some with uneven widths – occurs in the Caravanserai Fresco at Knossos (*PM* II, frontispiece), as I know from having worked with the original fragments (M. C. Shaw 2005).

20 The lines can be seen only in detailed photographs, as in *PM* III, 63, fig. 36.
21 Cameron 1975, 280–281.
22 *PM* II, col. pl. xviii.
23 Cameron 1975, 53, 281, cat. 48A; best visible in Warren 1975, 89. For the date, likely before LM II, see Immerwahr 1990, 173–174.
24 It appears that on the mainland the use of impressed and incised lines (the latter mainly used to draw circles using a compass) continue, but rather erratically. Their use is more common at Pylos and Tiryns.
25 Doumas 1992, 71–79, pls 36–38; 84–85, pls 44–48. In the Altar and Monkeys Fresco (Doumas 1992, 186, pl. 147), the cornice carrying the horns of consecration is delineated either by incision or the use of a very fine string impression, a distinction I am not able to make without inspecting the original.
26 Abramovitz 1980, 63, pls 3 a–d, 6 a–b.
27 *PM* II, 2, suppl. pl. xxvi, fig. 14.
28 Lang 1969, 83–89, cat. nos 49a H nws and 50 H nws, pls 31, 33, col. pl. D, reconstruction col. pl. N.
29 *PM* II, 725, fig. 452.
30 As noted by Cameron 1975, 281.
31 Unfortunately, I was unable to locate a published illustration that makes these lines clear. For the overall compositions, see *PM* IV, suppl. pl. lxvii a and b.
32 Marinatos and Hirmer 1960, col. pl. xxii. Once again, the impressed lines cannot be seen easily in the photograph.
33 Doumas 1992, 96–97, pls 63–64.
34 Cameron 1975, 283–284.
35 Cameron 1975, 282.
36 *PM* I, 780–781, fig. 508.
37 Christina Televantou (1987, 1994–1995) has written extensively about the use of incised sketches in the wall paintings of Thera. The sketches, however, were not preparatory to producing a painting. Rather they were scratched as graffiti on the already dry surface on the white background of a procession fresco of men in Xeste 4, which is still under study and restoration.
38 This is based on an illustration in Doumas 1992, 57, pl. 25.
39 Doumas 1992, 163, pl. 126.
40 Based on Doumas 1992, 92, pl. 57.
41 Doumas 1992, 133, pl. 94.
42 Birtacha and Zacharioudakis 2000. Though I was among the listeners who expressed doubt at the idea at the time of the oral presentation, I have had second thoughts on the issue since.
43 Doumas 1992, 132, pl. 95.
44 Cameron 1975, 281; Evely 1999, 156.
45 Doumas 1992, 38–39, pls 6–8.
46 This has been nicknamed the "Festoon Fresco." See Cameron 1975, 120, pl. 139B and C.
47 Doumas 1992, 175, pl. 137.
48 *PM* I, 773, fig. 504a.
49 Doumas 1992, 155, pl. 119, and 167, pl. 130.
50 Kritseli-Providi 1982, pl. Γ, α, β.
51 For a more detailed earlier study on the grid, see M. C. Shaw 2000.
52 Kaiser 1976, pl. 25.
53 Militello 1998, 104–107, fig. 27, pls 3A, 4D, E.
54 See *PT*, 314–330 for her definition of Minoan textile patterns and 328 specifically for the "interlock" varieties.
55 *PM* II, 2, col. pl. xii.
56 *PM* II, 2, suppl. pl. xxv, no. 7.
57 Furumark 1972, motif 62 (tricurved arch), fig. 68, and motif 12 (ivy), figs 35–36.
58 Betancourt 1985, fig. 22G, pl. 167, fig. 119L; Popham *et al.* 1984, pl. 152.2; Furumark 1972, 114.
59 *PM* II, 2, suppl. pl. xxvi, no. 20 and 729, fig. 456b; Sapouna-Sakellaraki 1971, pl. Bb.
60 Furumark 1972, motif 70, fig. 70.

61 For Karmares Ware, see Walberg 1976, 189, fig. 44, motif 16: 1, 2, 4; for a Postpalatial parallel, see Betancourt 1985, 167, fig. 119G.
62 *PM* II, 2, 679–681, fig. 430c; Sapouna-Sakellaraki 1971, 169, fig. 71a.
63 Sapouna-Sakaellaraki 1971, 169, fig. 71b–d, 171, fig. 72b.
64 Lamb 1919–1921, pl. viii 8–9.
65 Blegen and Rawson 1966, fig. 73.
66 *PM* II, 2, suppl. pl. xxvii, no. 21 and 729, fig. 456c, where it is incorrectly reproduced with ivies upside down; Sapouna-Sakellaraki 1971, pl. Ba.
67 Riefstahl 1944, 11–12.
68 Sapouna-Sakellaraki 1971, 193, and more recently Κωνσταντινίδη 1995.
69 Furumark 1972, motif 12, figs 35, 36; Niemeier 1985, fig. 22 (1) 16.
70 Fyfe 1902, 117 no. 40, Sapouna-Sakellaraki 1971, pl. Ea.
71 Cameron 1971, 37; *PM* II, 2, 731, fig. 457, there wrongly attributed to the "Ladies in Blue."
72 *PM* II, 2, 680, fig. 430d.
73 *PM* II, 2, suppl. pl. xxvii, no. 22, and 729, fig. 456d: outer crosses incorrectly rendered as curvilinear.
74 Sapouna-Sakellaraki 1971, 167, fig. 70a, d, 179, fig. 76g, col. pls Ea and Fb; Cameron 1975, vols II, pl. 182B and III, 146.
75 Rodenwaldt 1919, pl. 9.
76 *PM* II, 2, col. pl. xii, and 729, fig. 456e.
77 *PM* II, 2, 704–712, figs 443, 444.
78 Fyfe 1902, 117, fig. 40A; Cameron 1975, vol. II, 150, pl. 186C.
79 Blegen and Rawson 1966, fig. 73.
80 For an earlier copy prepared by Gilliéron, fils, see Rodenwaldt 1919, pl. 9. My examination and copy of the motif were carried out in 2001, with the kind permission of the then Director of the Museum, Nikolaos Kaltsas. My work was greatly facilitated by the then curator Eleni Papazoglou.
81 M. C. Shaw 1980, 169, ills 3–6. Barber (1998) reminds us of Minoan-like patterns appearing in wall painting depictions of cabins in Egyptian wall painting of ships, but the patterns are not as complex as the one surveyed above. Given that the Egyptians used the *tempera* technique, we cannot ascertain now whether an artist's grid was used there as well, but it is likely that it was, its lines drawn, rather than impressed, and now hidden under the final painting.
82 Omitted in the discussion here are Aegean gridded floors of the type illustrated in the logo of *METRON*. The subject is too large to be dealt with here, and, besides, the present author sees no clear antecedents in Minoan Crete in the period under examination here.
83 For a recent discussion of Egyptian techniques, see El Goresy 2000; Russman 2000.
84 Most of my comments on how Egyptian artists worked in the preliminary stages are based on Robins 1994, specifically 23–26, and 57.
85 For an exception, see M. C. Shaw 2000b.
86 For expert discussion on the aspects mentioned, including the use of Minoan textiles, see Barber 1991, 15, 49 n. 211.
87 See Robins 1994, 57–60, for difficulties encountered in her research for grids.
88 See Robins 1994, 70–83, for the grid's use in the Middle Kingdom, and 73, for its first appearance. Proportions were already of concern to the Egyptians before the Middle Kingdom. The system used then was one of successive horizontal lines drawn across the human figure at regular intervals – what Robins (1994, 4–69) calls "Guidelines."
89 See M. C. Shaw 1970, [on the Tomb of Hepzepha] that belongs also to the reign of Sesotris I. For further discussion of connections between ceilings and textiles, see Chapter 8.
90 In the Aegean there are no traces of a grid having been used in the rendering of human figures, in painting or in statuettes, but a number of scholars have inferred a canon based on their own perception of the inherent proportions that are measurable in the particular representations. Cameron, in Evely 1999, 157–160; Guralnick 2000; Weingarten 2000a, 103–112.
91 Bietak *et al.* 1994, 9–58, and more specifically, 44–58, pls 14–22.

92 For a special study of the grid, see Bietak, Marinatos, and Palyvou 2000.
93 Specifically, Bietak *et al.* 1994, pls 14D, 15B.
94 See, for instance, the reconstruction by Clairy Palyvou in Bietak, Marinatos, and Palyvou 2000, 84 and 86, fig. 7, which suggests a grid, but only by joining the short separate lines in a reconstruction. For an attempt to map out where the extant impressed lines actually appeared in the Maze painting, see M. C. Shaw 1998, 57.
95 M. C. Shaw 1995.
96 My thanks go to Nannó Marinatos and Sturt Manning for providing me with this information (pers. comm.). The new dating has in the meantime been published in Bietak, Dorner, and Janosi 2001, 27–119, particularly at 38–45.
97 On regional styles of Aegean painting, see Davis 1990; Morgan 1990.

8

Minoans, Mycenaeans, and Keftiu

*With a new introduction by Elizabeth J. W. Barber**

Introduction

Aegean textiles have proved to be a rich field of study – a field unrecognized 50 years ago, when I was first thinking about it. Blame the change on my mother. She loved to weave and she sewed all our clothes. Fibercraft was our childhood play: I still have (and use) something I sewed at 4 years of age, something I wove at 8, and a large sampler that I embroidered at 13 – this was not long after I fell in love with Aegean archaeology. When, finally, I took a course in Aegean archaeology in 1960, I was immediately struck by the wealth of gorgeous textile patterns on murals and the typical weaving patterns on pots. (I still have my elaborate tracing of the Ayia Triada woman dancing in her rapport-patterned flounces.) But everyone kept telling me, "They couldn't have made cloth that fancy back then: it's just artistic license. And since their textiles have perished, you can't study them."

Or blame it on my father. He always said, "If you can't get at an answer the usual way, find *other* ways." In 1974, I set aside two weeks to raid the Bryn Mawr library on the subject, hoping to put together a 10-page article proving that the technology – the fibers, spindles, looms, and dyes – had existed that would at least have made it possible for the Minoans to weave such textiles. Two things happened. I encountered there a friend, Dr Richard Ellis, also an archaeologist and a weaver, who felt the same about Mesopotamian textiles. And, with his help, I gathered so much material in two weeks that I then figured on a 60-page monograph. Thirteen years of research, travel, and experimentation later, it was a 500-page book, covering 20,000 years and 5,000,000 square miles of data: all to learn about no-longer-existing Aegean textiles!

One major key in my research had been the recognition that if a technique or resource was known to the east of the Aegean and then turned up mysteriously just to the west (in Europe), it almost certainly had to have been known by the people in the middle. That's why the book became so large. Another key was to attempt to weave recreations of the depicted patterns (and in the process, to second-guess their techniques for producing those patterns) and eventually full reproductions (as I began to encounter surviving fragments from elsewhere in Europe and the Mediterranean). Those experiments were extremely time-consuming, but as often as not produced "Aha!" moments of great value.

Many have now picked up the ball and they are running with it – the bibliography since 1987 (when I actually finished writing *Prehistoric Textiles*) has grown immensely. Maria Shaw has led the way in illuminating what the frescoes can tell us about ancient textiles (I would love to try weaving the Ayia Triada pattern for the third time, this time aligning it to the guide marks she found that the painter had laid down before painting the pattern). Marie-Louise Nosch and her research group have pursued the practical aspects of Aegean spinning and weaving. Paul Rehak, Abby Lillethun, and others have investigated clothing construction, while Nancy Hoskins has taken up the challenge of finding reasonable ways of weaving the patterns. I've been collecting examples illustrating the long tradition of women's ritual clothing manifest in the Aegean from the Neolithic to remarkably recent times. And so it continues.

Invited to work on Bronze Age cloth in Central Asia (Xinjiang) in 1995, I unexpectedly found the answer to a major puzzle of Aegean (and Hittite) men's clothing, namely the construction of the pattern-banded kilts so prominent among the Keftiu (the patterns of which, by the way, match LM IIIA pottery to a tee) (Figs 8.1, 8.2). As a weaver, I never could have imagined

Fig. 8.1. Keftiu man's kilt composed of patterned bands with fringes at their ends (best seen along the left man's back leg) rather than along the long side, where expected if loom-woven as stripes; tomb of Rekhmire (Th 100) (after Vercoutter 1956, pl. 19.156)

making a striped cloth by sewing together narrow bands; but that is just what these ancient Central Asian herders had done. I finally realized that the nomadic lifestyle didn't lend itself to hauling a large loom about, but that plaiting one's yarn into long narrow bands, plain or fancy, and then sewing them edge to edge into a garment would be both relatively easy and portable. These early horsemen were clearly Caucasoid and almost certainly Indo-Europeans (Tocharians and/or Iranians), thus close cousins of the horse-taming Mycenaean Greeks and Hittites. No surprise then that their textile traditions should be similar.[1] And I had already shown in *Prehistoric Textiles* that the Greek textile vocabulary was layered: the terms needed for band-weaving and plaiting were Indo-European, whereas the terms for weaving on a large loom were later borrowings – some of them (recognizable by their word-shapes) clearly from the Minoans.

Another aspect of Aegean textiles that continues to intrigue me, although I have not yet reached a critical mass to publish it, is the use of Aegean textiles on a certain class of Egyptian boats. These long, slim 18th Dynasty ships, of which we have many paintings and a few models, are characterized by a large rectangular cabin amidships, the sides of which are often covered with red, white, and blue spiral designs of various well-known Aegean types (Figs 8.3, 8.4). I keep wondering if these are the Egyptians' "Keftiu boats." However that may be, the creation of such cabins on what seem to be pleasure boats must have entailed a fairly brisk trade in large and elaborately patterned Aegean wool rugs. The alternative is that the cabin sides were wood (making for a hot, less airy cabin), painted to *resemble* Aegean rugs. Either way, Egyptians connected boat cabins with Aegean textiles. Evidence that Aegean people themselves used cloth to cover the tops and sides of pavilions and rooms includes, of course, the maritime fresco from Thera, the side chamber of the tholos at Orchomenos, and the ikria fresco at Mycenae (see Chapter 6) – not to mention the later tradition of temporary pavilions constructed predominantly of figured cloth that were set up for special occasions at Classical Greek temples (the *locus classicus* being Euripides' description in *Ion*, as discussed at length in *Prehistoric Textiles*). In this area too, Maria Shaw

Fig. 8.2. Late Bronze Age Central Asian pattern-banded clothing, made by plaiting narrow bands from colored wool, then sewing the bands together edge to edge; from Zaghunluq (photo I. Good)

Fig. 8.3. Nile boat with a cabin covered in designs typical of Aegean textiles; tomb of Pere (Th 139) (Davies 1936, pl. 56; courtesy of the Oriental Institute of the University of Chicago)

Fig. 8.4. Designs from four other 18th Dynasty Egyptian boat cabins; left to right: on model boat (after Daressy 1902, pl. li, no. 5091); mural, Tomb of Neferhotep (Th 50; after Davies 1933, pl. xliii); on model boat from Thebes (after Huxtable, in Glanville 1972, fig. 5, cf. pl. xi); painting, Deir el-Bahri (after Kantor 1947, pl. xi.A)

has taken the lead in suggesting various other fresco fragments from Knossos as depicting wall and ceiling hangings (see Chapter 4).

All in all, the study of Aegean textiles is thriving. It is to be hoped that by putting together, at intervals, the findings of all the separate researchers, we will inspire each other to yet new insights and discoveries about this remarkably complex and beautiful textile tradition.

Minoans, Mycenaeans, and Keftiu
PT, 311–312, 330–357

If, as you wander about in the Theban tombs, you look up from the typical Egyptian wall scenes of feasting and daily life, you are likely to see in the ceiling panels above you a startling array of Aegean-looking scrolls and spirals and even bulls' heads. Archaeologists have long exclaimed over the appearance, without visible means of transport, of these strongly Aegean artistic motifs on Egyptian ceilings; for Aegean pottery – which we know was imported – does not show many of these design elements, let alone the large, all-over patterns. How did the Egyptian artists come by them, and why did they choose to put them on their ceilings?

A few scholars have also noted that the designs in question are of a typical textile variety, and have hazarded the suggestion that the Egyptians had been importing Aegean cloth, now perished, from which they had copied these designs. The patterns are thus of considerable interest to us.

Nor are these the only places in which Egyptian artists have been suspected of depicting Aegean textiles. The small but famous series of tombs in which the Aegean-looking gifts are brought to the Pharaoh by emissaries sporting long Minoan curls, fancy kilts or loincloths, and elaborate sandals (of a sort still used to this day in Anatolia and the Balkans), under the labels of "people of Keftiu" and/or "people of the Isles in the midst of the Great Green (Sea)," stirred decades of heated debate over the accuracy of the artists and the geographical origins of the people depicted – a debate finally resolved to the satisfaction of most by Jean Vercoutter's monumental study of the entire series.[2]

Starting from the premise that different tomb owners had different reasons for depicting the foreigners, Vercoutter was able to demonstrate that some of the owners – the viziers – had been concerned with, and would have seen for themselves, both the objects brought from abroad and the people who brought them, whereas others – the temple treasurers and the military men – would probably have seen only the objects and so had copied the bearers somewhat freely from the tombs of the viziers.[3] The accuracy of the details of the tomb paintings, then, had to be assessed anew for each tomb; and even the viziers may not have been above copying from each other sometimes. But where objects of the same class survive today for comparison (e.g., pottery and some metalwork), it has become clear that the top Egyptian artists were excellent observers of foreign people and things when they wished to be. And Aegean men and goods were among the observed.

Given the material about ancient cloth that we have accumulated here, what can we say about these topics now? What more can we deduce safely about Aegean cloth from the evidence of the representational art – Aegean and/or Egyptian – by building on our direct knowledge of the looms, dyes, fibers, and techniques of the 2nd millennium? Can we assess the degree of Egyptian familiarity with Aegean textiles and how the Egyptians acquired that familiarity? And what else of more general archaeological interest might the study of these textiles and of this international contact add to our understanding of the ancient world?

The Egyptian evidence

In order to provide some sort of understandable framework for the detailed arguments to follow, I shall begin with a quick overview of Egyptian-Aegean relations as they appear to us at the moment.

We know from the combined archaeological and literary records[4] that Minoan contact with Egypt, however much earlier it may have begun, flourished off and on from the Middle Kingdom, early in the 2nd millennium, down into the New Kingdom, and most especially during the reign of Queen Hatshepsut (c. 1504–1482 BC), who apparently received official Minoan embassies at court. Although the detailed correlation to Minoan archaeological periods is vehemently disputed, we are safe in saying that this large period corresponds approximately to the Middle Minoan period plus a little bit of Late Minoan. No indisputably LM II pottery has been found in Egypt, although LM IB probably has, if it has been reliably distinguished from the LH IIA sherds.[5]

Late in the reign of Hatshepsut's stepson and successor, Thutmose III (1482–1450 BC), we see evidence, discussed below, that the Mycenaeans had either joined or replaced the Minoans in a major expedition to the Egyptian court at Thebes. The pottery problems just mentioned imply that this ought to correspond to LH II and/or LM IB periods. For from then on, Minoans and Minoan artifacts effectively disappear from the Egyptian scene.

After a short hiatus of maybe 30 years, during which time no one at the Egyptian court, at least, seems to have laid eyes on Aegean people of either sort, trade with the Mycenaeans flourishes again, reaching its peak during the reigns of Amenhotep III (1417–1379 BC) and of his son Akhenaton (1379–1362 BC). The Aegean influence at both of their courts is considerable, and the imported pottery excavated at their sites is chiefly LH IIIA (at Amarna, 98% late IIIA2).[6]

During the next two dynasties, the Aegean trade clearly continues, for we have LH IIIB pottery throughout the long and stable reign of Rameses II (1304–1238 BC). But Aegean influence at court was minimal, no embassies are recorded, and the last we hear of Aegean people is in the rather one-sided accounts of the battle of Rameses III, in his Year 8 (c. 1190 BC), against the infamous Sea Peoples, who seem to have included some folk of Mycenaean culture among them. Then we learn nothing more.

Surviving Egyptian representations of official Aegean emissaries occur entirely within the reigns of Hatshepsut and Thutmose III, a short space of 54 years. We will ignore all the other "possibly Aegean" figures – mostly either generic prisoners or representatives of the four corners of the world – that Vercoutter seems to have demonstrated to be either copies of the emissary paintings or fairly free inventions (in the tombs of Puyemre, Amenemopet [= Imenemipet], Kenamun, Amenemheb, and Ineni [= Anena]).[7] That leaves us with five tombs.

The earliest of these is the tomb of Senmut (Th 71), officially priest of Amon and chief steward under Queen Hatshepsut. In modern parlance he was the queen's finance minister and the chief architect of her public works (her temple at Deir el-Bahri being one of the most felicitous pieces of architecture in all of ancient Egypt), as well as the person who, more than anyone else, it seems, helped Hatshepsut manage her power and keep her hold on the throne. As such, Senmut was a prime victim of the unpleasant attentions of Thutmose III, who defaced every trace he could find of his hated stepmother and her flunkies as soon as he finally regained his throne in 1482 BC. Senmut had died in Year 19 of Hatshepsut's reign, about 1485 BC.

Senmut's tomb contained two things of interest to us. One is a painting of an embassy of Aegean people who wear Minoan-style loincloths, codpieces, cinchbelts, and hairdos, and who are bringing typically Aegean-looking vessels as gifts, presumably to Hatshepsut (Fig. 8.5). Unfortunately, Thutmose's destruction crew and other more recent accidents of fate have destroyed most of the precious scene: parts of three Aegean ambassadors remain today, and parts of two more were recorded in the last century.[8] In two or three cases the loincloth was plain (one of them was bright red) with a fancy edging. In another case the belt and/or loincloth has bands of sawtooth alternating with bands of dots or bars, all done in red, white, and blue. As Hatshepsut's favorite, Senmut was presumably present at the reception of the Aegean embassy, and he seems to have counted it among the more memorable sights of his life at court.

The other article available for our attention is the handsomely adorned cloth found tied upside down on the back of a mummified horse.[9] The band that decorates it has the honor of being the earliest pattern-woven cloth from Egypt after the all-white towel looped in a zigzag design and the scrap with little blue stripes at the edge, both from the 11th Dynasty, centuries earlier.[10] The geometric pattern here is quite complicated, worked out in red, brown, and yellow with a warp-float technique.

8. *Minoans, Mycenaeans, and Keftiu*

Fig. 8.5. Minoan ambassadors to the Egyptian court at Thebes, from the tomb of Senmut (Th 71); early 15th century BC. Note codpieces, long hair, and yo-yo and barred-band patterns (Davies 1926, fig. 2; courtesy the Metropolitan Museum of Art, New York)

Chronologically, the next tomb to show Aegean emissaries is that of Antef (Th 155), who was Chief Herald under Hatshepsut, and who was politically adroit or innocuous enough to keep that post into the early reign of Thutmose III. Unfortunately, this register of the scene is mostly gone (although the register of Syrian ambassadors just below is well preserved), so we see only the remains of one prostrate emissary wearing a Minoan-style loincloth, and behind him the distinctive sandals of a second Minoan.[11] As Chief Herald, Antef also apparently assisted at the reception of these embassies. It would be wonderful to know if the Aegean presentation was the selfsame event as that shown by Senmut, but that is beyond us. In Antef's tomb, too, there is a second thing to catch our interest: the painted ceiling (see Fig. 8.15, below). That will be described presently.

Third comes the tomb of Useramon (or Amenuser; Th 131), who was vizier – in effect, prime minister – during the early years of Thutmose III. His tomb was finished shortly before Year 28 (that is, 1476 BC, since Thutmose counted his regnal years from the date when he was officially designated pharaoh as a very young child, with Hatshepsut as his regent). This scene is far better preserved, and we can count

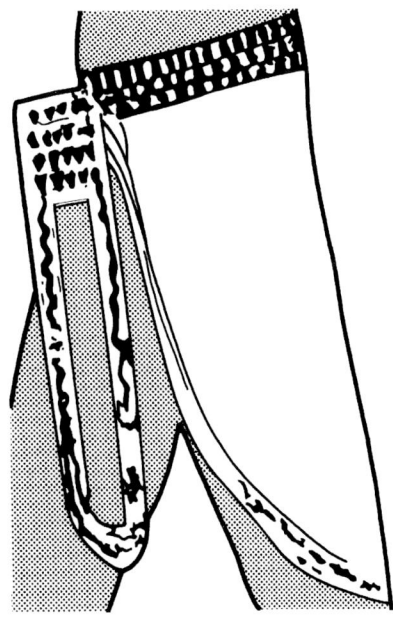

Fig. 8.6. Loincloth on the 12th porter, tomb of Useramon (Th 131); early 15th century BC. Textile pattern is red on white (after Vercoutter 1956, pl. 16.137)

a total of 16 fully or partly visible Aegean visitors. Like those in Senmut's tomb, they wear Minoan loincloths (Fig. 8.6) with prominent codpieces and carry Minoan-style vessels. The fabrics are shown either as plain or with small, busy designs on them that can no longer be made out with any accuracy. As in Antef's tomb, the men wear elaborate sandals with brightly colored, prominent belts, but not so obviously cinched in as those in Senmut's tomb. Either the wasp waists of the Minoan men did not impress this artist so much, or he felt more bound by the "proper" human figure. Once again, the clothing is shown as exclusively red, white, and blue.

The fourth comes the celebrated tomb of Rekhmire (Th 100), nephew of Useramon and the next holder of the office of vizier. Rekhmire kept that office through the rest of Thutmose's long and energetic reign, and briefly into that of the following pharaoh, Amenhotep II (1450–1425 BC). Like most officials, he began his tomb almost immediately upon becoming important, so as to be as ready as possible for the next world at all times. Among the scenes he ordered painted was one of the reception of the foreign embassies, 16 Aegeans among them clad in Minoan-style loincloths and codpieces (Fig. 8.7). A number of years later, however – and we wish we knew how many – the vizier went to the trouble of having this group of figures partially repainted. What was altered was the style of clothing; each and every loincloth was erased and a pointed kilt painted over it. Fortunately for us, the expungers did not do a very complete job and we can make out some of the original paint. The old cloth patterns again seem to have been mostly small, all-over designs, whereas the new ones are invariably in stripes or zones, some vertical, some horizontal. Again the predominant colors are red, white, and blue, with only two exceptions (other than the pair wearing yellow and black-spotted leopard skins, nos. 5 and 14 on Fig. 9.7). Man 9 seems to have a mostly yellow kilt, and Man 4 has one small yellow panel among the hodgepodge of designs on his.

This is the first of the tombs to provide a linguistic label for the Aegean visitors: they are said to be princes from the Keftiu country – consistently a designation for Crete – and the "Islands in the Midst of the Great Green (Sea)."[12] It has been cogently argued that the people of the islands were indeed specifically the Mycenaeans – as one would expect from all considerations other than the term "Islanders" – but that the Egyptians first encountered northerners of this culture during the initial period of Mycenaean expansion, when such islands as Rhodes and then Cyprus were the jumping-off points for further travel.[13] That most of the crew and even the entrepreneurs should have been the islanders, born and bred to life in a boat, rather than the mainlanders ("tamers of horses"), seems fairly reasonable.

The last of the tombs showing Aegean emissaries is that of Menkheperraseneb (Th 86). This man, eldest son of Rekhmire, was High Priest of Amon under Thutmose III and Amenhotep II – that is, he administered the treasury where the tribute was stored. As such, he would have seen as much as he wanted of the objects brought by such embassies, but may or may not have been present at the reception itself. At any rate, Vercoutter shows that most of the 12 Aegean people shown in Menkheperraseneb's tomb (Fig. 8.8) seem to be free copies and rearrangements of parts of the figures in Rekhmire's tomb, to which he would have had access, whereas the objects portrayed are often new and different – including textiles! – but are quite believably Aegean.[14] Clearly, the visit depicted is the same one recorded by Rekhmire in the repainted version.

The following structure emerges. During the reigns of Hatshepsut and Thutmose III, the court at Thebes received at least two official embassies of Minoans, who wore the usual Minoan loincloth with cinchbelt and fancy boots. The Egyptian artists who recorded the scenes were enormously

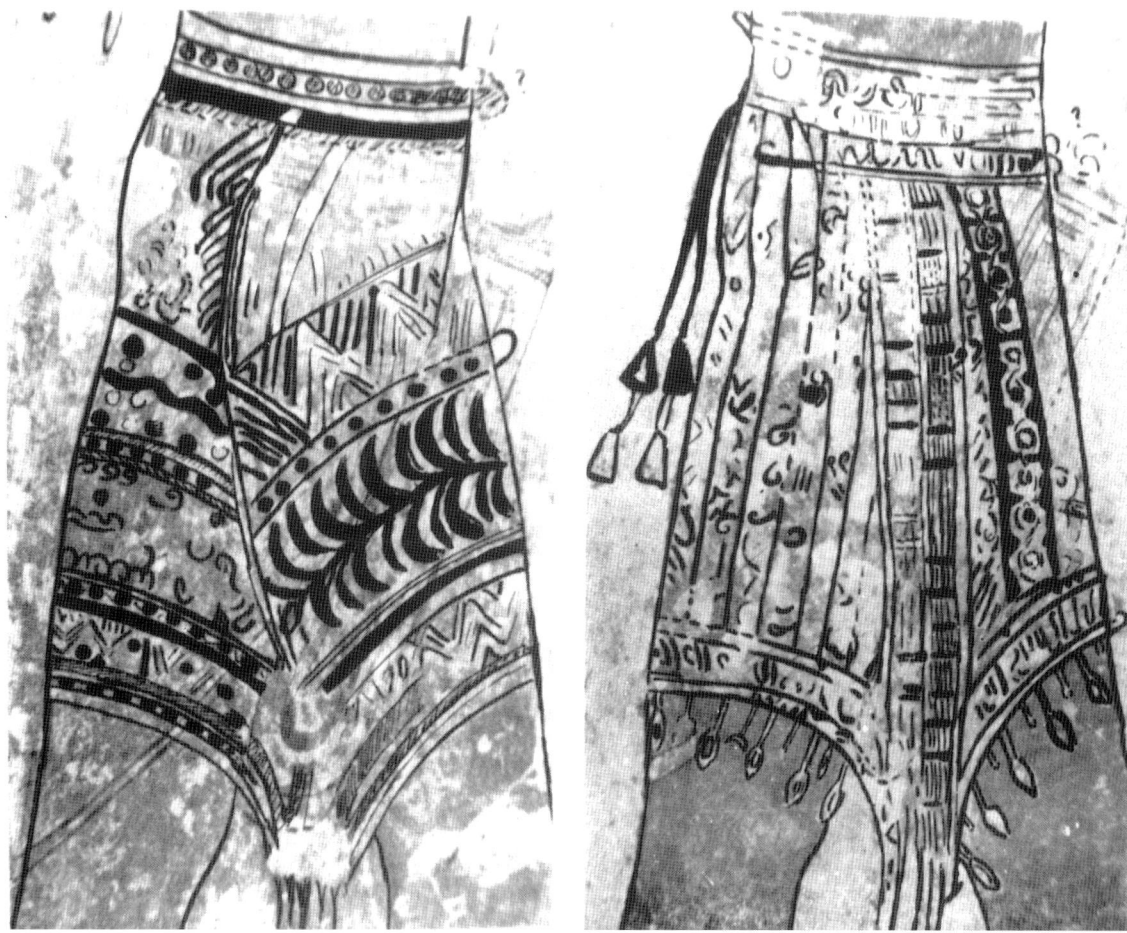

Fig. 8.7. Kilts of two Aegean ambassadors to the Egyptian court at Thebes: leader on right, fourth emissary on left; tomb of Rekhmire (Th 100); early 15th century BC. Note traces of the original loincloth and codpiece (cf. Figs 8.5, 8.6), repainted with kilts (Vercoutter 1956, pls 21.162, 19.156)

impressed by the prominent "codpieces" that hung down in front;[15] they also perceived the Minoans as wearing exclusively red, white, and blue, and generally depicted the cloth as having a small, curly, all-over pattern and a fancy edging. At some time during Rekhmire's term of office, however, the Egyptians received another official Aegean embassy at court, but these men were wearing elaborate paneled, pointed, and tasseled kilts, along with the usual belts, booties, and long curly hair. Then there were apparently no more embassies from the Aegean.

The cloth in these kilts was perceived as quite different from that in the earlier loincloths. In several cases it consists of patterned stripes running vertically, in a few cases horizontally, the end stripe down the overlapping edge at the front of the kilts being a little wider, fancier, stiffer, and longer than the other stripes, with a tassel – if any – emerging from the bottom end of this stripe. Such an arrangement accords remarkably well with what we know of making cloth on the warp-weighted loom: the heading band is necessarily different from the body, and stiffer; and

Fig. 8.8. First five Aegean ambassadors to the Theban court, from the tomb of Menkheperraseneb (Th 86); mid-15th century BC. Note the kilts, and the stylized "bolt of cloth" over the arms of the 1st and 4th porters (Davies and Davies 1933, pl. 5; courtesy the Committee of the Egyptian Exploration Society)

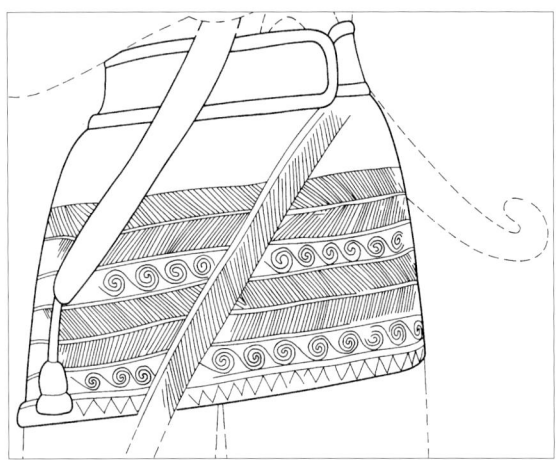

Fig. 8.9. Kilt worn by a male in martial dress, from a Hittite relief on the King's Gate at Boğazköy, in central Turkey; 3rd quarter of the 2nd millennium BC. Compare the elongated front corner and figure-striped cloth with the kilts in Figs 8.7 and 8.8 (courtesy Museum of Anatolian Civilizations, Ankara)

it results in a fringe of its own narrow warp at one or both ends, which are corners of the final cloth. The stiffness (and tassel) could also be accounted for by assuming that some of the bands along the front overlap and the bottom edges of the kilts had been sewn onto the kilt afterwards. The other general type of Aegean kilt depicted by Rekhmire's artists is made of panels that virtually had to have been sewn together to form the garment, if we are to believe the artist at all, since on a single kilt some panels run vertically, some horizontally, and each has a wildly different pattern from the others – a real feat to weave all in one piece, but easy to weave separately and sew together. The very first emissary in Rekhmire's group has an additional embellishment: a fringe all the way around the bottom of his kilt, which suggests that this kilt was woven the other way – from top to bottom rather than side to side. If the fringe truly belongs with that kilt pattern, it means that an elaborated side-selvedge formed the overlap edge, and the figured stripes were woven vertically on the loom rather than horizontally (not my choice of how to do it most efficiently!). In this, however, it corresponds to the way in which a very similar kilt elsewhere must have been woven: the pointed kilt on the King's Gate figure at Boğazköy (Fig. 8.9). Here the figured stripes run horizontally on the kilt, but the fringe runs down the overlapping edge. In other words, here too the patterned stripes must have been woven vertically on the loom – or the fringed edge was woven separately and sewn on.

Mycenaean men wore either the kilt or the tunic, never the loincloth. The Minoans on Crete seldom wore anything but the loincloth (or a great, stiff robe) until just the time when the Mycenaeans arrived on their island. Is it coincidence that the only other ethnic group to turn up wearing a kilt in the middle of the 2nd millennium is the Hittites, who are the only other Indo-Europeans around at this point besides the Mycenaeans? I think not. I strongly suspect that the kilt came to the Aegean with the Mycenaeans as an Indo-European garment,[16] and that it traveled to Egypt with Mycenaean or heavily Mycenaeanized emissaries.[17]

Before we leap, however, to the tempting conclusion that the change of dress in Rekhmire's tomb is the same as that recorded in the Knossos frescoes, and thus is nicely synchronous, we must stop and look carefully: the textile match is not so neat. Why don't either the Minoan or Mycenaean frescoes of kilts show these elaborate stripes and panels? The Egyptian artists may not have had the time to record in explicit detail the pattern on each and every kilt, as the visitors paraded past. But they clearly did have enough time to record a reasonably accurate impression. Their perception that these emissaries wore striped and paneled kilts, entirely different from those of any other race they knew, came from observation. So who was wearing them? Are we to take the fresco's label, Keftiu (Minoans) and Islanders, at face value? Mycenae is not on an island. Later the label of "Islanders" clearly came to refer to the Mycenaeans in general; but this is one of the earliest uses,[18] and therefore ought to be more literal. Was this the particular costume of Rhodes and perhaps Cyprus, the islands from which Mycenaean maritime expansion into the East Mediterranean pushed off? We shall find reason to return to this question presently.

The question of colors is also puzzling. Egyptian artists normally used red, yellow, blue, green, black, and white freely in their paintings. That they chose never to use green for their portrayal of Aegean clothing matches exactly what we found in the Aegean frescoes of clothing. But the persistent lack of yellow on the textiles until Rekhmire's tomb, where yellow is used even then on two of the 13 kilts (although freely used on the two leopard skins, as well as elsewhere in the scenes), does not correspond. Certainly by LM I, at Thera and on Crete, yellow was in plentiful supply. But it is also clear that the Egyptians perceived the earlier contingents of Aegean people as wearing only red, white, and blue; and their observations are otherwise sufficiently accurate – especially for overall characteristics like this – that I have to believe that for some reason they saw only red, white, and blue cloth. This reduced color scheme will turn up again to haunt us.[19]

Thus armed, we can now turn our attention back to the Egyptian ceiling designs, which, as we have mentioned, sometimes include motifs that look both very Aegean and very "textile" in nature.

The motif most often cited under this heading is the spiral, characteristic of the Aegean since the Early Bronze Age at least. The chronological list of occurrences of spirals on tomb ceilings, given in Table 8.1, begins with a few of the Middle Kingdom, then jumps to the 18th Dynasty, ending with a few in the 19th, 20th, and 21st.[20] All these have spirals – but no two ceilings are exactly the same, and a few are wildly different from any others. Another motif that is cited as Aegean-looking is the design of a bull's head with a rosette between its horns, much as we see it

Table 8.1; Egyptian Tombs with Spiral Décor

Tomb	Owner's Name and Chief Title	Period
Assiut	Hepzefa (Hapdjefai), nomarch (see Fig. 8.11)	12th Dynasty
Qau 18	Wahka II (B)	12th Dynasty
Th 21	User, royal scribe and steward	Thutmose I–II
Th 81	Ineni (Anena), overseer of granary	Amenhotep I–Thutmose III
Th 67	Hapuseneb, high priest of Amon (see Fig. 8.10)	Hatshepsut
Th 39	Puyemre, priest of Amon	Hatshepsut–Thutmose III
Th 155	Antef, great herald (see Fig. 8.15)	Hatshepsut–Thutmose III
Th 82	Amenemhet, scribe to Usseramon, overseer of weavers of Amon (see Fig. 8.15)	Thutmose III
Th 87	Minnakht, overseer of granaries	Thutmose III
Th 251	Amenmose, royal scribe	Thutmose III
Th 262	[name lost], overseer of fields	Thutmose III?
Th 17	Nebamun, royal scribe and physician	Amenhotep II?
Th 78	Haremhab, royal scribe	Thutmose IV–Amenhotep III
Th 90	Nebamun, standard-bearer, captain of police	Thutmose IV–Amenhotep III
Th 226	[name lost], royal scribe	Amenhotep III
Th 40	Huy, vizier	Tutankhamon
Th 181	Nebamun and Ipuky, sculptors	late 18th Dynasty
Amarna 49	Ay, first vizier, then pharaoh	Akhenaton–Ay
Th 50	Neferhotep, "divine father"	Ay–Haremhab
Th 19	Amenmose, priest	Rameses I–Seti I
Th 51	Userhet, mortuary priest	Seti I
Th 16	Panehesy, priest	Rameses II
Th 45	Dhutemhab, overseer of weavers of Amon (tomb usurped from Dhout: time of Amenhotep II)	Rameses II
Th 359	Inherkha (Anherkhawi, Khai-inheret), chief of public works	Rameses III
Th 65	Imiseba, chief of temple-scribes	Rameses IX–X
Th 68	Nesipanoferher, priest of Amon	Herihor

in three dimensions in the famous bull's head rhyton from Mycenae. As it happens, the Egyptians for some reason always chose to combine these rosetted bull's heads with spiral motifs, wherever we have them; so listing the tombs with spiral ceiling-patterns nets the bulls too – in the tombs of Neferhotep, Inherkha, and Imiseba.

Spirals occur so freely on the Aegean pottery, which we know was imported, that one might at first wonder if textiles need to be dragged into the explanation. And bull's heads were sometimes used as an artistic motif in both Syria and Egypt quite independently of the Aegean, it seems. So neither of these designs will clinch our textile case immediately.

But the third motif cited occurs only once in all of Egyptian paintings: a quatrefoil interlock almost identical in form to that on the skirt of the Ayia Triada dancer. It differs only in its colors (using the usual Egyptian six-color scheme instead of being limited to red, white, and blue) and in the inclusion of a skinny quatrefoil inside every, rather than every other, large quatrefoil. These interlock patterns show a long and natural development in Crete in the Middle Minoan period, where they are entirely at home with other aspects of the local decorative arts, whereas in Egypt this design is isolated. It occurs on the soffit or "ceiling" of a doorway in the tomb of one Amenemhet, a high functionary during the early reign of Thutmose III, the tomb having been completed about Year 28, i.e., *c.* 1476 BC (Fig. 8.15).[21] Pottery correlations indicate that the early reign of Thutmose III should fall at about the end of LM IB,[22] that is, about the time the Ayia Triada fresco was being destroyed, not painted initially. So every consideration brings us to the conclusion that the Egyptian artist borrowed the pattern from the Aegean world, and almost certainly from the world of textiles, since the design occurs on Crete mostly in frescoes depicting textiles.

The plot thickens when we investigate Amenemhet's background. According to the inscriptions in his tomb, he was scribe and chief steward to Useramon, that same powerful vizier who officiated at the reception of an embassy of Minoans to the court of Thutmose III; moreover, he had also inherited the important and honorable post of "head of the weavers of Amon."[23] As Useramon's scribe, Amenemhet was thus in a prime position to see and record whatever Useramon was doing, including entertaining Minoans, and as head of the weaving shops that made luxury cloth for the highest priests of the realm, he certainly had technical knowledge and an interest in textiles – probably far beyond any other high official in the land. If anyone had both interest in and access to fancy imported textiles, it would have been Amenemhet. Maybe it is not so surprising, then, to find such an exact replica of a typical Minoan textile pattern, and one of the loveliest they ever invented, reproduced in his tomb. And it is not the only typically Aegean design there: another of his eight ceiling patterns is a handsome diagonally running spiral, not quite like any other found in the tombs, and hence not likely to have been copied from them. Are we to imagine Useramon rewarding his faithful helper with the particularly suitable gift of a couple of magnificent and curiously wrought textiles, perhaps out of the largesse that we see the pharaoh showering onto his vizier?

Having established to near certainty that at least one Minoan textile found its way to Egypt and was copied onto a ceiling, we now must go back and take the other ceiling designs more seriously. For, as Smith's Law says, "What did happen, can happen." Let us begin by investigating the nature of Egyptian treatment of ceilings in general.

From the time that tomb ceilings begin regularly to receive decoration at all, in the 12th Dynasty, until the late 19th or early 20th Dynasty (when scenes of importance to the funeral rites finally creep up to that heretofore unimportant surface), the ceiling designs fall into four categories. They show sky, with or without stars (e.g., Th 362;[24] or the great star chart in Senmut's unfinished tomb under the temple of Deir el-Bahri[25]); or birds in flight, as if startled up from a swamp thicket (e.g., Th 50, tomb of Neferhotep;[26] or Th 31, that of Khons, with crickets and ducklings all in a flap[27]); or a grape arbor, viewed as if from directly underneath, with the leaves and clusters of grapes hanging down among the grid of wooden staves (e.g., Th 359, tomb of Inherkha;[28] or Th 16 Panehesy); or a repetitive polychrome design framed by architectural strips painted yellow to imitate wood. It is the origin of this last group that interests us. But note, meanwhile, that all of the other categories of ceiling designs represent things that one normally sees overhead.

The earliest and most common repetitive designs throughout the entire period are checkers, zigzags, and lozenges: the normal Egyptian mat patterns. One sees handsome representations of these mats elsewhere in the tombs, where their function as mats is clear: for example, forming the shade-giving wall of a pavilion behind Urarna, a nobleman of the 5th Dynasty,[29] or covering the magnificent wall and awnings around a sacred door painted on a 12th Dynasty wooden coffin.[30] Most of these same designs occur already as painted wall decoration in a 1st Dynasty tomb at Saqqara (no. 3121), where they presumably also represent mats.[31] That is, the making of elaborately patterned, polychrome mats was a native Egyptian art going back to the earliest dynasties and probably much earlier.

Why mat patterns on the ceiling, then? One reason is that mats were often laid on poles across rafters in the houses, in order to prevent the mud of the roof from crumbling down onto the occupants.[32] In these cases, when one looked up one would indeed see mats. Fancy mats are painted in just this position on the expensive plaster ceilings of some of the rooms of Amenhotep III's palace at Malkata, in imitation of the cheaper roofing. Since the tombs were fashioned of stone or brick, the tomb architects, too, took up this strategy of painting decoration.

Another type of structure in which mats were commonly used overhead was the outdoor pavilion erected to provide shade from the hot tropical sun. We see such pavilions, with checkered mat coverings over a flat or gently sloping grid-like framework of yellow wooden beams, both on land and on boats – for example, in the 12th Dynasty tomb of Antefoker (Th 60).[33] The constructional details of these pavilions can be seen especially clearly on a 12th Dynasty model boat now in the British Museum (no. 9525).[34]

In 1929, Ludwig Borchardt pointed out evidence of another type of pavilion, following an excavation at Luxor. He noticed, high up along a wall that belonged simultaneously to the Medinet Habu temple and to the palace of Rameses III, a series of beam holes in arcs of 3, 5, or 7.[35] If one placed into them the ends of rafters that were suitably supported at the other end, and threw some sort of floppy covering over each group, one would produce a series of barrel-vaulted tents or pavilions of varying sizes, the largest in the middle. Borchardt went on to show that this arrangement answers very closely to a ceiling, for example, in the small Ramesside tomb of Irinefer, which consisted of a barrel vault with rafters indicated as running both lengthwise at the apex and crosswise in the middle, with a pillar to support both at the crossing.[36] In the tombs, all of this is sculpted in mud, plaster, or stone, of course, and then painted yellow to imitate wood with blue-painted hieroglyphs carved into it. The empty panels between are painted with repetitive polychrome patterns, many of which are the old traditional mat patterns that we have been discussing, but some of which are not. (We shall discuss those in a moment.) It seems entirely reasonable that, in real life, large and elegantly woven mats should have been used to cover such pavilions, so arranged that the guests would be able to view and enjoy the bright decorations. But one could also imagine other coverings being used – rugs, for example, or even leather. (Leather would have the disadvantage of not letting any air through; yet we possess the fragmentary remains of a colorful leather canopy that belonged to the 21st Dynasty princess Isimkheb.[37])

One of the most dazzling arrays of decorated ceiling panels in a barrel-vaulted tomb occurs in that of Inherkha (or Anherkhawi or Khai-inheret, depending on how one reads his name), no. 359 at Deir el-Medineh.[38] Here we find the ceiling divided into eight compartments: four on either side of a central yellow rafter, with three yellow cross-rafters making the remaining divisions. Within its frame of yellow rafters, each panel had a border that looks like a fringe, or perhaps a set of lashings, on all four sides (compare this to the fringe hanging down from the top of the lower canopy on the funeral barque of Neferhotep[39]). Within that border the panels contain (as one moves down one side of the room and back up the other):

1) concentric rectangles of rosettes (there is only one other similar ceiling, in Th 68 Nesipanoferher;[40] but compare the ceiling of the side chamber of the tholos tomb at Orchomenos in Greece);[41]
2) an S- and C-spiral rapport enclosing rosetted bulls heads (for the bull plus spiral motif, see also five other ceilings discussed below);
3) a bead network (or four-petal rapport) with rosettes in the interstices (an increasingly common pattern in the 18th Dynasty, first found without rosettes in the Middle Kingdom, e.g., at el-Bersheh);[42]
4) alternating wide and narrow stripes, the wide stripes filled alternately with rosettes or running spirals (compare several similar but not identical panels in Th 68 Nesipanoferher);[43]
5) a stylized grape-arbor pattern;
6) a diagonally running spiral rapport, with glyphs in the interstices;
7) a bead network with rosettes, virtually identical to no. 3;
8) a C-spiral rapport (Helene Kantor gives numerous Aegean examples in various media, and Egyptian examples, all on scarabs, except one that is apparently a ceiling design and published by Champollion without provenience[44]).

All of these designs give the immediate impression of rugs; and with the exception of the grape arbor, all are more easily explained as textile, leatherwork, or perhaps beadwork patterns than anything else. But those, too, are reasonable things to hang overhead on a pavilion for decoration as well as for shade.

Having demonstrated that every category of design shown on Egyptian tomb ceilings until very late is derived from things that Egyptians could see overhead, and that the repetitive designs among them derived from mats, cloth, and whatever else one might cover an outdoor pavilion with, we are in an even stronger position than before in maintaining that Amenemhet, the scribe of Useramon and overseer of the temple weavers, was copying a textile when he put a distinctly Minoan textile pattern on his ceiling.

The entire line of argument suggests, in fact, that all of the spiral and other such designs that are not easily ascribed to the matting should have had their origin in textiles and similar arts (that is, arts that produce large, floppy coverings). Yet our textile studies have given us every reason to believe that the ancient Egyptians never wove such things – at least not until the late 18th Dynasty at the very earliest, since we see their exploratory first attempts under Amenhotep II and Thutmose IV.[45] It would be very easy at this point, then, to jump to the conclusion that all spiral ceilings before the late 18th Dynasty were therefore copies of specifically Aegean textiles. But the case is not so simple.

When Kantor drew together the material known in 1947 for Aegean-Egyptian relations, she pointed out five occurrences in Egypt of a very distinctive type of pattern, in which the curl of a spiral ends in a vertical bud or palmette (as in Fig. 8.10). The motif is perfectly at home in the Aegean, but again isolated as a tiny group in Egypt. Three of the Egyptian examples are from tomb ceilings: Th 67 Hapuseneb (Hatshepsut's reign), Th 251 Amenmose (Thutmose III's reign), and a fragment from Th 262 (Thutmose III's reign).[46] A fourth example is on the cabin of Queen Hatshepsut's boat, as painted on the wall of her temple at Deir el-Bahri; and the fifth is tooled on the leather of the chariot found in the tomb of Maherpra, a high official under Amenhotep II.[47]

Here, then, we have proof positive that some of the Aegean-type designs had to do with leather – another commodity that survives not at all in the Aegean. As a weaver, such evidence relieved my mind; for although I could readily see weaving the version of the pattern with a simple bud in the center, as shown by Hatshepsut and her priest Hapuseneb, I would hate to have to weave (or repeatedly embroider) the detailed palmettes of the fancier version. The design is not impossible to produce on a loom, it just is not the sort of pattern that evolves naturally and easily in that medium, at least not without a drawloom. Were the Minoans and/or Mycenaeans also exporting tooled leather?[48]

Once we entertain the notion of leather having been imported, it behooves us to notice objects of appliquéd leather such as the canopy of Princess Isimkheb, and the horsecloth shown among the New Year's gifts in the tomb of Kenamun (Th 93), chief steward of Amenhotep II.[49] The overall pattern of rosettes and stars on the latter, if not identical to anything on the ceilings, nonetheless suggests another possible source of at least a few of these designs. And there is nothing to indicate that the Egyptians did not make the leather horsecloth themselves. In short, we may have to reckon with ceiling patterns from leather covers for canopies and pavilions, in addition to those from mats and textiles. And they too may be either imported or native.

Fig. 8.10. Ceiling pattern from the tomb of Hapuseneb (Th 67) composed of spirals ending in a vertical "bud"; early 15th century BC (after Jéquier 1911, pl. 28.43, and Kantor 1947, pl. 11B)

We must also reckon with Hatshepsut's boat. It is only the first of a considerable series of boat cabins with spiral designs on the sides. The chief group that I have encountered includes: two model boats from the tomb of Amenhotep II (nos. 5089, 5091),[50] and wall paintings in the tombs of Kenamun (Th 93),[51] Nebamun and Ipukhy (Th 181),[52] and Neferhotep (Th 50).[53] All of these show close variants of a design that is slightly different form Hatshepsut's: a simple spiral rapport with a rosette in the enclosed quadrilateral, and a polychrome barred edging around the whole. (In general, the spiral is yellow, its center green, the diagonal rows of quadrilateral spaces alternately red and blue with contrasting rosettes.) Apparently the pattern became a favorite, for we then find versions of it on the tomb ceilings of Huy, vizier to Tutankhamon (Th 40), of Userhet (Th 51, Rameses I), and of Dhutemhab (Th 45, late Rameses II); and on the ceiling of a chapel of Rameses II at Deir el-Medineh.[54]

Since it was readily visible to the public eye on boats and chapels, we are not forced to postulate continual imports of such a pattern. But we may still wonder how the design got onto the boats of Hatshepsut and Amenhotep II in the first place. Are we supposed to be reminded of the colorful deck-shelters, or ikria (Fig. 6.12), apparently of textiles, leather, and perhaps felt, that we see depicted on boats as well as separately in frescoes at Thera and Mycenae?[55] Did the Aegean sailors give the Egyptians a new idea for a particularly handsome and elegant form of cabin, originally of furlable textiles, but perhaps eventually of painted wood? Is this a part of what we are to understand by the expression "Keftiu boats" in the Egyptian shipyard lists?[56]

Whatever the precise answers to these questions, we are finding that the probability of importations of perishable manufactured goods from the Aegean is multiplying in all directions: not just textiles, now, but leather and maybe even boat parts were imported. But we are also refining our understanding to the point where we can distinguish the reflections of textiles among them with some confidence. It is time, then, to go back and pick out the most obvious and interesting Aegean textiles among the ceilings and see what we can add to our knowledge of the craft.

There is an important change, however, in the kind of representation that we find around the reign of Amenhotep III, which will also affect our interpretations. The earlier pieces seem to stand out from each other as few in number and quite distinct in character; but later, the designs begin to smear together, as though either a lot of rough copying was going on, or else a great many imports of approximately but not exactly the same design were flooding the market. Strictly within the evidence of the paintings, I find it hard to make a choice, but the great increase in imported Aegean pottery during the reigns of Amenhotep III and Akhenaton suggests that increased trade – of textiles along with the pottery – is at least partly responsible.

The later group, then, in its homogeneity, is probably telling us what the average bolt of imported Aegean cloth looked like during that period. (Here we seem to be getting hints of the Mycenaean tendency towards a simplified and uniform "mass" production.) The earlier group, however, is probably telling us what the cream of the crop looked like: the pieces selected by the emissaries as most likely to win Pharaoh's favor, or selected by the early merchants as most likely to win customers abroad. Some of the earlier ceilings, in fact, are so idiosyncratic that it is almost impossible to believe that the artists were not sitting there copying the master's favorite pavilions. In one very early case, the tomb of Hepzefa at Assiut (Fig. 8.11), the artist has crammed in six patterns in a way that makes a hash of the architecture, as if he had to fit them all in at any cost. But viewed on its own, the ceiling looks like six odd-sized rectangles of cloth were stitched together to make a larger piece. One senses the same pride and pleasure in personal ownership that is evident in the pets sometimes pictured under the master's and mistress's chairs – here a pet monkey, there a pet goose. The pets were clearly a real and treasured part of daily life; handsome canopy tops, a conspicuous symbol of status, luxury, and taste, may have been just as real and just as carefully copied for the next world.

We have one other curious bit of evidence for the cloth- and mat-covered pavilion, in a set of wooden linen-chests from the 18th Dynasty tomb of Kha that are painted with all-over designs remarkably similar to those on some of the ceilings.[57] Each of these four oblong boxes has a lid like a low-pitched, gabled roof (with different designs on each of the faces), making it look like a small house or pavilion. On three of the boxes, the back side is painted with a simple checker pattern, the short sides with either that pattern or other all-over designs, and the front with a scene of the lord and lady dining with the aid of a servant. The illusion is quite strong that the people are sitting inside the box/pavilion. The fourth chest has textile-type patterns on all four sides. Using patterns from textiles for the outside of a linen-chest seems peculiarly appropriate.

The earliest ceilings to show Aegean-type designs are actually from the height of the Middle Kingdom, including that from the tomb of the nomarch Hepzefa[58] – from an era of trade rather

Fig. 8.11. Part of the painted ceiling in the 12th Dynasty tomb of Hepzefa, at Assiut; early 2nd millennium BC. Note the "wrought-iron fence" motif of interlocked hearts in the center panel (photo of an early drawing by Baroness von Bissing; courtesy of H.-W. Müller)

than of official embassies. The spirals occur in a heart shape that is mirrored again at the apex to form what I nickname the "wrought-iron fence" motif. This motif is then offset, and interlocked at the spirals (how Minoan a trick!) to spring another row of fencework, and so forth. The space within the heart is occupied by a palmette, while the space between groups is a solid diamond framed with dots. Two other ceiling designs in the tomb are composed of squared spirals, or meanders. Maria Shaw has cleverly shown that one of these is the squared counterpart of the spiral heart, hence the two are closely related,[59] while the other contains as an element the dot-surrounded diamond. Since meanders were as uncommon as spirals in Egypt, one must suspect them also of being imports,[60] whether they piggybacked on cloth or on mats.

Mats? Did the Minoans also make fancy mats? Or had they simply acquired a technique of weaving other than weft-float, perhaps one like double-cloth that would make "squaring the circle" a natural transformation of the design?

The "wrought-iron fence" motif turns up again – diamonds, palmettes, and all – on a doorway in the tomb of Ukhhotep at Meir (12th Dynasty, under Amenemhet II); on the ceiling of the tomb of Antef (Fig. 8.15), the Great Herald of Hatshepsut who depicted Aegean emissaries on his walls (see above); and on the kilt of the third Aegean visitor in the tomb of Menkheperraseneb, who served under Thutmose III and Amenhotep II (see Fig. 8.8). A very similar motif (with rosettes instead of diamonds, and no palmettes) was painted on the ceiling of a barrel-vaulted room among a group of private houses at Thebes – perhaps a small chapel – from the time of Thutmose IV.[61] The latest example that I know of is an unfinished version on the ceiling of the tomb of Amenmose (Th 19), a priest who must have died during the reign of Seti I, early in the 19th Dynasty. Here the diamonds have been made over into the sort of concentric lozenges typical of Egyptian tomb ceilings, and the palmettes have disintegrated into equally typical Egyptian dotted quatrefoils, such as those that decorate the next ceiling panel in the same tomb.

The simplest hypothesis is that one or more textiles with this design came into Egypt in the 12th Dynasty, presumably from Crete (MM I/II?), where spirals, rapports, and interlocks are at home,[62] and that a second importation occurred from the same general source at the time of Antef (LM IB?).[63] Menkheperraseneb's depiction presumably represents yet a third sighting of this motif (LM IIIA?), this time on a stranger's clothing. The chapel from the time of Thutmose IV must show yet another importation, co-occurring, as we shall see, with the earliest known representation of the bull-and-spiral motif, whereas the 19th Dynasty example would seem to be an Egyptianized rendering of one of the earlier examples.

There is something else of note on the ceiling of Menkheperraseneb's tomb (Fig. 8.15). There are no spirals here; in fact, the design might be passed over very easily as being the familiar old mat pattern of lozenges and zigzags, except for three things. First, the design is the only ceiling pattern I know of in Egypt that is exclusively red, white, and blue – exactly like the Keftiu kilts. Second, the bands forming the zigzags and lozenges look unusually thin: the proportions are quite different from the usual Egyptian ones. Third, this main design has an edging (unusual at this period, though not unheard of) that is also entirely in red, white, and blue; the edging is also composed of lozenges, right until those lozenges unceremoniously turn into zigzags. A most un-Egyptian irregularity!

There is only one other example that I know of in which such a change of design occurs in midstream: the belt band from Lefkandi,[64] where zigzags become diamonds by way of chevrons.

There the shift is undoubtedly the happy result of boredom interacting with the particular weaving technique; but in paint one is not restricted by what the warp will do, nor does the work progress so slowly. So why that particular change? The zigzags on the ceiling edge are also of an unusual type for Egypt: a single line zigzags continuously, with little nests of tents inside each resulting triangle. This is exactly the form we see suddenly filling up LM IIIA pottery.[65] It also occurs on the kilt of the fifth Aegean porter in Menkheperraseneb's tomb, alternating with a band of chevrons (see Fig. 8.8),[66] and twice in closely related forms on the fourth kilt in Rekhmire's tomb (see Fig. 8.7, left; and see the sixth in Menkheperraseneb's, which is a rough equivalent thereof). In fact, a comparison of Popham's newly excavated horde of LM IIIA pottery with the kilts in the tombs of Rekhmire and Menkheperraseneb reveals an astonishing number of parallel motifs.[67]

Menkheperraseneb's Aegean visitors are a problem in themselves. Many of them, as Vercoutter has pointed out, look as though they are merely reversed copies or amalgamations of the porters in his father's tomb.[68] Now, as overseer of the Treasury of Amon, Menkheperraseneb was in a good position to study the objects brought to Thebes by the embassies and left behind to be stored in the temple, even if he was perhaps not invited to the reception itself. But it is also just conceivable that the reason his pictorial record is so similar to that of his father, yet also notably different, is that the son was in fact at the reception, that each of the two men recorded the same event independently, and that the different scribes and artists in their respective retinues simply noticed and emphasized different things. The kilt with the "wrought-iron fence" pattern and the red, white, and blue ceiling, both absent from Rekhmire's tomb, would seem to resurrect the younger man's credibility somewhat. Menkheperraseneb's tomb is also the only one to bother to show the Aegean visitors bringing textiles over their arms (along with the celebrated metal vases). His paintings show four such porters.

Fig. 8.12. Ceiling fresco from the robing room of the palace of Amenhotep III at Malkata, Thebes, showing a spiral rapport and bulls' heads; early 14th century BC (photo courtesy the Metropolitan Museum of Art, New York; Rogers Fund, 1911: no. 11.215.451)

Leaving the era of the embassies and picking our way with care into the later group of ceilings, we find that the most interesting design is that in which a bull's head with a rosette between its horns occurs among running spirals. The earliest example we have is on the ceiling of that same barrel-vaulted chapel that contained the "wrought-iron fence" motif, from the time of Thutmose IV.[69] The next example comes from the ceiling of the robing room of Amenhotep III, in his palace at Malkata, near Thebes (Fig. 8.12).[70] (This palace has long been a source of discussion on account of its many apparent ties with Aegean art.) And finally we have the ceiling designs from three rather late Theban tombs: Th 50 Neferhotep, Th 359 Inherkha, and Th 65 Imiseba.

Each design is significantly different – there is no question here of copying directly either from each other or from the same single source. The chapel ceiling has a horizontal row of mournful-looking bulls' heads between vertically running S-spirals, each bull having a disc or rosette filling the space between the incurving horns. Each bull's head occupies the space of two spirals. In Amenhotep's robing room, however, each bull's head (with a rosette filling the space between horns) is set diagonally within the quadrilateral spaces left by a spiral rapport. As with the spiral rapport patterns on the boats, these spaces are contrastively colored: in other words, it is the same pattern we discussed earlier on the boats, only with rosetted bulls' heads in place of plain rosettes.

Neferhotep's design is wildly different. C-, S-, heart-, and eyeglass-spirals are complexly interlocked into a square grid; the rosetted bulls' heads are set square in the large red mushroom-shaped interstitial spaces, and grasshoppers crouch in the slim, blue, diagonal spaces. The bulls, unlike before, have their eyes on the sides of their skulls (rather than the front), and they have blotchy hide-marks down their noses, but no nostrils are shown. (Another elaborate ceiling design in the same tomb also is based on a rather similar spiral rapport, but with rosettes and hieroglyphs in the interspaces.)

Inherkha's pattern, on the other hand, harks back to the form in the chapel, with rows of bulls' heads set between vertically running S-spirals. But here the spirals are reversed each time (so that they mirror their neighbors), the horns are vestigial supports for great discs above, and a peculiar fan is painted below each nose – whether to represent a great snort of breath or a mythical beard of the sort that bulls wear in Mesopotamian monuments, or just to fill up the space utterly. Imiseba's pattern is different again, with a proportionally rather small bull's head with a small disc between the horns, set in a field framed by interlocked spirals with a banana-like element in the small interspaces.

Where are these coming from? Bulls' heads with rosettes on the forehead occur in Aegean art earlier than anything here (e.g., the silver-and-gold rhyton from Shaft Grave IV at Mycenae, 16th century BC).[71] Smith has pointed out that an even closer parallel occurs on the two 14th century silver wishbone-handled bowls from Enkomi in Cyprus and Dendra in Greece.[72] On both cups the horns curve down, and on the Enkomi cup the rosettes occur below rather than above the bulls, but that may be largely a function of the shape of the space to be decorated. The forms of the heads are extremely similar. (On the Dendra cup the "rosettes" appear, as on one of the rhyta, as a swirl of hair on the bull's forehead – between the horns in that sense.) Closer still is a cup with a row of bulls' heads with rosettes between the horns, depicted among the precious "gifts" brought to Egypt by the Keftiu themselves at the time of Senmut (Fig. 8.5).[73] Smith also points out a rather less similar Ugaritic Hathor-head with spiral curls and a small disc between the horns, as well as a constellation vaguely similar to our bulls and spirals in a mural at Nuzi – a bull's head motif with a little crossed circle between the horns and some curly plant-like designs on either side.[74] The artists here seem to have been drawing from an international fund of motifs, which makes our design rather harder to trace.

It is noteworthy that all of the Egyptian bull-and-spiral ceilings occur relatively late in the Egyptian series: from Thutmose IV on – that is, after the court embassies are over and the Minoans are no longer on the scene. So if they are Aegean, these patterns must be Mycenaean or "Mycenaeanized," and they may even be Syrian. They also occur at precisely the time when we see little in the way of fancy textile motifs depicted in the Greek mainland palaces. Again we must ask: if the Mycenaeans themselves were not making such things, then who was? The people of Rhodes, or Cyprus, or some other Mycenaeanized outpost?

We receive a fresh shower of representations of Aegean-type designs at the close of the Bronze Age, with such tombs as Th 359 Inherkha and Th 68 Nesipanoferher. We have already discussed at length the eight-rug paneled ceiling of Inherkha, who was chief architect to Rameses III. The ceiling of Nesipanoferher, priest of Amon under Herihor of the 21st Dynasty, is no less varied and no less full of spirals. If anything, these panels appear the most rug-like of the lot. After that, there is nothing more, other than isolated late copies and reworkings.

Our ultimate deduction concerning the Egyptians' knowledge of Aegean textiles is that they knew quite a lot. They seem to have received a variety of handsome Minoan textiles during the 12th and early 18th Dynasties in particular, and to have viewed the wearers and bearers of such textiles at least twice at the Theban court in the early 15th century BC, under Hatshepsut and Thutmose III. The Egyptian observers recorded a Minoan preference for red and blue dyes, and for complicated all-over patterns. We can even use the Egyptian evidence to name some favorites: the running and rapport spiral – including the "wrought-iron fence" variant – and the quatrefoil interlock.

Then, as Minoan pottery gives way to Mycenaean, we find that Minoan-style cloth gives way to some other sorts of Aegean-looking textiles. We have called them "Mycenaean" by convention, but we have also seen some reasons to doubt the precision of this term. At any rate, loincloths are replaced by kilts in the last Aegean embassy to Thebes, and new patterns of cloth turn up: figured panels and stripes (including the "tented" zigzag) on the kilts, bull's head patterns, and increasingly baroque spiral rapports. These textile imports appear to continue down to the era of the Sea Peoples, early in the 12th century BC.

Our final task will be to integrate into this picture a few other sources bearing on textiles while we deal with a few last problems.

Some other evidence

From approximately the 12th through the 18th Dynasties in the Faiyum, the Egyptians seem to have been in contact with the Minoans in a way quite different in character from momentary bouts of trade or tribute. It is not at all clear what is happening there, despite the massive work of Kemp and Merrillees in reviewing all the previous excavations in the area and all the Minoan and possibly Minoan pottery there.[75] The center of activity seems to be el-Lahun (ancient Rehone), which lies on one of the Faiyum waterways and was probably an important docking site.[76] Close around it lie the other sites with a high incidence of Minoan artifacts: Gurob, Kahun, and el-Haraga.

It is in a modest grave at Gurob that Brunton and Engelbach found the low-whorl spindle of European – presumably Aegean – form, but made of local Egyptian materials.[77] The only simple explanation for its presence is that some European women were living in Gurob more or less permanently, and doing their own spinning. If they were also weaving according to their own custom, their products may have been the source of some of the Aegean designs we see in Egypt. That is, it is not necessarily the case that everything was imported from across the Mediterranean. Some of the people may have moved too.

Next door in 12th Dynasty Kahun, meanwhile, someone had left behind something equally un-Egyptian for Petrie and his team to find: a "handful of weaver's waste" of spun wool in three different colors.⁷⁸ We have no reason to believe that the Egyptians were skilled in either weaving or dyeing wool at this time, or that they were even raising the appropriate types of sheep (those depicted are all of the hairy varieties); yet the presence of cut-off ends from a loom virtually proves that the weaving had taken place right there in Kahun. So now we can be virtually certain that foreigners, presumably women and probably Aegean, were busy weaving in the Faiyum in the 12th Dynasty according to their own foreign customs and with some of their own foreign materials.⁷⁹ We get a surprise, however, when we look at the colors: they are red, blue, and green! So *someone* knew how to make green dye.

We find more evidence later of these northern-style textile crafts being plied in Egypt: specifically, some more spun and unspun, colored and uncolored wools found at Amarna.⁸⁰ Indeed, we see all sorts of hints that Akhenaton and his family enjoyed bright and cheery textiles quite different from the traditional white Egyptian linen: the geometrically figured sashes that the king wears as he drives his chariot or leans out of the window,⁸¹ or the fat red cushions, patterned with tiny blue and yellow diamonds, that the family lounges on in the boudoir (in the famous Princess Fresco, now in the Ashmolean Museum)⁸² or leans on in the Window of Appearances.⁸³ Akhenaton's successor(s) used the same window cushion, according to Neferhotep;⁸⁴ and we possess a long, tapering sash similar to Akhenaton's that belonged to Rameses III.⁸⁵

There is no guarantee that these particular textiles were Aegean, however. The Syrians, and many others as well, are shown wearing elaborate cloth by this time; three splendid examples of ornate Syrians are those presenting tribute to Tutenkhamon in the tomb of Huy (Th 40),⁸⁶ those groveling at the feet of Amenhotep III in Tomb Th 58,⁸⁷ and the faience inlays of prisoners discussed below. As with the Minoans, we tend to see tiny all-over designs on the cloth, but mostly dots, rosettes, and skinny quatrefoils.

On the other hand, we have some guarantees that Syrian textiles were well known in Egypt by the mid-18th Dynasty. We have already discussed at length the gala tunic of Tutankhamon,⁸⁸ with its wide embroidered band of mixed Egyptian and Syrian motifs, clearly produced specifically for that pharaoh (his cartouche is part of the décor), but by foreign craftspeople. We have mentioned Thutmose III's records of bringing home foreign textiles and workers as part of the booty from the sack of Megiddo.⁸⁹ And we have discussed the introduction into Egypt of both the tapestry loom and tapestry technique: both apparently from Syria, and both presumably at about this time.⁹⁰ Tapestry technique caught on and embroidery was largely rejected in Egypt; but we learn something valuable about how the Syrians were producing their textiles around 1450 BC. At the moment it seems unlikely that southern European women practiced either tapestry or embroidery yet (though they may have been learning these arts from the same sources at about the same time). As for the other weaving method that turns up in Egypt in this millennium – namely, warp-patterning of narrow bands – the source is harder to pinpoint. We know, at least, that the Europeans had the technique long before Senmut's horsecloth was made;⁹¹ and we can see that by the time of Rameses' girdle, such work was being done in Egypt itself, since the pattern on the latter consists of ankh signs.⁹²

By the time of the Ramesside pharaohs, too, the inhabitants of Syria and Palestine had enlarged their repertoire of cloth made for wearing. A series of brightly colored faience tiles from the Egyptian palaces at Tell el-Yahudiyeh (Fig. 8.13), Kantir (see Fig. 8.14), and Medinet Habu shows us what these new fashions were,⁹³ and they include decorated versions of both the old Syrian spirally wrapped gown and an elaborate kilt, often paneled or friezed with amazing beasts, and often sporting a tassel at the extra-long (i.e., pointed) front corner.

I have complained before about the paneled kilt. The first hint of trouble came when we discovered that the kilts of the clearly Aegean emissaries shown by Rekhmire and Menkheperraseneb, with their elaborately figured panels and stripes, did not match the textiles of the kilts shown in the Mycenaean frescoes. The kilts shown in the last frescoes at Knossos certainly have an elaborate tassel in the right spot, and dip down in the front; but the textile motifs are the little, curly, all-over designs characteristic of the Minoans, and although they may have a wide bottom border, the kilts are not composed of panels. The Mycenaean mainlanders, on the other hand, show themselves wearing tunics as often as kilts, and the cloth in both cases is exceedingly plain: frequently edged and occasionally fringed, but never constructed of fancily designed stripes or panels. We pointed a finger inquiringly at Rhodes and Cyprus, the Mycenaean

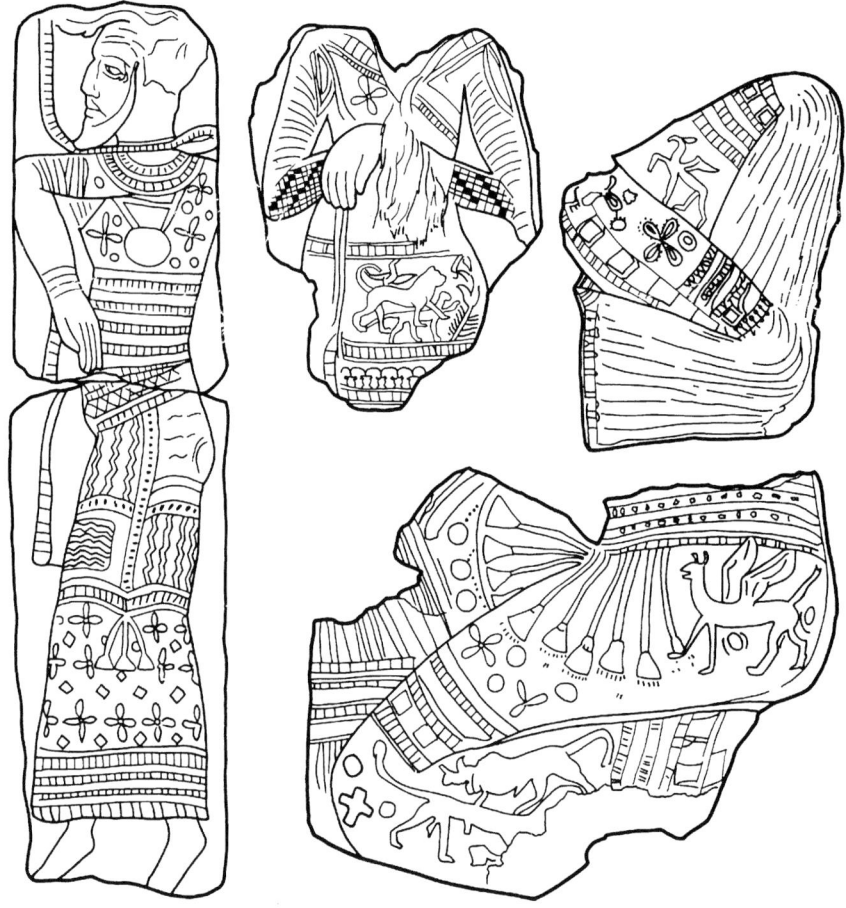

Fig. 8.13. Fragments of faience tiles of foreigners from a 20th Dynasty Egyptian palace at Tell el-Yahudiyeh; 12th century BC. Note panels and beasts on kilts; cf. Fig. 8.14 and PT, 321, 334, 366, figs. 15.7, 15.18, 16.5 (after Wallis 1900, pls 5, 6)

stepping-stones to the east Mediterranean, on the basis of Rekhmire's inscription. Now we have another pointer to the east for such kilts, in the faience tiles. What is going on?

First, Rhodes and Cyprus. We have little direct evidence from either of these islands about cloth or clothing, except for some non-detailed bronze figurines of kilted males, and two ivory mirror handles from Cyprus (Late Bronze III) showing heroic male combatants wearing knee-length kilts decorated with wide ornamental bands.[94] They look not unlike some of those in the tombs of Rekhmire and Menkheperraseneb. For the rest we must content ourselves with the general archaeological picture. That means pottery.

Pottery in Bronze Age Europe and the Near East seldom looks so much as though it was copied off of textiles as does the Mycenaean pottery of Rhodes, Cyprus, and the other Mycenaean or "Mycenaeanized" islands and seaports. The friezes of chariot riders, friezes of wild animals, panels of ornate birds and geometric motifs, and the little fillers are all typically "textile" in shape and busy treatment. Remember that in a textile you can only deal

Fig. 8.14 Fragments of faience tile depicting a red-skinned captive wearing a friezed kilt, from the palace of Ramses II at Kantir; 13th century BC (Hayes 1937, pl. 8; courtesy the Metropolitan Museum of Art, New York)

with one row of weft at a time; so construction in strips or panels is a necessity. You can choose, of course, to ignore this natural division in the overall design you wish to create: you can make a huge tapestry of a single tableau, as with the French Gobelins. But the path of least resistance in weaving is to plan the decoration in manageable strips or friezes, especially if you are attempting "representational" art rather than merely trying to cover the cloth with a pleasing pattern. In other words – to sort this out – it looks as though some weavers in the Mycenaean world had begun to depict real objects (e.g., birds and animals) and even scenes (e.g., chariot processions), making them quite naturally into friezes and panels; and the potters had followed suit, taking forms directly from the weavers. For remember, too, that a pot-painter is *not* tied to strip-shaped space the way the weaver is, and in fact is normally confronted with complexly curved and nonlinear surfaces that may be covered in any order whatsoever.

The filler ornaments, too, are the sorts of things a weaver working with floating weft will come up with in order to reduce the length of the floats. (Long floats snag easily, and add nothing to either the beauty or the practicality of the fabric.) Fillers, of course, can have other sources in art; but in weaving they are sometimes a necessity.[95]

Technical analyses have demonstrated that all the Mycenaean-looking pottery on Rhodes in the LH IIIA and IIIB periods was in fact imported from the mainland, and specifically from the

Argolid,[96] and that in Cyprus at least the chariot kraters, along with much else, were imports from there too.[97] So if we are looking to Rhodes and Cyprus for such fancy textiles as may have inspired this pottery, we will need to look right back to the mainland, to the Argolid, where we thought all the textiles were plain. So once again: who was wearing the paneled and tasseled kilt?

When it comes right down to it, we really don't have a very complete picture of Argolid male fashions in LH III period. The frescoes that we happen to have from the Argolid show men hunting or peaceably driving chariots. If only a particular class wore the kilt, perhaps the right scenes didn't happen to be preserved. Where we see the kilt is on the warriors at Pylos, and on the men portrayed in the Knossos processions and the Egyptian tribute-bearing scenes – on men who may have worn them as "messengers" or "sailors" or both.[98]

As a matter of fact, the Egyptians show us large quantities of warriors and sailors wearing the pointed kilt at the very end of this period. In the reliefs depicting Rameses III's battles against the Libyans and the Sea Peoples, it seems as though most of the enemies, especially among the Sea Peoples, as well as half the Egyptian allies are wearing it.[99] The kilt seems, in fact, to have become the basic garb for war and/or travel throughout the Mediterranean – outside Egypt – and the different ethnic types of kilt-wearers can be distinguished only by their hats and hairdos, noses, beards, and jewelry, when at all. The reliefs themselves do not show enough detail for us to see what patterns, if any, occurred on the garments; we see only the division of the kilts into panels or registers, the pointed front, and sometimes the stylized triple tassel. For such patterns we have to go back to the much more elaborate and purposely colorful representations of prisoners and generic enemies on the faience tiles.

What we see are friezes of creatures. Some are of bulls or of wild goats(?) among flowery fillers (Fig. 8.14), remarkably like those on the pottery of Argos and its export markets in the LH IIIB period, of the whole sub-Mycenaean world in the LH IIIC period, and on into Archaic and Classical times in Ionia.[100] One beast, however, seems to be a rhinoceros, and others are Syrian-looking griffins (see Fig. 8.13). So the Syrians themselves have had a hand in this. But the owner of the "wild goat" kilt is shown with red skin (Fig. 8.13),[101] a convention used by the Egyptians only for themselves (males) and for the people from Keftiu and the islands of the north – not for Syrians, Palestinians, Hittites, or any others. So despite the fact that this man wears his kilt over a long white tunic, in the fashion of so many people of the Levant, he is evidently perceived as being of Aegean blood.

The only sense I can make of all these facts is that the Mycenaeans and the Syrians had been very busy mixing up new textile ideas together during the time between the deaths of Thutmose III and Rameses II (certainly the traceable spread of Mycenaean pottery makes this probable), and that in the course of swapping techniques, fashions, motifs, and even people back and forth, they had all by various routes come to the point of making and sometimes wearing friezed and/or paneled kilts, some of which were handsomely decorated in a new Animal Style. To judge from the styles, it would even seem possible that tapestry technique had finally made its way to the Aegean, alongside the native weft-float methods. Looking back at the frieze on the bottom of Tutankhamon's tunic, one can suspect the Syrians, at least, of making some of their figures in embroidery as well as tapestry and that technique too may have reached the Aegean, as we saw from the frescoes of miniature ornaments.

Fig. 8.15. Egyptian ceiling patterns from the 18th Dynasty, probably copied from contemporary Aegean weavings (drawing M. Stone). Top left: Quatrefoil interlock pattern decorating the soffit of a doorway in the tomb of Amenemhet (Th 82); early 15th century BC (after Davies and Gardiner 1915, pl. 32D; courtesy the Committee of the Egypt Exploration Society); Right: "Wrought-iron fence" motif and two meander-based patterns from the ceiling of the tomb of Hatshepsut's herald Antef (Th 155); early 15th century BC (after Säve-Söderbergh 1957, pl. 19); Bottom left: Ceiling pattern from the tomb of Menkheperraseneb (Th 86); mid-15th century BC. Note the unusually skinny lines and shifting border pattern (after Davies and Davies 1933, pl. 30B; courtesy the Committee of the Egypt Exploration Society)

We have one more strong hint at Mycenae that cloths friezed with representational subjects were indeed being manufactured for Mycenaean use: one of the ikria or deck-shelters (see Fig. 6.10) painted in a room on the Citadel consists of friezes of very Aegean nautilus shells between familiar bands of lozenges.[102] Maria Shaw shows by her parallels that such figured screens were almost certainly woven of cloth originally.

Within this context of figured cloth, the late ceiling patterns that show representations of bulls' heads, and the striped patterns that involve running spirals, seem even more plausibly to have come out of an Aegean/Syrian textile *koine*, at least, if not strictly, from the Aegean.

Conclusions

No matter which way we have turned, we have found the evidence strong for a lively Bronze Age Aegean textile industry and a lively trade with Egypt during most of the 2nd millennium BC. We saw that the Aegean representations of cloth, found chiefly in the frescoes, accorded well with what we knew of European cloth from direct archaeological evidence; and we saw that Minoan cloth in particular was highly ornate. By the time we were done, we had even isolated several of the favorite Minoan patterns: the spiral band and rapport, the four-lobed interlock, and the "wrought-iron fence" motif with its filler of dot-edged diamonds.

We concluded from the evidence that the Egyptians were well acquainted with this cloth. They observed it during the 18th Dynasty on the persons of the Aegean ambassadors to the Theban court, and among the "gifts" brought by those emissaries. We also deduced step-by-step that wealthy Egyptian noblemen were using Aegean rugs as brightly decorative and even ostentatious canopy-covers from Middle Minoan/Middle Kingdom times (early 2nd millennium) to perhaps as late as the collapse of Bronze Age societies *c.* 1200 BC. It is for this reason, apparently, that we find so many Aegean textile-type motifs on the tomb ceilings of particular classes of the Egyptian nobility. The Egyptian representations, once recognized, even added to our stock of information about the Aegean textiles – concerning motifs, colors, uses, and dating.

Aegean merchant-sailors spread these textiles liberally about the East Mediterranean while bringing fresh ideas and techniques back to Greece and Crete. We can see the return-cargo of ideas in the frescoes in particular. We found we could follow the sequence of Aegean carriers both by the changes in style of clothing among the emissaries and captives shown by the Egyptians, and by the continuing parallels to the textile patterns in the Aegean pottery motifs. Finally, toward the close of the 2nd millennium the entire Aegean culture plunged deep into a Dark Age, along with Egypt, the Levant, and the East Mediterranean trade. What may have happened then is a puzzle for the next chapter to be taken in its own right and on its own evidence.

Notes

*   Aside from the new introduction, this chapter is from *Prehistoric Textiles* (1991, 311–312, 330–357), adapted for republication by Anne P. Chapin. To the greatest extent possible, this text preserves its original published form. Notes and illustrations are renumbered according to the new arrangement; in-text citations are reformatted as endnotes and edited for consistency; and spelling is adjusted as needed. Texts reproduced by permission.
1   See Barber 1997 for Aegean banded kilts; and for Central Asian bands, see Barber 1999, 54–60, figs 3.7–3.11, and color pl. 6, which includes references to ancient Aegean weights similar to those used in East Asia for making such bands.
2   Vercoutter 1956.
3   Vercoutter 1956, 188–198, etc.
4   The chief studies from which this framework has been compiled are Kemp and Merrillees 1980, which sorts out the Minoan pottery in Egypt; Hankey and Warren 1974, which pulls together the Aegean and Near Eastern imported artifacts relevant to cross-dating; and Vercoutter 1956, which sorts out both the literary and the graphic evidence for Minoans and Mycenaeans in Egypt. The reader is hereby directed to these sources for details. Many aspects of the difficult problems involved in Egyptian-Aegean relations are irrelevant to an analysis of the textile trade, such as, for example, the year-dates. I have used some current "standard" dates, those of the *Cambridge Ancient History* (1970–), not because I have anything to say on the subject of absolute chronology, but in order to make the framework more graspable to the uninitiated. (For

the same reason I ignore the fact that Thutmose III technically ruled with Hatshepsut during the 20 years in which she alone wielded the power, and treat him for rhetorical purposes as having followed her.) On the other hand, I believe that the textile material has some independent information to add on the subject of cross-dating in the Late Bronze Age, as will appear below.

5   Kemp and Merrillees 1980, 226, 232, 245.
6   Hankey and Warren 1974, 147.
7   Vercoutter 1956.
8   By Hay: Hall 1909–1910, pl. 14.
9   *PT,* 157, fig. 5.9.
10 See *PT,* chapter 5.
11 Säve-Söderbergh 1957, pl. 13.
12 Vercoutter 1956, 57, 133–134.
13 Vercoutter 1956, 154–156.
14 Vercoutter 1956, 283–284.
15 These appendages have been much discussed: they may indeed represent codpieces, or more simply, as Vercoutter plausibly suggests (1956, 253–256, fig. 33), the end-flap of the loincloth that has been brought forward between the legs and then caught up under the belt so as to fall down in a showy cascade in the front. That explanation is the only one that satisfactorily accounts for the fancy edging, which is always the same as that on the main body of the cloth. The artists of Senmut's tomb obviously didn't quite understand what they were trying to draw, whereas Useramon's artists understood it much better. This point, incidentally, demonstrates clearly that Useramon was not copying from Senmut, and that another opportunity to scrutinize the Minoans in person had occurred in the meantime. For a detailed discussion of the Minoan loincloth, see Sapouna-Sakellaraki 1971.
16 Barber 1975.
17 There is at least one early representation of a kilt, and a tasseled one at that, from within the purely Minoan world: on the sword hilt from Mallia (see *PT,* 320, fig. 15.5). Perhaps this surprisingly short-haired acrobat is a foreigner, like the seven youths imported to Knossos from Athens in the Theseus legend. On the other hand, the design on his kilt is a typically Minoan one. Perhaps the Minoans were already picking up the kilt from across the sea.
    Note too that a series of textile ideograms in Linear B has been identified as referring to cloth intended for kilts (Duhoux 1974). Some of the variants show a warp fringe, others show a double line (heavy or applied border?) at one or both ends or along one long side. Some also show the "ingot" shape that would produce the long points in front, while others have a median line lengthwise, suggesting folding at the waist. (See Duhoux 1974, 117 for all relevant diagrams.) The ideograms are frequently surcharged with *WE*, written out once as *we-a$_2$-no* (*weanós*: Duhoux 1974, 119–121), which comes from the basic Indo-European root for clothing (Lat. *vestis*, etc.). It suggests that to the Mycenaeans the kilt was somehow the most basic form of clothing, with other types (like the tunic) having other, special names.
18 Vercoutter 1956, 127–134.
19 I have been quite puzzled by this discrepancy. Two fairly simple possibilities are (1) that the Minoan yellow dyes were too fugitive to withstand the voyage, and (2) that the Minoans hadn't yet begun to use yellow dye at the time of the earliest embassies. But I feel quite unsatisfied by both.
    There is a third possibility, which I had suppressed in my head until I read the recent cogent analyses (from new data) of the Theran fresco of the Saffron Gatherers as being a depiction of a specifically female ritual, probably connected with puberty (N. Marinatos 1984, 62–72; Immerwahr 1990, ch. 4). Saffron is above all else a potent and beautiful yellow dye, and here it is connected specifically with women (it is also a medicine for menstrual ills). Did the original Minoan culture – before it became diluted with Mycenaeans – reserve yellow cloth specifically for a particular class of women and their associates? Is that why the Egyptians didn't see Aegean men wearing yellow textiles until the very end? The restriction of the later Homeric and Classical epithets and descriptions involving yellow clothing to women, especially young women like Iphigenia, Athena, and the Muses, would also seem to go back into this Bronze Age stratum. And the Knossos Cupbearer, with his bright yellow kilt, may have been serving the goddess of such affairs, or he may, as a

20 Table 8.1 is only as comprehensive as I could make it from the literature available to me. There may well be more examples. "Th" refers to Thebes, as numbered by Gardiner and Weigall 1913 (q.v. for titles and dates), and, for Th 359, by Bruyère 1933. For a list of the sources I used for each of these tombs, as well as for all the other tombs cited in this chapter as having probable Aegean connections, see *PT*, 396, appendix D.
21 Davies and Gardiner 1915, 1.
22 Hankey and Warren 1974, 146.
23 Davies and Gardiner 1915, 8.
24 Davies 1933, 9; cf. Frankfort 1929, 59.
25 Hayes 1959, 112.
26 Davies 1933, pl. 56.
27 Davies 1948, pl. 19.
28 Bruyère 1933, pl. 5.
29 Davies 1901, pl. 15.
30 Terrace 1968, pls 2, 3.
31 Emery 1949, pl. 50.
32 Smith 1958, 171, fig. 60.
33 Davies 1920, pls 20, 30.
34 Glanville 1972, 15, fig. 14.
35 Borchardt 1929, 111, fig. 1.
36 Borchardt 1929, 111, fig. 2.
37 Brugsch 1889.
38 Bruyère 1933, 32–35, pls 3–5.
39 Davies 1933, pl. 22.
40 Jéquier 1911, pl. 32; Borchardt 1929, fig. 9.
41 Marinatos and Hirmer 1960, pl. 161; Stubbings 1973, fig. 38.
42 Newberry n.d., 11, fig. 1.
43 Jéquier 1911, pls 34, 35, 37.
44 Kantor 1947, pls 4, 5.
45 See *PT*, chapter 5.
46 Norman Davies (1922, 51) says he found the pieces of this last fragment on the floor of "a certain tomb," along with a number of other handmade but heterogeneous fragments, and "found, after industrious inquiry among the older guards, that they did not come from this tomb at all, or one of its period, but had been laid there temporarily by an inspector and forgotten on his removal from the district." In a footnote he adds: "I have since found that this shell-like design (from tomb 162) occurs again in the contemporary tomb, No. 160." (Tomb 160 is Saite, so I have not considered it in this chapter.) One would assume from these statements either that Davies later determined that the fragments with the shell or spiral pattern derived originally from the nameless tomb Th 162, or that Th 162 was the tomb on whose floor he found the fragments originally.

But apparently Davies got his own notes a bit mixed, as well, as I learned upon pursuing the attribution. According to information kindly supplied me by Marsha Hill of the Metropolitan Museum of Art (pers. comm. 1986), "following recent work on the Davies tracings and notebooks he housed in the Griffiths Institute, it seems that the pattern fragment was found by Davies stored in tomb 16 and considered by him as probably from tomb 262 (he himself had confused the numbers of tombs 162 and 262)." I will therefore refer to it as from Th 262, which Engelbach (1924, 18) and Porter and Moss (1960–, 344) attribute to the reign of Thutmose III.
47 Daressy 1902, pl. 22, no. 24147.
48 Spirals and rosettes also occur all over the chariot of Yuaa and Thuiu, the parents of Queen Tiy (see Quibell 1908, 65–67, pls 51–54, no. 51188). The walls of this chariot are made mostly of molded and gilded plaster; the chariot also shows no sign of wear and therefore was presumably only a funeral gift (Quibell 1908, 65–67).

A well-used little chair from the same tomb also has a spiral pattern on the edges (no. 51112: Quibell 1908, 52–53, pl. 36). Both objects may have had leather prototypes.

49 Davies 1930, pl. 22.
50 Daressy 1902, 258–260, pl. 51.
51 Davies 1930, pl. 42.
52 Davies 1936, pl. 63.
53 Davies 1933, pl. 22 and – different boat – pl. 43.
54 Bruyère 1948, 123, fig. 67. The pattern also has a forerunner in the design that occurs first in the tomb of User, a royal scribe and steward under Thutmose I and II (Th 21), then again in the tomb of Nebamun, a high functionary who served principally under Thutmose IV (Th 90). In both cases, the curved quadrilateral space between the spirals has been broken up into four triangular corner pieces surrounding the central rosette. A version in which the spiral rapport has disintegrated into disconnected spirals in a square grid occurs in the tomb of Puyemre, a priest of Amon under Thutmose III (Th 39). In all three tombs, the triangles filling in the quadrilaterals show the same alternation of color in diagonal rows between red and green that we see later between red and blue; the spirals are yellow, and their centers green. Such idiosyncratic similarities are not due to chance.
55 See Chapter 6.
56 Vercoutter 1956, 54.
57 Schiaparelli 1927, figs 109–112.
58 Shaw 1970.
59 Shaw 1970, 29, pl. 6.18, 6.19.
60 See Kantor 1947, pl. 6, for other examples.
61 Anthes 1943, 15–16, pl. 6.e.
62 Arthur Evans (*PM* II, 2, 745, figs 480, 481) shows several scarabs, from the 12th Dynasty on, with spirals in mirrored heart-shapes, but no other details correspond. He also shows a similar motif labeled "on kilt of man of Keftiu, Rekhmara Tomb" – but there is nothing like it in Rekhmire's tomb today. It is evidentially the Menkheperraseneb kilt pattern turned upside-down.

   Evans, in fact, would see the "wrought-iron fence" motif as Egyptian, whereas I see the design as relatively isolated in Egypt and find in the Bronze Age Aegean a whole host of family members, in the form of heart-spirals, if not the exact design itself. Walberg, for instance, shows many such hearts on Middle Minoan pottery, including occasional mirrored hearts (Walberg 1983, pl. 30 – note especially 3(i)6, 3(iii)7; Walberg 1976, 181, fig. 36 – note especially 3.5, 3.6 – and 193, fig. 48.i.11); Popham (1970, pl. 7.c; Popham 1984, pl. 166, nos. 56, 58, 59) shows it on LM II pottery. The heart-shaped spirals are also depicted on textiles, for instance on the skirt of the seventh lady in the Knossos Procession Fresco (*PM* II, 2, suppl. pl. 25). And for interesting later Aegean evidence of specifically blue point-to-point hearts, see *PT*, 371–372, fig. 16.11.
63 Antef's tomb contains not only this ceiling motif, but also, the reader will recall, one of the depictions of Aegean foreigners, and two more suspiciously Aegean-looking ceiling motifs: a heart-shaped meander and a unique labyrinth pattern.
    A further curious tie among the earlier "wrought-iron" motifs is that the heart-shaped spirals are apparently blue in each case: Ukhotep, Wahka, Antef, and Hepzefa. (My warm thanks to H. W. Müller for providing me with a color picture of this last.) Menkheperraseneb's is black, and no colors are given for the chapel. In most of the other spiral patterns in Egypt, the spirals are yellow or white.
64 *PT*, 197.
65 Popham 1984, pl. 171.4 plus many closely related variants.
66 Vercoutter 1956, 267 fig. 61, pl. 22.169.
67 Popham 1984.
68 Vercoutter 1956, 266 etc.
69 Anthes 1943, pl. 6.
70 Frankfort 1929, pl. 13; Smith 1958, 169, pl. 121A.
71 Marinatos and Hirmer 1960, pl. 175; cf., more subtly, a steatite rhyton from the Rhyton Well at Mycenae; Wace 1919–1921, pl. 13.1D.

72 Smith 1965, 32–33, pl. 52; Buchholz and Karageorghis 1973, 459, pl. 4: no. 1684; Persson 1931, pls. 1, 12–15.
73 Vercoutter 1956, pl. 35.231.
74 Smith 1965, 33, pl. 51.
75 Kemp and Merrillees 1980.
76 Kemp and Merrillees 1980, 15.
77 Brunton and Engelbach 1927, pl. 13.8; *PT,* 65, fig. 2.32.
78 Petrie, Griffith, and Newberry 1890, 28.
79 I have long been puzzled as to why the Minoans should turn up in such force precisely in the Faiyum. One would think that the Minoans, being of necessity ship-travelers, would end up with a "colony" along the Nile somewhere, or on the Mediterranean shore. (By "colony" I envisage only something similar to the Greek "colony" we have today in Chicago or the Spanish Basque "colony" in Idaho: when anyone else back home decides to go off to the New World, he heads for Uncle Nick in Chicago or Uncle Jaime in Boise simply because they are already there and can help them get started, and so the crowd grows.) Recently a student of mine, John Yohannes, was doing some serious map-study of Egypt with this question in mind and noticed that there is a trench running southeast from the Mediterranean coast all the way to the ridge immediately northwest of the Faiyum. Is it possible that this was still an open waterway around 2000 BC, and that the Minoans, sailing eastward along the coast after crossing directly to Libya from Crete, found their way up it to a point much closer to the Faiyum than to any other part of Egypt, so that they considered this their "home base" from then on? A small amount of judicious geological fieldwork of the sort done recently at Troy (Kraft, Kayan, and Erol 1980) could answer this question.
80 Frankfort and Pendlebury 1933, 18; Pendlebury *et al.* 1951, 109.
81 Davies 1903, pl. 17; 1905, pl. 35.
82 Frankfort 1929, 16, fig. 12.
83 Davies 1905, pl. 35.
84 Davies 1933, pl. 9.
85 *PT,* 120, fig. 3.34, 156–157.
86 Davies and Gardiner 1926, pls 19, 20.
87 Davies 1936, III, pl. 60.
88 *PT,* 160–161, figs 5.10, 5.11.
89 *PT,* chapter 5.
90 *PT,* chapter 5, col. pl. 1. In assessing the textiles from the tomb of Thutmose IV, it is important to notice the number of heirlooms there. One cloth bears the cartouche of Amenhotep II – and so do at least five of the vases; one cloth bears the name of Thutmose III, and so does one of the vases (Carter and Newberry 1904, 18–19, 143–44). All of this gives me the feeling that the big textile innovations had more or less stopped by the time of Thutmose IV, and now people were coasting, and consolidating the gains in technique. Thutmose IV had nothing better to show than what his father and grandfather had left him. The piece with Thutmose III's name shows that its weavers had not yet decided how best to negotiate the edge of a tapestry color-field (see *PT,* chapter 5, col. pl. 1), and were still experimenting. By the time they wove the cloth for Amenhotep II, they had chosen the slit technique and were becoming rather ambitious in their designs. Tutankhamon's tapestries many years later show no further innovations of technique – but they are done with extreme skill. This is one reason why I put the "watershed" line dividing Egyptian attitudes towards textiles somewhere in this part of the 18th Dynasty.
91 *PT,* 323, fig. 5.9.
92 *PT,* 120, fig. 3.34.
93 Wallis 1900; Daressy 1910; Hayes 1937.
94 Buchholz and Karageorghis 1973, pls 480, 481: nos 1747, 1748.
95 Note that only in weaving does the craftsperson have to create the "blank" spaces at the same time as the "filled" ones. The weaver's considerations of sequence and existence are crucial aspects of any discussion of textile art, and are often overlooked by historians of ancient art.
96 Mee 1982, 81–89.
97 Catling and Millett 1965; *PT,* 369, fig. 16.10.

98 Fashions in kilts may be worth paying attention to. Both Evans (*PM* II, 2, 745) and Vercoutter (1956, 64–67) complain that the silly Egyptians labeled as "king of the Keftiu" a figure who looks much more like a Syrian in both coiffure, skin color, and style of kilt: to Vercoutter this is one more sign that Menkheperraseneb's scene was a freely and carelessly invented pastiche. I'm not so sure. The kilted Keftiu in the lower row (where the "king" is) alternate with people in clearly typical Syrian dress, while their own kilts mostly show special features associated with the Levant and Anatolia (the half-circle at the waist, the peculiar placement of the border, the slanted fringe, the color scheme). In fact, there are very close parallels to some of these details on the ivories from Ugarit (Smith 1965, figs 55, 57).

    I am coming to feel, therefore, that Menkheperraseneb's second row represents a different boatload of strangers – Keftiu and coastal Semites traveling together, who may even have arrived at the Theban court at a rather later date than the people of the upper row (since the types in this lower row do not occur in Rekhmire's tomb). That is, perhaps the combining of Cretan, Anatolian, and Levantine cultural traits was actually occurring at this time in Ugarit, on Cyprus, and/or on the south coast of Anatolia (e.g., in Pamphylia, which means "All Races"), rather than only in the heads of "careless" Egyptian artists. (Compare the cargo mixed from very similar sources in the 14th century BC shipwreck found at Ulu Burun, off the south coast of Turkey; Bass 1987.) Perhaps some group of Minoans, displaced eastward by the geological and political catastrophes of which we have ample evidence, had chosen a useful easterner to be their leader in exile. Having resurrected this king's claims to some authenticity by means of his followers' kilts, however, I will leave him as prey for the historians.

99 See Sandars 1985, figs 14, 74–77, 79–84, 86, 87, 90 right, 93.
100 See *PT*, 366, fig. 16.5.
101 Hayes 1937, 26.
102 Shaw 1980, fig. 4 and pl. 26 fig. 1.

9

Observations, Summaries, and Conclusions

Anne P. Chapin

What, then, is to be learned from this investigation of Aegean patterned textiles? First, a comment on the obvious: the peoples of the Aegean placed a high value on textiles – on the making of textiles, on the wearing of textiles, and on the use of textiles in various aspects of life, from the functional to the symbolic. Textile production was a major component of Bronze Age economies in both the Minoan and Mycenaean periods, and the vast majority of human figures depicted in Aegean art wear clothing made from textiles: loincloths, kilts and tunics, cloaks and mantles, and dresses (chitons), skirts, and blouses are all fashioned from woven fabrics.[1] Indeed, as noted in the first chapter, an attitude of cultural superiority over those who do not wear cloth can be detected in both the Shipwreck scene of the Theran miniature fresco, where naked warriors drift in the sea, defeated and drowned (Fig. 1.3), and in the Battle Fresco from Pylos, where helmeted Mycenaeans dressed in kilts and lappet armor fight a fierce but culturally backward enemy wearing only raw sheepskin coverings (Fig. 1.4). The message embedded in these frescoes is clear: civilized and victorious people wear cloth; a lack of cloth signals cultural and military inferiority.

Creating these textiles required a substantial investment in time and materials, even for utilitarian cloth with low thread counts and/or loose weaves. As discussed in Chapter 2, the raw materials had to be grown or raised, collected, and processed in preparation for spinning. Then, as modern replication experiments demonstrate, it took about eight hours with a drop spindle to produce enough thread to weave on a warp-weighted loom for just one hour, and numerous hours on the loom to weave a length of cloth.[2] For many today it is difficult to imagine the time and labor required to produce what can now be spun and woven in mechanized plants in the blink of an eye. Historically, this investment of time and resources meant that textiles functioned as repositories of wealth that could be collected or traded as an early form of currency. The premium cloth depicted in Aegean art, woven with high thread counts and colorful decorative patterns, thus represents a far more significant investment in time and resources than may at first be recognized.

In the social and historical contexts of preindustrial production, then, the finest textiles depicted in Aegean art were true luxury items. As Thomas Campbell explains in his introduction to the Metropolitan Museum's extensive collection of historical textiles, "the prominent role that luxury textiles played in the secular and religious life of many societies ensured that no cost or effort was spared in making them as beautiful and artful as possible. As a result, they often represent the highest creative achievement of a society or of a particular milieu within it, combining inspired design, fine quality materials, and impeccable craftsmanship."[3] The pictorial evidence surveyed in this volume indicates that the Minoans and Mycenaeans alike placed a particularly high value on the production of fine textiles. These fabrics were sewn into the elaborate costumes depicted in Aegean art, used in ritual, presented as gifts to the divine, displayed as luxurious room decoration, and set up as screens and sunshades on ships in ceremonious maritime events. But, as discussed below, the Minoans and Mycenaeans put these textiles to subtly different purposes. On Crete and in the Cycladic islands during the Neopalatial period, members of the elite appear to have "cornered the market" for luxury textiles by claiming the finest materials and decorated fabrics for themselves and displaying them prominently in social and ritual events. Indeed, this conspicuous consumption of textiles announces a taste for luxury living that is also evident in the widespread popularity of villa architecture. In contrast, the Mycenaeans (including the priestly and warrior elite) preferred clothing made from plain cloth or fabrics woven with simple patterns. Highly decorated textiles were fashioned into ritualized objects, such as flounced costumes, and ikria and shields made of cloth. These were used ceremonially, probably for purposes related to building a sense of communal Mycenaean identity and maintaining existing power structures.

The Minoan period: luxury textiles and social display

Artistic and archaeological evidence indicates that the cultural and economic elite of both the Protopalatial and Neopalatial periods were conspicuous consumers of luxury textiles. These fine fabrics, and the striking costumes and lavish household furnishings into which they were fashioned, appear prominently in Neopalatial art. Indeed, many of the most famous works of Minoan art, such as the faience Snake Goddess figurine from Knossos (Fig. 3.3) and the frescoes from Xeste 3 at Akrotiri (Figs 3.21–3.30), preserve detailed depictions of the elaborate textiles favored by the Minoans and those living within their circle of influence. As discussed in Chapter 3, these were woven with bright colors and intricate patterns, and sewn into the form-fitting, figure-flattering flounced costumes preferred by women of privilege. Further embellishments included gathering the warp (vertical) threads into ornamental fringes and tassels, and reinforcing points of stress (such as seams and hems) with decorative bands woven with additional patterns and motifs. Artistic depictions also reveal that a variety of luxury garments were produced, from light and sheer chemises and veils to fleecy mantles and skirts decorated (embroidered or appliquéd?) with intricate pictorial motifs. Together with an abundance of jewelry and accessorizing head bands and hair ribbons, the images of affluence in frescoes of the Neopalatial era underscore the impression of great wealth lavished on individualized personal adornment – the *haute couture* of the Bronze Age.

Elite costumes made from luxury textiles thus characterize the affluence, sophistication, and high social standing of the self-assured women who appear in Neopalatial art. It is surely significant that the most complicated textile patterns are found in Minoan paintings from Crete. The Lady in Red (Fig. 3.2) and the Ladies in Blue from Knossos (Fig. 9.1),[4] the women of the shrine at Pseira (Figs 3.5, 3.6), and the Ayia Triada "Goddess" (Fig. 3.9) preserve some of the most intricate textile patterns known from images of women's dress. Frescoes of men in elaborate costumes are fewer in number, but at Knossos, the Cupbearer and his cohort of companions in the Procession Fresco indicate that the display of luxury fabrics with intricate rapport patterns continued into the LM II–IIIA period (Figs 3.15, 3.18, 3.19). These textile patterns are so involved that fresco painters incised grids into the surface of the prepared wet plaster, and then use the lines to guide their renderings of the complex patterns (Frontispiece; Fig. 9.2). That Minoan patrons preferred detailed renderings of luxury textiles, and that artists went to great lengths to reproduce them with care, is indicative of the importance placed on elite textile production in Neopalatial society.

Less familiar in the fragmentary archaeological record, but no less impressive, are the images of luxurious wall hangings in frescoes decorating the walls of villas and affluent houses. Best preserved are paintings of wall hangings from the town of Akrotiri on Thera. The House of the Ladies features an eye-catching design with a red-dotted net pattern and four-pointed stars (lozenges), painted as if above or behind a ritual scene that possibly depicts the presentation of a flounced garment by one female figure to another (Fig. 3.32). From the upper floor of Xeste 3 come fragments of a relief fresco that looks rather like a quilt sewn with star-like rosettes in

Fig. 9.1. *The Ladies in Blue, from the palace at Knossos, with an inset detail of the tricurved arch textile pattern and impressed grid (photos A. Chapin. Archaeological Museum of Heraklion, Hellenic Ministry of Culture and Sports – Archaeological Receipts Fund)*

a tricurved ribbon network (Fig. 4.6); other rooms of the upper floor were painted with large, psychedelic spirals. These compositions seem infused with celestial symbolism, and their presence on the walls of these buildings may have set the stage for ceremonial activity.[5] Frescoes from Ayia Triada (Fig. 4.1), Epano Zakros (Fig. 4.2), Galatas (Fig. 4.3) and Poros-Katsambas (Figs 4.4, 4.5) are likewise identified as wall hangings that would have similarly transformed interior spaces into extraordinary settings for special events. Historical data demonstrates that luxurious wall hangings (e.g., tapestries) are exceptionally expensive,[6] so perhaps the frescoes functioned as less expensive imitations. If so, it is fortunate that bits of the wall paintings preserve the likenesses of these otherwise lost textiles.

Evidence for floor coverings comes from surviving fragments of mats and floor frescoes. A fragmentary mat made from pressed pine needles was found at Phaistos, as described in Chapter 2, and floor frescoes from both Phaistos and Knossos utilize the inlaid (*incavo*) technique for quatrefoil and labyrinth designs that evoke mats or rugs, perhaps plaited, woven, or knotted (Figs 4.14–4.17). The Zebra Fresco from the Royal Road near Knossos also uses the *incavo* technique to reproduce a floor covering suggestive of exotic animal skins, or perhaps textile designs inspired by animal skins (Figs 4.18, 4.19). Frescoed imitations of perishable floor coverings probably made such luxurious interior décor appear permanent and lasting.

Fig. 9.2. Kilt of male figure (no. 22) showing the incised artist's grid, from the Procession Fresco, Knossos (photo A. Chapin. Archaeological Museum of Heraklion, Hellenic Ministry of Culture and Sports – Archaeological Receipts Fund)

Ceiling cloths were likely produced to protect household residents from the bits of debris that fell from their ceilings, which were made of wood, reeds, and mud. The evidence for such textiles is scarce, but can be inferred from the Quadruple Spiral Relief Fresco from Knossos (Fig. 4.11), which recalls the ceiling decoration of the Mycenaean tholos tomb at Orchomenos (Figs 4.12, 4.13), and various Egyptian tomb paintings that depict ceiling cloths with characteristically Aegean patterns. Elizabeth Barber has previously suggested that as early as EM II, the well-developed textile industry at Myrtos: Fournou-Koriphi, on the southeastern coast of Crete, could have produced textiles woven with trademark Aegean polychrome patterns and exported them to Egypt.[7] The ceiling painting of the 12th Dynasty Tomb of Hepzefa at Assiut (Fig. 8.11) is one of the earliest to imitate Aegean-style spiral designs and a "wrought-iron fence" motif in a patchwork of patterns suggestive of a large ceiling cloth.

Egyptian interest in the Aegean can also be documented for the 18th Dynasty, when direct contact between the two regions is chronicled by New Kingdom tomb paintings depicting embassies of gift-bearing "Keftiu" wearing Aegean-style costumes. The Tomb of Menkheperraseneb (Th 86), for example, depicts men in patterned kilts holding lengths of cloth over their arms (Fig. 8.8); these textiles were evidently popular enough with Egyptians that copies were painted as permanent ceiling decoration in their tombs. During the New Kingdom, spiral designs and bulls' heads with rosettes between the horns may show Aegean inspiration (as do diamond patterns and rosette designs), but one of the closest imitations of a Minoan textile is a quatrefoil interlock design from the ceiling of the Tomb of Amenemhet (Th 82), dating to the reign of Tuthmoses III (1482–1450 BC); its resemblance to the fabric worn by the Ayia Triada goddess, of contemporary date, is uncanny (Figs 3.10, 8.15).

Maritime textiles comprise the last category of elite Aegean fabrics surviving in pictorial form in Neopalatial art, and these are likewise depicted in contemporary Egyptian art of the early New Kingdom. Textiles serving as canopies and wind screens were set up on the decks of large ships for ceremonial events in order to protect high-ranking passengers from the natural elements. In the Flotilla Fresco of the West House at Akrotiri, these cloths are decorated with wavy and striped patterns, and resemble awnings supported on poles, but, as discussed in Chapter 6, close examination of the painted details suggest that the "awnings" may actually be wind screens that were rolled up and stowed away as the ships entered the protected harbor (Fig. 6.1). Comparisons can be found in Egyptian art of the early 18th Dynasty, when certain Egyptian luxury watercraft were equipped with large rectangular cabins set amidships; some were brightly decorated with Aegean-style red, white, and blue spiral designs (Fig. 8.3). Whether these cabins were made of woolen cloths stretched on wooden armatures, or were wooden cabins painted to look like imported cloth, is not as important as the observation that wealthy Egyptians associated boat cabins on their luxury ships with imported Aegean textiles.

Archaeological evidence for luxury textiles

Artistic evidence for how the Minoans themselves defined the *luxury* in luxury textiles is now confirmed by recent archaeological discoveries surveyed in Chapter 2. Artists working in the Neopalatial era depicted their best fabrics as finely woven with colorful and complex patterns and embellished with decorative borders, fringes, and tassels; these pictorial details are consistent with the bits of surviving textiles. Most importantly, from the Building of the Benches at Akrotiri come pieces of finely woven linen embellished with an ornamental hem, an embroidered design, and various fringes and tassels. This cloth, which was found in association with a golden ibex figurine and a pile of goat horns, demonstrates that artistic representations of fine textiles are faithful to cloth that was actually produced.[8] Furthermore, the ritual context of this cloth, though not yet fully analyzed, seems to support artistic evidence for textiles used as votive or ritual offerings.

Other textile fragments reviewed in Chapter 2 underscore the diversity of materials used by Minoan weavers. Dark-colored goat hair and light-colored nettle fiber were identified in a narrow textile band found at Chania; the band itself confirms abundant pictorial evidence for bands in costume design. Fragments of wool thread were identified recently at Akrotiri. Dye vats found at Alatzomouri-Pefka, near Gournia in east Crete, were likely used for the washing and dyeing of wool; the discovery of murex shells there suggests that at least one dye color was purple. The use of silk is suggested by the discovery in the House of the Ladies at Akrotiri of a cocoon from the silk moth *Pachypasa otusi*, and it is possible, although not proven, that sea silk from the fan mussel *Pinna nobilis* was used for cloth manufacture. Pine needles, sedges, and perhaps grasses were used for matted items at Phaistos and Akrotiri. Altogether, the evidence points to a distinctive variety of materials being worked into textiles and related materials.

As discussed in Chapter 2, the distribution patterns of the tools of textile production, together with documentary evidence, offer support for the hypothesis that the members of the Minoan palatial elite were the primary consumers of luxury textiles. In the Early Bronze Age, entire communities were involved in a wide range of textile activities, from growing the raw material to manufacturing the finished fabric, but with the rise of the palaces on Crete, textile production changed. Evidence for spinning (in the form of spindle whorls) largely disappears – either spinning took place in undiscovered locations, or wooden spindle whorls were used but do not survive. In both scenarios, thread was produced differently from before. Protopalatial deposits of loom weights confirm that weaving took place in the palaces at Knossos and Phaistos, and in Quartier Mu at Mallia. Further, a distinctive type of three-sided sealstone from Mallia carved with images identified as warp-weighted looms suggests administrative control of weaving activities (Fig. 2.9).[9] Pietro Militello suggests that palatial interest in textile production was probably limited to their own needs, rather than focused on producing a surplus for trade or export.[10]

In the Neopalatial period, textile production – particularly weaving – was concentrated in certain houses, villas, and smaller palaces. Not every household owned a loom, and presumably, weaving had become a specialized craft. The largest palaces, Knossos and Phaistos, offer no evidence for weaving, but that is not to say that palace officials were uninterested in textile production. To the contrary, administrators living and working in villas throughout Crete maintained connections with palace officials, and it is likely that the palatial elite acquired textiles from these and other sources. At the Royal Villa of Ayia Triada, for example, loom weights, a Linear A tablet inscribed with a quantity of wool, and a group of noduli indicate official interest in the final stages of textile production, particularly weaving. The frescoes of the villa correspondingly highlight luxury textiles. One fresco reconstructed as a wall hanging has a lively fabric pattern (Fig. 4.1), and the "goddess" of Room 14 wears a flounced costume of intricate and delicate design (Figs 3.9–3.10).

Towards Minoan hierarchies of cloth

The privileged access to the finest textiles enjoyed by members of the Minoan cultural and economic elite is underscored by the social hierarchies depicted in miniature frescoes dating to the Neopalatial era. These frescoes, typically painted with figures only 6–10 cm in height, present panoramic views of the Aegean world, from the palaces to the hinterlands. The better-preserved miniature frescoes were painted with dozens or even hundreds of figures, and, as suggested in Chapter 1 for the miniature fresco from the West House at Akrotiri, suggest that the elaborate costumes depicted in some of the most famous frescoes of Crete and the Aegean may actually have been rather rare, and that most people wore simpler clothing. If true, then sumptuous dress would have immediately distinguished Minoan high society from the masses, and would have announced an individual's elite social standing, access to wealth, and control of economic resources.

Social hierarchies are distinguishable in LM I miniature frescoes depicting the environs of Knossos. The Sacred Grove and Dance Fresco (Fig. 9.3) depicts a group of women in flounced

9. *Observations, Summaries, and Conclusions* 245

Fig. 9.3. The Sacred Grove and Dance Fresco from the palace at Knossos (photo A. Chapin. Archaeological Museum of Heraklion, Hellenic Ministry of Culture and Sports – Archaeological Receipts Fund)

skirts performing before a crowd of men and women gathered around an open area, perhaps the West Court at Knossos.[11] The audience is painted *en masse* in the shorthand technique (in which faces but not bodies are painted on backgrounds of red or white) while a few distinguished female guests in flounced skirts watch from front-row seats. This fresco offers good evidence for the close connections between court costume, public performance, and elite display before the Minoan population. The fresco's companion piece, the Grandstand Fresco, depicts a crowd gathered in a palatial setting, perhaps the central court of the palace itself.[12] Seated female figures are painted at a larger scale than all other figures, and they are dressed in the most elaborate costumes with brightly colored, striped flounces (Fig. 9.4). Nearby, but painted on a somewhat smaller scale, groups of standing women wear flounced skirts and plainer, bell-shaped skirts made of striped cloth. Lowest in rank are the men and women of the crowd – anonymous and costumeless members of the multitude indicated only by their heads. The fresco thus depicts a Knossian social hierarchy with three levels, with the best-dressed, seated figures being the most prominent. As the Minoans themselves describe and define their social hierarchies, the privileged few enjoy the finest textiles; the rest watch the show.

Male social hierarchies are also revealed through a study of costume. Large-scale male figures painted in frescoes at Knossos (discussed in Chapters 3 and 7) wear loincloths and kilts ornamented with complex rapport patterns (Figs 3.1, 3.15, 3.18, 3.19, 7.7–7.11). But outside the palatial center, male garments fashioned from patterned cloth are harder to find. White kilts ornamented with large red running spirals are worn by two male figures in a procession fresco from Xeste 4 in Akrotiri;[13] and in Xeste 3, a youth holds strips of cloth decorated with undulating bands – possibly the cloth was to be made into the young man's new garment for a robing ritual.[14] The Theran

Fig. 9.4. Seated Ladies from the Grandstand Fresco, Knossos (drawing A. Chapin after PM III, col. pl. xvii, center)

fabric patterns are less complex than those recorded in Knossian painting and presumably reflect the non-Cretan location of Akrotiri. Further afield, a painted plaster offering table recently discovered at Ialysos on Rhodes in an LM IA context depicts acrobats in colorful flounced kilts,[15] and some contemporary art objects from the Greek mainland, such as the LH I Lion Hunt dagger from Shaft Grave IV at Mycenae,[16] represent male hunters and warriors in decorative flounced garments. But most male figures in Neopalatial painting wear plain costumes with little or no decoration,[17] as evidenced by the animal handlers and mature man of Xeste 3, all clad in white kilts,[18] and the multitude of male figures wearing plain white loincloths and kilts in miniature frescoes from Knossos, Tylissos, and the West House at Akrotiri. Unadorned capes, tunics, and one-shouldered garments are also found in miniature frescoes from Ayia Irini on Keos.[19] Patterned textiles were thus true luxury items, and clothing made from them seems to have been restricted to the social and economic elite.

The raiment of power and authority

Wealth and social status were thus communicated through preferential access to luxury textiles. It is worthwhile noting that some frescoes, most notably the Grandstand Fresco and the Sacred Grove and Dance Fresco from Knossos, indicate that the Minoan elite appeared before crowds wearing their *haute couture* clothing. In this context it may be significant that people across the world, and throughout time, have often gathered in crowds and appreciated special viewings of their own "rich and famous" – their political and religious leaders, star athletes, celebrities, and others. Today, for example, the Catholic Pope regularly greets huge masses of devotees from a balcony above St Peter's Square in Rome while dressed in his ecclesiastical robes. Well-dressed politicians around the developed world give speeches before large audiences to rally the people. In cities and towns, crowds gather for parades and other public events that celebrate the accomplishments of their athletic teams and local luminaries. Weddings and funerals, holiday festivities, and religious observances often feature large public gatherings. In the Classical era, Greeks traveled from across the Mediterranean to compete in (and watch) the athletic contests held at the Panhellenic sanctuaries at Olympia, Delphi, Isthmia, and Nemea; and in Rome, huge crowds attended military triumphs, chariot races, and gladiatorial contests. In both the Greek and Roman societies, religious rituals were attended by whole communities, and each individual must have felt he was part of the spectacle that was being led and performed by members of the elite. From these few examples, it is evident that crowds are and have always been powerful forces in human society, binding the many into one and shaping a sense of belonging to a community that is greater than one's own immediate family, clan, or kinship group.

In these contexts, the study of collective behavior and collective action (which lie within the purviews of sociology and psychology) are of interest. Though crowds can be violent and are sometimes prone to rioting, under the right circumstances, crowds become festive and even euphoric. Social boundaries that typically separate individuals break down, and a sense of social unity emerges.[20] It seems probable that the Minoan elite recognized the positive power of crowds to fuel emotion and to build group loyalty, and that they staged public events to build a particularly *Minoan* sense of communal identity, one that connected individual men and women of all social ranks to a distinctively "Minoan" vision for collective society. As suggested by the miniature frescoes, both elite and non-elite all participated, whether on stage as ceremonial actors or as viewers in the audience, and in so doing, they would have built a common ground based on shared experience and public memory.

One does not have to look far to find the stages for these performances – west courts, theatral areas, and central courts are all basic components of Minoan palatial architecture, and each offers prime space for performance.[21] Indeed, recent scholarship characterizes Minoan palaces as places for communal gatherings rather than as seats of hierarchical power or centers of a redistributive economy (though this author feels that these roles are not mutually exclusive).[22] At both Knossos and Phaistos, pottery associated with large-scale drinking and eating ceremonies can be traced back in time from the palatial era to the beginning of the Bronze Age, and even into the Final Neolithic period, more than 1000 years before the palaces were constructed. More specifically, these deposits are found in areas that were later transformed by architecture into western and central courts. One might infer from this that the palaces (as built structures) offered architectural frames for open-air gatherings in spaces long used for large-scale communal rituals and ceremonies.[23]

And, significantly for this argument, with the construction of palaces comes clear evidence that a social, religious and political elite were in power, both inside the palaces and without, and that these people were consumers of luxury goods, including textiles.[24] By the Neopalatial era, one sees members of this elite class painted in fresco, participating in ritual and appearing before crowds, dressed in their finery and looking the part they needed to play: rich and powerful, and confident in their ability to lead. The success of these events and elite public performances is evident in how a "Minoan" sense of identity was forged from disparate social groups, as suggested by the "Minoanization" of outlying territories, first on Crete itself in the Prepalatial and Protopalatial periods, and later, in the Neopalatial era, across the islands of the Aegean.

Some artworks offer additional description of these performances. Elite costumes were worn by individuals, particularly women, who acted in priestly roles and presumably served as intercessors between mortal and divine realms. Luxury textiles were carried as offerings and presented in ritual events, presumably to appease the god(s) and maintain a sense of order and balance in society. And, since Minoan deities were depicted in art in anthropomorphic terms as well-dressed people (see, for example, the seated goddess of Xeste 3 (Fig. 3.27), the line between mortal and divine was blurred. Indeed, the social status of any elite mortal, whether man or woman, was no doubt enhanced by the godlike appearance of his or her fine dress. In addition, other types of textiles marked events as ceremonious. Most obviously, the Flotilla Fresco from the West House at Aktoriti shows how decoratively patterned textiles were used as wind screens and/or awnings on the decks of seagoing vessels, and emphasized the social rank of those they protected (Fig. 6.1).

While intricately designed cloth and clothing was used to great effect in public displays before large crowds, it was also used as a signifier in more secluded settings. Within individual households, luxurious wall hangings made from the finest textiles set the stage for private ceremony by transforming interior rooms into sacred spaces, as suggested in particular by the frescoes from the House of the Ladies at Akrotiri (Figs. 3.31, 3.32). And thus public and private ritual seems linked by the incorporation of luxury textiles into ceremonial action.

Luxury textiles played important roles throughout the fabric of Minoan society. Their production both defined and maintained social hierarchies within an economic and political system led by a ruling elite and administered by palace officials. All levels of society contributed to the production of luxury textiles, from the lowly farmers and herders to the palatial weavers, but it was the social elite who commanded them – who wore them and were identified by them, who used them in public and private display, and who presented them to the divine. Luxury textiles thus distinguished those with the most social and economic power from everyone else. And even though the specific nature of Minoan rulership continues to elude precise definition, the surviving artworks present good evidence for the significant roles that luxury textiles played in what are arguably the three top social priorities of the Minoan ruling class: elite display, public performance, and fostering communal identity.

The Mycenaean period: tradition and innovation in patterned textiles

Mycenaeans adopted both Minoan weaving technologies and their ideas about elite consumption of luxury textiles, and adapted these inherited traditions for their own purposes. From a Mycenaean point of view, patterned textile production was a centuries-old practice that signified a powerful connection with the past. The command of the resources necessary to produce such fine fabrics still communicated high social status, but the use of luxury textiles for contemporary social display seems to have been tempered by Mycenaean concepts of cultural heritage, identity, and social memory. These shifting priorities led the Mycenaean political and social elite to put time-honored forms of textile decoration to new and innovative uses.

Archaeological and documentary evidence for Mycenaean patterned textiles

The few fragments of cloth – all linen – that survive from Mycenaean contexts support the impression from art that Mycenaean textiles were plainer and less decorative than textiles of the

Minoan era. Linen fragments found in Grave Circle B at Mycenae and in a tomb at Ayia Kryiaki on the island of Salamis, for example, were tabby-woven (a plain weave) and undecorated. But, as discussed in Chapter 2, Linear B tablets found in Mycenaean palaces (including Knossos) and nonpalatial regional centers such as Iklaina and Ayios Vassilios provide documentary evidence for a well-organized textile industry characterized by palatial administration, regional involvement, and an emphasis on linen and wool production, from raw material to finished textile. Tablets record the numbers of sheep in various flocks and document the amounts of wool and flax collected from villages and towns. A variety of ideograms and terms differentiate types of cloth, and craft specialists – such as spinners, weavers, fullers, and finishers – are listed among the dependents receiving food rations from the palaces. Moreover, the documents reveal something of regional specialization – the Pylos tablets seems focused on linen production, whereas those of Knossos deal with sheep and wool. As little physical evidence for textile manufacturing has been found in the palaces, it is likely that cloth was made in outlying communities.

Patterned textiles in Mycenaean dress

As mentioned above, the undecorated scraps of linen found in archaeological contexts support the impression gained from art that Mycenaean aesthetic taste favored simpler and plainer fabrics. Some Minoan-style forms of costume, particularly festal attire with flounced skirts, continued to appear in Mycenaean art, particularly in procession frescoes, but the fanciest rapport patterns of earlier generations were replaced by simple striped designs or by fabrics woven with uncomplicated all-over scatter patterns. Decorative bands that reinforced edges, seams, and hems on bodices, skirts, and tunics were still made by the Mycenaean weavers, but even these were plainer than before.[25] The transition to simpler fabrics can be detected in the later frescoes from Crete.

Under Mycenaean authority, Knossos dominated Crete well into the LM IIIA era of the 14th century BC, when the walls of this enduring palace still bore the famous Procession Fresco (Figs 3.15, 3.18, 3.19, Frontispiece). As Suzanne Murray observes in Chapter 3, this fresco's air of festive pageantry is heightened by the display of valuable ritual vessels and offerings, each carried by bearers dressed in ornate costumes made from Minoan-style luxury fabrics woven with interlocking rapport patterns and finished with sumptuous trim and decorative belts. Intricate patterns enjoyed a brief popularity in mainland textile production, as indicated by the interlocking crosses, scale patterns, and tricurved arches found on fresco fragments from earlier phases at Mycenae (Figs 3.39, 3.40). These patterns have close parallels with those depicted in the Knossos Procession Fresco, but they are not painted with the assistance of an incised grid and their execution is accordingly less tidy. Since these motifs disappear from later Mycenaean depictions of costume, it would appear that the earlier taste for complexity was replaced by a preference for easier-to-weave scatter patterns (dots, dashes, crescents), striped or wavy ("ripple" or "marbled") patterns, and plain fabrics ornamented with decorative trim. Complicated patterns were evidently less dominant as visual markers of elite social status in Mycenaean society.

The trend towards plainer fabrics is already evident in the "unisex" costumes worn by male and female figures on certain monuments dating to Final Palatial Crete. The LM IIIA Ayia Triada sarcophagus, for example, depicts both men and women in long tunics colored blue or yellow and trimmed with decorative bands (Fig. 9.5).[26] Similarly, both men and women wear hairy skirts, though the women cover their upper bodies with short-sleeved bodices. Men in LM IIIA paintings from Knossos, and women and men in the Final Palatial frescoes of Ayia Triada, are clad in long robes with diagonal stripes, presumably to communicate their priestly roles in ritual activity.[27] Only a few figures stand out in this Final Palatial crowd – La Parisienne from Knossos with her gaily striped costume being one, perhaps correctly identified as the image of a deity (Fig. 3.50). Although some of these costume types appear earlier in the Neopalatial era, they become popular in Final Palatial painting. One might conclude that, with the fall of Minoan hegemony and the installation of Mycenaean overlords on Crete, less emphasis was placed on the sex and social class distinctions that had been important within the Minoan culture. Rather, newly emerging social identities that blended Minoan and Mycenaean cultural traditions are reflected in the new habits of dress.

Yet, the processional theme survived into the Mycenaean era, and with it, Minoan-inspired festal attire took its place in the Mycenaean ritual wardrobe. Women in bare-breasted, flounced outfits populate procession frescoes at all major mainland palatial centers – Thebes, Tiryns, Mycenae, and Pylos – but the compositions are mechanical and stiff, with repetitive poses and standardized garb. Each figure carries an offering – typically floral, but vases, pyxides, figurines, necklaces, and bolts of cloth also appear. These votives are carried towards an unpreserved goal, perhaps a seated figure (a goddess or her priestess) or perhaps an altar. The frescoes themselves,

Fig. 9.5. Offering scene on the painted limestone sarcophagus from Ayia Triada (photo A. Chapin. Archaeological Museum of Heraklion, Hellenic Ministry of Culture and Sports – Archaeological Receipts Fund)

however, seem so formulaic as to raise the question as to whether flounced costumes were still being produced in the Mycenaean era, and whether such ceremonies were enacted in the 13th century BC – a proposition impossible to answer definitively without additional evidence. But the depiction of simpler fabric patterns is consistent with Mycenaean textile production in general, and suggests that the frescoes represent events familiar to contemporary painters. The continued appearance of Minoan-style festal costume in Mycenaean contexts thus speaks to the power of tradition enacted through ritual. A comparison can perhaps be made to academic regalia – manufactured today but inspired by centuries-old medieval garb, and worn only in ceremonial contexts.

Costumes made of plain fabrics are also much in evidence in male dress. In the Mycenaean era, the Minoan loincloth and kilt were superseded by the ubiquitous thigh-length tunic worn by hunters, warriors, and charioteers alike (Fig. 9.6). That certain white-skinned figures (e.g., the charioteers of Tiryns[28] and the archer of Pylos[29]) also wear these garments suggests that tunics were worn by men and women alike, as occasion required.[30] These all-purpose garments are typically solid in color (white, yellow, and blue are popular) and embellished with decorative colored bands sewn at the neckline, seams, and hems. Patterned fabrics seldom appear in male garb, and when they do (as with the priestly men painted in the vestibule of the Pylos megaron, Fig. 3.49), modest all-over scatter patterns predominate.

Fig. 9.6. Horse, hunter, and hound from the Boar Hunt Fresco, Tiryns (National Archaeological Museum, Athens; photo A. Chapin)

Dressing the Mycenaean palace: patterned textiles on walls, ceilings, and floors

The concepts of tradition and innovation are also readily apparent in Mycenaean depictions of interior furnishings made from patterned textiles and displayed in elite architectural settings. Wall hangings and ceiling coverings are represented by three catalogued frescoes and one stone relief, all of which feature motifs strongly reminiscent of textile design: the Quadruple Ivy Leaf Fresco (Fig. 4.7) and the Rockery Fresco (Fig. 4.9), both from Mycenae, the fresco from Pylos depicting a male figure below a possible wall hanging (Fig. 4.10), and the ornate stone ceiling in the burial chamber of the Treasury of Minyas at Orchomenos (Figs 4.12, 4.13). The frescoes are painted with repetitive patterns bordered by an undulating white band that probably signified the hem of the represented wall hanging. The ivy pattern of the Quadruple Ivy Fresco represents a continued (but increasingly rare) Mycenaean interest in rapport patterns and is reminiscent of the complicated designs of the Minoan era, prior to the fall of Knossos. Could this painting depict an antique textile? In contrast to clothing, which suffers from regular wear and tear, wall hangings (such as tapestries from the historical era) can survive for centuries, often in good condition, which raises the possibility that Mycenaeans did not just borrow the idea of decorative wall hangings from previous generations of Minoans, but acquired them too. If so, then the value of wall hangings as signifiers of wealth and prestige is emphasized, as is their connection with history and tradition.

Some of the depicted designs, such as the net pattern of the Pylos fresco (Fig. 4.10) and the spirals and rosettes of the Orchomenos ceiling (Figs 4.12, 4.13), are familiar from earlier representations of textiles from the Minoan era, but the ripple pattern is newly popular in the Mycenaean period. Wavy parallel lines appear both as woven designs on the flounced skirts of Mycenaean women (e.g., the processional figures from Pylos, Fig. 3.46) and as pictorial devices signifying the veining of rockwork and stone (e.g., the stone vase from Pylos, Fig. 5.6). The ripple motif in the Rockery Fresco from Mycenae (Fig. 4.9) thus connotes textile *and* stone, while the pieced-together appearance of the restored composition suggests the imitation of a patchwork quilt. The Pylos fresco with its ripple and net patterns, and the various panels of the relief ceiling in the Treasury of Minyas at Orchomenos, likewise suggest quilts or cloths sewn from smaller pieces of fabric.

In contrast to the continuous tradition for wall hangings and ceiling cloths, the Mycenaean evidence for floor coverings, in the form of floor frescoes found in the palaces at Pylos, Tiryns, and Mycenae (discussed in Chapter 5), shows distinct innovation both in concept and design. First, it should be observed that nothing quite like these painted plaster floors is known from Crete, though parallels can be drawn to Minoan slab floors embellished with plaster strips painted red.[31] A plaster floor, gridded and painted with imitation stone motifs, was recently discovered in Middle Bronze (MB) Age contexts at Ialysos on Rhodes, and offers a significant comparison to the Mycenaean floors, though one that is centuries earlier in date.[32] Additionally, a gridded plaster floor painted with a variety of motifs, including floral, was found in the MB Levantine palace at Tell Kabri, in modern Israel, and offers another early parallel.[33] Presumably future archaeological discoveries will fill in the gaps of this brief chronology.

The Mycenaean floors were laid in plaster, gridded, and painted with designs imitating expensive stonework (Mycenae) or textiles (Tiryns); at Pylos, the designs suggest both textiles and cut stone. Tricurved arches, scale patterns, cross-hatching, and interlocked quatrefoils are time-honored textile patterns, while circles, zigzags, wavy lines, and concentric arcs suggest slabs of mottled and veined stone. Together, this mix of textile and stone motifs creates a novel hybrid design, one that recalls the mix of patterns in the images of Mycenaean wall hangings. Emily C. Egan suggests that the intention was to dazzle the viewer with the richness of the implied materials. The textile patterns are well known from earlier generations of costume decoration, but in the megaron at Pylos they lose their connection with human vestments and are transformed, monumentalized, and made permanent in their new situation at the heart of the Mycenaean palace, seemingly to dress the palace. Indeed, the "mistake" in the gridding of the floor may actually have been an intentional but somewhat subliminal effort to guide the visitor's gaze towards the *wanax*, who was seated to the right of the entrance to the Throne Room (Fig. 5.13). The elaborate floor designs, then, were not mere imitations of expensive materials, but functioned as potent visual statements of the power of humankind to transcend the materiality of nature and approach the sublime realm of the divine – altogether, an appropriate setting for the godlike figure of the *wanax*, the Mycenaean king.

Innovative use of patterned textiles to dress the Mycenaean palaces extended to the incorporation of important symbols of military authority, both on land and sea, in the form of cloth versions of figure-eight shields and ikria. Artistic imagery shows that figure-eight shields, together with rectangular-shaped tower shields, were carried by Mycenaean warriors into hunt and battle as early as LH I, *c.* 1625–1525 BC (according to traditional chronology), but by the 13th century BC, these armaments were antique and there is little evidence that they were still being used. Large-scale images of figure-eight shields, however, appear in LH III frescoes at Mycenae, Tiryns, and Knossos, where they served as emblems of military strength; their spotted oxhide coverings would have made bold and powerful visual statements as wall decoration. The LH IIIB Shield Fresco from the palace at Pylos, however, is painted as if it depicts shields of cloth hanging from cloth-covered pegs. As discussed in Chapter 4, the textile designs are unmistakable: barred bands and lozenge net ("diaper") patterns, painted blue and black, and yellow and red (Fig. 4.21). But cloth coverings would have rendered the shields useless in battle, so a ceremonial function seems likely. Shields of cloth could have been carried in military parades and formal processions, and displayed on walls, and presumably the sight of them would have evoked memories of heroic warriors who fought in military campaigns of the Mycenaeans' own celebrated past.

The LH IIIB Ikria Fresco of Mycenae likewise illustrates longstanding Aegean symbols of authority. Unroofed, lightweight, and portable, ikria (as depicted in art and discussed in Chapter 6) were stern screens set up on the rear decks of seagoing vessels and probably functioned as captains' seats. They were made from wooden frames, open at the front, and covered with oxhide or cloth to protect the occupant from the natural elements. In both the Flotilla and Ikria Frescoes of the West House at Akrotiri, dating to the Neopalatial Minoan period, ikria are sheathed in oxhide, but at Mycenae in LH IIIB, the ikria are painted as if covered in cloth woven with textile motifs, including scale and ripple patterns, lozenge (net) patterns, stylized leaves or foliate bands, and nautilus friezes. Iconographic study of ikria in Minoan and Mycenaean contexts suggests they were erected for maritime ceremonies and symbolized naval power. Now that nautical subjects (specifically, ship processions) are better recognized in Mycenaean painting, the large-scale frescoes of ikria made from patterned textiles at Mycenae, discovered in rooms adjacent to the palace megaron, underscore the connections between military strength, palatial authority, and the social prestige accompanying the display of luxury textiles in a ceremonious contexts.

And thus, patterned textiles and their frescoed imitations were used to "dress" Mycenaean palaces and other significant structures. Frescoes depicting wall hangings and ceiling cloths served as luxurious interior décor and, as inferred from their ritual associations, probably set the stage for ritual and ceremony. The frescoes of ikria and figure-eight shields made from patterned cloth shaped social and historical memory through official palace decoration. Even the painted plaster floors of Mycenaean megarons communicated visual messages of enduring wealth and power. As such, palace decoration with textile themes connected the contemporary Mycenaean viewer with a continuing vision of a gloried past, as presented in an equally magnificent Mycenaean present. The embellishment of Mycenaean palaces with textile-themed decoration thus seems intended to enhance, monumentalize, and concretize important cultural memories and beliefs.[34]

A closing note on the Aegean artist's grid

Throughout this volume observations are made about the use, or lack thereof, of an artist's grid comprised of thin lines incised or impressed into wet plaster and used as a guide to paint complicated textile patterns (for overall discussion, see Chapter 7; in this chapter, Figs 9.1, 9.2). Of all the observable drafting aids used by Aegean artists, the grid is one of the most sophisticated,[35] as it was individualized for each composition by the size of the "squares" (which were roughly square or somewhat rectangular in shape) and angled as needed for each pattern. The grids range in size from quite small (about 7 mm on one side) to significantly larger, at well over 3 cm per side. As an artist's tool, the usefulness of the grid ended when the textile pattern was rendered, but no attempt was made to erase the impressed lines; rather, they were simply covered over by the painted colors. The Aegean grid is significantly different from the more famous grid associated with the Egyptian canon of proportion, generally defined as 18 squares for standing figures and 15 squares for seated figures.[36] Whereas the Egyptian grid governed the proportions of entire figures, and did in fact influence the rendering of some Aegean figures,[37] the Aegean artist's grid was used selectively for rendering textile patterns. Furthermore, the Egyptian grid was painted on large wall surfaces, whereas the Aegean grid was impressed into wet plaster, but only where it was needed.

The Aegean grid appears overwhelmingly in frescoes from sites on Crete; elsewhere, the preference for simpler textile patterns did not require its use. Its general disappearance from Aegean (Mycenaean) fresco painting by the 13th century BC might suggest that knowledge of the artist's grid had been lost to Aegean painters, but the grid's unexpected appearance in individual squares of the painted plaster floor in the Mycenaean megaron at Pylos, where it was again used to render complex textile patterns, confirms that this particular bit of artistic knowledge had indeed survived the centuries. In all, the peculiarities and particularities of the Aegean grid demonstrate that the textile patterns rendered in Aegean painting are the result of specific and intentional choices made by the painters, presumably in consultation with their patrons, as to the types of textile patterns to be depicted. As such, the existence – and persistence – of the grid underscores the importance of patterned textiles to the peoples of the prehistoric Aegean.

Woven threads and the fabric of society

The evidence for woven and patterned textiles surveyed in this volume indicates that textiles, and textile production, were central to Aegean economic and social life throughout the Bronze Age. Archaeological excavations have uncovered the tools of textile production and, occasionally, bits of cloth, while documentary evidence provides information on the palatial administration of textile production. But it is the artwork, particularly the frescoes, that preserves something of the appearance of these ancient textiles. A variety of fabrics can be distinguished in the paintings, from sheer and lightweight to heavy and opaque. Colors of thread, tassels, and cloth are carefully described in the paintings, as are the decorative patterns that were expertly woven into the cloth. Even possible sewing techniques can be recognized, from the stitching of seams, to embroidered designs and appliqué ornaments, to the stuffing and cording of quilted materials. Different uses for these textiles are recognizable, too, from the numerous types of garments used for Aegean dress; to various household furnishings (wall hangings, ceiling cloths, and floor coverings) found in palaces, villas, and houses; to the maritime and military textiles made into sails, awnings, ikria, and ceremonial shields. And thus it is the artwork that offers the most detailed information for what Aegean textiles looked like, and for how they were used in Minoan and Mycenaean society.

The artistic depictions of textiles investigated in this volume also serve as mirrors of Aegean society. Textiles around the world, and across time, are used in nearly all aspects of human life, and because of this, they communicate complex and multifaceted messages about the values of any given society. Textiles were, and continue to be, indicators of social status, ethnicity, age, and gender. Textiles communicate messages of power, wealth, and political authority, on individual, interpersonal, and collective levels. In the Aegean Bronze Age, the most elaborate patterned textiles combined an astute sense of aesthetic design with the best textile technologies available at that time. For the Minoans, patterned textiles were true luxury items, owned and consumed by the social elite, displayed in rituals and pageants, and traded abroad; for the Mycenaeans, when most textiles were plainer in decoration, elaborately patterned textiles (and renderings of them in art) were infused with cultural memory and tradition. For both dominant cultures of the Aegean Bronze Age, patterned textiles were central to communicating ideas about identity, both at the individual and communal levels. As such, it seems clear that Thomas Carlyle was quite correct, and that Aegean society, like that of the modern era, was indeed "founded upon cloth."

Notes

1. Nudity is observed on occasion in Aegean art. In frescoes, it is generally restricted to images of male youths who have yet to pass into adulthood, some athletes (e.g., the runners of the West House miniature frescoes), and those who find themselves in or near water (e.g., fishermen, swimmers, drowning men). On seals, some female figures who appear nude are actually wearing clothing. On nudity in in seal iconography, see Kyriakidis 1997; Crowley 2013, 352, 355; in fresco painting, see Chapin 2012.
2. Barber 1997, 515.
3. Campbell 1995–1996, 10.
4. *PM* I, 545–547, figs 397–398; Immerwahr 1990, 58–59, 172.
5. Blakolmer 2004, 22–27.
6. Campbell 2002.
7. Barber 1994, 109; 1997, 516, pl. cxciiia, b; 1998, 14.
8. Spantidaki and Moulherat 2012.
9. Burke 2010, 43–48.
10. Militello 2007, 44.
11. *PM* III, 66–68, col. pl. xviii; Immerwahr 1990, 65–66, 173, pl. 23; Chapin 2011, with relevant bibliography.

12 *PM* III, 46–65, col. pls xvi–xvii, figs 28–34, 36; Immerwahr 1990, 63–65, 173, pl. 22; Chapin 2011, with relevant bibliography.
13 Doumas 1992, pl. 138.
14 Doumas 1992, pls 109, 113.
15 Marketou 2010, 782, fig. 58.2.
16 Marinatos and Hirmer 1960, pl. xxxvi bottom; Xenaki-Sakellariou and Chatziliou 1989, 24–25, pl. 1.
17 Glyptic art offers such a small pictorial field that fabric patterns are schematic at best, but they are sometimes discernable. See Crowley 2012, 234–235, pl. liii; 2013, 173–175.
18 The kilt of the mature man has a black border (Doumas 1992, pls 110, 114). The animal handlers of Room 5 are not yet published in detail, but photographs suggest that they wear white garments; see Vlachopoulos 2008, 451, figs 41.3–41.6.
19 Abramovitz 1980, 57–59, pl. 4c.
20 On collective behavior, see Miller 2013; on crowds, see Borch 2013.
21 West courts, as intermediary spaces between palaces and towns, would have been accessible to greater numbers of people than central courts, which were situated inside palaces and had restricted access. There is a large bibliography on Minoan architecture, and recent interest in emphasizing the social roles of courts by referring to palaces as "court-centered buildings," "court compounds," or "court complexes". On Minoan architecture, see Graham 1962; Hitchcock 2000; McEnroe 2010; J. Shaw 2015; on courts, see Vansteenhuyse 2002; Driessen 2004; Letesson and Vansteenhuyse 2006. Spaces near Minoan tholos tombs were also used for funerary ceremony, the tombs themselves being communal in form and function.
22 See, for example, Driessen 2002; MacGillivray 2002; Day and Relaki 2002; Letesson and Vansteenhuyse 2006. For the view that the palaces were arenas for factional competition, see Hamilakis 2002.
23 Day and Wilson 2002; Schoep 2006; Tomkins 2008, 2012; Todaro and Di Tonto 2008; Todaro 2013.
24 The economic impact of palaces must not be forgotten. Large-scale storage facilities, together with plentiful evidence for an active administration, indicates that MM II Phaistos and its officials played important, privileged roles in the local economy (Militello 2012a).
25 *PT*, 325.
26 Long 1974; Immerwahr 1990, 100–101, 180–181 (AT no. 2), pls 50–53; Burke 2005; Chapin 2012, 301–302, pl. lxviiib.
27 See, for example, the red-skinned figures in diagonally banded robes in the Palanquin-Chariot Fresco and the Camp Stool/La Parisienne Fresco from Knossos (Immerwahr 1990, 92–95, 175–176, Kn nos 25–26, pl. fig. 27; and the white-skinned figure in a similar costume in the Fresco with Women and Deer at Altar from Ayia Triada (Immerwahr 1990, 102, 181 [AT no. 4]; Militello 1998, 139–142, 287–290, 291, col. pls I, L).
28 On the Boar Hunt Fresco, see *Tiryns* II, nos 113–193; Immerwahr 1990, 129–130, pls 68–70.
29 Brecoulaki *et al*. 2008.
30 Chapin 2012, 302.
31 J. Shaw 2009, 151–152.
32 Marketou 2013, 105, fig. 2.
33 Niemeier and Niemeier 2002, 254–266, pls iv–xviii, fig. 6.1.
34 While the relationship between memory and material culture has been the subject of much recent archaeological study (e.g., Van Dyke and Alcock 2003; A. Jones 2007; Yoffee 2007; Barbiera, Choyke and Rasson 2009; Georgiadis and Gallou 2009), the role of collective memory in the decoration of Mycenaean palaces has not yet been the subject of significant investigation. For the role of memory in the Mycenaean megarons at Mycenae and Pylos, as evidenced by battle frescoes, see Chapin forthcoming. For memory in palatial Mycenaean society as understood from mortuary evidence, see Button 2007; for memory and Middle Helladic tumuli, see Whittaker 2009; for the related sphere of ancestor worship, see Lupack 2014.
35 Recent research into how Theran artists created the flowing lines of their compositions suggests the use of a device somewhat akin to a template or a French curve, used to create the sinuous curving outlines of figures. See Birtacha and Zacharioudakis 2000; Papaodysseus *et al*. 2008.
36 For a detailed study of the Egyptian canon, see Robins and Fowler 2010.
37 Weingarten 1997; 1999; Guralnick 2000.

Abbreviations

AEGEAN WALL PAINTING Morgan, Lyvia, ed. 2005. *Aegean Wall Painting: A Tribute to Mark Cameron* (*British School at Athens Studies* 13). London: British School at Athens.

ANCIENT TEXTILES Gillis, Carole and Marie-Louise B. Nosch, eds. 2007. *Ancient Textiles: Production, Craft and Society. Proceedings of the First International Conference on Ancient Textiles, Held at Lund, Sweden and Copenhagen, Denmark, on March 19-23, 2003.* Oxford: Oxbow Books.

CMS *Corpus der minoischen und mykenischen Siegel.* Mainz: Akademie der Wissenschaften und der Literatur.

GORILA Godart, Louis and Jean-Pierre Olivier. 1976–1985. *Recueil des inscriptions en Linéaire A*: Études Crétoises 21, vols. 1–5. Paris: P. Geuthner.

KNOSSOS: PALACE, CITY, STATE Cadogan, Gerald, Eleni Hatzaki and Adonis Vasilakis, eds. 2004. *Knossos: Palace, City, State. Proceedings of the Conference in Herakleion Organised by the British School at Athens and the 23rd Ephoreia of Prehistoric and Classical Antiquities of Herakleion, in November 2000, for the Centenary of Sir Arthur Evans's Excavations at Knossos* (*British School at Athens Studies* 12). London: British School at Athens.

KOSMOS Nosch, Marie-Louise and Robert Laffineur, eds. 2012. *KOSMOS: Jewellery, Adornment and Textiles in the Aegean Bronze Age. Proceedings of the 13th International Aegean Conference, University of Copenhagen, Danish National Research Foundation's Center for Textile Research, 21-26 April 2010* (*Aegaeum* 33). Leuven: Peeters.

MELETEMATA Betancourt, Philip P., Vassos Karageorghis, Robert Laffineur and Wolf-Dietrich Niemeier, eds. 1999. *Meletemata: Studies in Aegean Archaeology Presented to Malcolm H. Wiener as He Enters His 65th Year* (*Aegaeum* 20). Liège: Université de Liège Histoire de l'art et archèologie de la Grèce antique; Austin: University of Texas at Austin, Program in Aegean Scripts and Prehistory.

METRON Laffineur, Robert and Karen P. Foster, eds. 2003. *METRON: Measuring the Aegean Bronze Age. Proceedings of the 9th International Aegean Conference, New Haven, Yale University, 18–21 April 2002* (*Aegaeum* 24). Liège: Université de Liège, Histoire de l'art et archèologie de la Grèce antique; Austin: University of Texas at Austin, Program in Aegean Scripts and Prehistory.

MONUMENTS OF MINOS Driessen, Jan, Ilse Schoep and Robert Laffineur, eds. 2002. *Monuments of Minos: Rethinking the Minoan Palaces. Proceedings of the International Workshop "Crete of the Hundred Palaces?" Held at the Université Catholique de Louvain, Louvain-la-Neuve, 14-15 December 2001* (*Aegaeum* 23). Liège: Université de Liège, Histoire de l'art et archèologie de la Grèce antique; Austin: University of Texas at Austin, Program in Aegean Scripts and Prehistory.

PM I–IV Evans, Arthur J. [1921–1935] 1964. *The Palace of Minos at Knossos,* vols. I–IV. London: Macmillan, repr., New York: Biblo and Tannen.

POTNIA Laffineur, Robert and Robin Hägg, eds. 2001. *POTNIA: Deities and Religion in the Aegean Bronze Age. Proceedings of the 8th International Aegean Conference, Göteborg, Göteborg University, 12-15 April 2000* (Aegaeum 22), Liège: Université de Liège, Histoire de l'art et archèologie de la Grèce antique; Austin: University of Texas at Austin, Program in Aegean Scripts and Prehistory.

PT Barber, Elizabeth J. W. 1991. *Prehistoric Textiles: The Development of Cloth in the Neolithic and Bronze Ages.* Princeton: Princeton University Press.

P*YLOS* I Blegen, Carl W. and Marion Rawson. 1966. *The Palace of Nestor at Pylos in Western Messenia,* vol. I. *The Buildings and Their Contents.* Princeton: Princeton University Press.

P*YLOS* II Lang, Mabel. 1969. *The Palace of Nestor at Pylos in Western Messenia,* vol. II. *The Frescoes.* Princeton: Princeton University Press.

TAW I–III *Thera and the Aegean World,* vols. I–III. I: Doumas, Christos, ed. 1978. London: Thera and the Aegean World. II: Doumas, Christos, ed. 1980. London: Thera Foundation. III: Hardy, D. A., Christos Doumas, Jannis A. Sakellarakis and Peter M. Warren, eds. 1990. *Thera and the Aegean World III: Proceedings of the Third International Congress, Santorini, Greece, 3-9 September 1989.* London: Thera Foundation.

TEXNH Laffineur, Robert and Philip P. Betancourt, eds. 1997. *TEXNH: Craftsmen, Craftswomen, and Craftsmanship in the Aegean Bronze Age. Proceedings of the 6th International Aegean Conference, Philadelphia, Temple University, 18-21 April 1996* (Aegaeum 16). Liège: Université de Liège, Histoire de l'art et archèologie de la Grèce antique; Austin: University of Texas at Austin, Program in Aegean Scripts and Prehistory.

T*EXTILES* Gleba, Margarita and Ulla Mannering, eds. 2012. *Textiles and Textile Production in Europe from Prehistory to AD 400* (Ancient Textile Series 11). Oxford: Oxbow.

T*HERA* I–VII Marinatos, Spyridon. 1968–1976. *Excavations at Thera,* vols. I–VII. Athens: Athens Archaeological Society.

T*IRYNS* II Rodenwaldt, G. [1912] 1976. *Tiryns II: Die Fresken des Palastes,* repr. Mainz am Rhein: Philipp von Zabern.

WPT Sherratt, Susan, ed. 2000. *The Wall Paintings of Thera. Proceedings of the First International Symposium, Petros M. Nomikos Conference Centre, Thera, Hellas, 30 August-4 September 1997,* vol. I. Athens, Thera Foundation – Petros M. Nomikos and Thera Foundation.

ΧΑΡΙΣ Chapin, Anne P., ed. 2004. *ΧΑΡΙΣ: Essays in Honor of Sara A. Immerwahr* (*Hesperia: The Journal of the American School of Classical Studies at Athens* Suppl. 33). Princeton: American School of Classical Studies at Athens.

Bibliography

Abramovitz, Katherine. 1980. "Frescoes from Ayia Irini, Keos. Parts II–IV." *Hesperia: The Journal of the American School of Classical Studies at Athens* 49, 57–85.

Adovasio, James. 1983. "Notes on the Textile and Basketry Impressions from Jarmo." In *Prehistoric Archaeology along the Zagros Flanks* (Oriental Institute Publications 105), eds. Linda S. Braidwood, Robert J. Braidwood, Bruce Howe, Charles Reed and P. J. Watson. Chicago: Oriental Institute of the University of Chicago, 425–426.

Adovasio, James M., Olga Soffer and Bohuslav Klíma. 1996. "Upper Paleolithic Fibre Technology from Pavlov I, Czech Republic." *Antiquity* 70, 526–534.

Aldrete, Gregory S., Scott M. Bartell and Alicia Aldrete. 2013. *Reconstructing Ancient Linen Body Armor: Unraveling the Linothorax Mystery.* Baltimore: Johns Hopkins University Press.

Aloupi, E., Y. Maniatis, T. Paradalis and L. Karali-Yannacopoulou. 1990. "Analysis of Purple Material Found at Akrotiri." In *TAW* III, 488–490.

Anderson, Eva and Marie-Louise Nosch. 2003. "With a Little Help from My Friends: Investigating Mycenaean Textiles with Help from Scandinavian Experimental Archaeology." In *METRON*, 197–205.

Anthes, Rudolf. 1943. "Die deutschen Grabungen auf der Westseite von Theban in den Jahren 1911 und 1913." Mitteilungen des deutschen archäologischen Instituts, Abteilung Kairo 12, 166.

Apostolakou, Vili, Thomas M. Brogan and Philip P. Betancourt. 2012. "The Minoan Settlement on Chryssi and Its Murex Dye Industry." In *KOSMOS*, 179–182.

Atkinson, Thomas D., R. C. Bosanquet, C. C. Edgar, A. J. Evans, D. G. Hogarth, D. MacKenzie, C. Smith and F. B. Welch. 1904. *Excavations at Phylakopi in Melos Conducted by the British School at Athens* (Society for the Promotion of Hellenic Studies, Suppl. Paper 4). London: Macmillan.

Baines, John. 1996. "Contextualizing Egyptian Representations of Society and Ethnicity." In *The Study of the Ancient Near East in the Twenty-First Century*, eds. Jerrold S. Cooper and Glenn M. Schartz. Winnona Lake, Ind. Eisenbrauns, 339–384.

Balfanz, Kathrin. 1995. "Bronzezeitliche Spinnwirtel aus Troia." *Studia Troica* V, 117–144.

Barber, Elizabeth J. W. 1975. "The Proto-Indo-European Notion of Cloth and Clothing." *Journal of Indo-European Studies* 3, 294–320.

Barber, Elizabeth J. W. 1994. *Women's Work: The First 20,000 Years: Women, Cloth and Society in Early Times.* New York and London: W. W. Norton.

Barber, Elizabeth J. W. 1997. "Minoan Women and the Challenges of Weaving for Home, Trade, and Shrine." In *TEXNH*, 515–519.

Barber, Elizabeth J. W. 1998. "Aegean Ornaments and Designs." In *The Aegean and the Orient in the Second Millennium* (Aegaeum 18), eds. Eric H. Cline and Diane Harris-Cline. Liège: Université de Liège, Histoire de l'art et archèologie de la Grèce antique; Austin: University of Texas at Austin, Program in Aegean Scripts and Prehistory, 13–17.

Barber, Elizabeth J. W. 1999. *The Mummies of Ürümchi.* New York: W. W. Norton.

Barber, Elizabeth J. W. 2005. "Half-Clad Minoan Women, Revisited." *Kadmos* 44, 40–42.

Barber, Elizabeth J. W. 2007. "Weaving the Social Fabric." In *Ancient Textiles*, 173–178.

Barbiera, Irene, Alice M. Choyke and Judith A. Rasson, eds. 2009. *Materializing Memory: Archaeological Material Culture and the Semantics of the Past* (British Archaeological Reports 1977). Oxford: Archaeopress.

Bard, Kathryn A. 1996. "Ancient Egyptians and the Issue of Race." In *Black Athena Revisited*, eds. Mary R. Lefkowitz and Guy MacLean Rogers. Chapel Hill, N.C.: The University of North Carolina Press, 103–111.

Bass, George F. 1961. "The Cape Gelidonya Wreck: Preliminary Report." *American Journal of Archaeology* 65, 267–276.

Bass, George F. 1967. *Cape Gelidonya: A Bronze Age Shipwreck* (Transactions of the American Philosophical Society 57.8). Philadelphia: American Philosophical Society.

Bass, George F. 1985. "The Ulu Burun Shipwreck." In *VII. Kazı Sonuçları Toplantısı*. Ankara: Kültür Turizm Bakanlığı, Eski Eserler ve Müzeler Genel Müdürlüğü, 619–635.

Bennett, Emmett, L. 1958. *The Olive Oil Tablets of Pylos. Texts of Inscriptions Found, 1955*. Salamanca: Universidad de Salamanca.

Bennet, John. 1989. "Outside in the Distance: Problems in Understanding the Economic Geography of Mycenaean Palatial Territories." In *Texts, Tablets, and Scribes: Studies in Mycenaean Epigraphy and Economy Offered to Emmett L. Bennett Jr.* (Minos Suppl. 10), eds. Jean-Pierre Olivier and Thomas Palaima. Salamanca: Universidad de Salamanca, 19–42.

Bennet, John. 1990. "Knossos in Context: Perspectives on the Linear B Administration of LM II–III Crete." *American Journal of Archaeology* 94, 193–211.

Bennet, John. 1992. " 'Collectors' or 'Owners'? An Examination of Their Possible Functions within the Palatial Economy of LM III Crete." In *Mykenaïka: Actes du IXe Colloque international sur les textes mycéniens et égéens organisé par le Centre de l'Antiquité Grecque et Romaine de la Fondation Hellénique des Recherches Scientifiques et l'École française d'Athènes, Athènes, 2-6 octobre 1990* (Bulletin De Correspondance Hellénique Suppl. 25), ed. Jean-Pierre Olivier. Athens: L'École française d'Athènes, 65–101.

Bennet, John and Jack L. Davis. 1999. "Making Mycenaeans: Warfare, Territorial Expansion, and Representations of the Other in the Pylian Kingdom." In *POLEMOS: Le contexte guerrier en Égée á l'âge du Bronze. Actes de la 7e Rencontre égéenne internationale Université de Liège, 14-17 avril 1998* (Aegaeum 19), ed. Robert Laffineur. Liège: *Université de Liège*, Histoire de l'art et archéologie de la Grèce antique; Austin: University of Texas at Austin, Program in Aegean Scripts and Prehistory, 105–120.

Betancourt, Philip P. 1985. *The History of Minoan Pottery*. Princeton: Princeton University Press.

Betancourt, Philip P. 2007a. *Introduction to Aegean Art*. Philadelphia: Institute for Aegean Prehistory Academic Press.

Betancourt, Philip P. 2007b. "Textile Production at Pseira: The Knotted Net." In *Ancient Textiles*, 185–189.

Betancourt, Philip P. and Costis Davaras, eds. 1998. *Pseira II: Building AC (the "Shrine") and Other Buildings in Area A*. Philadelphia: University of Pennsylvania Museum.

Betancourt, Philip P., Vili Apostolakou and Thomas M. Brogan. 2012. "The Workshop for Making Dyes at Pefka, Crete." In *KOSMOS*, 183–186.

Betancourt, Philip P., Michael C. Nelson and Hector Williams, eds. 2007. *Krinoi kai Limenes: Studies in Honor of Joseph and Maria Shaw*. Philadelphia: Institute for Aegean Prehistory Academic Press.

Bietak, Manfred, Josef Dorner, Irmgard Hein and Peter Jánosi. 1994. "Neue Grabungsergebnisse aus Tell el-Dab'a und Ezbet Helmi im östlichen Nildelta, 1989–1991."*Ägypten und Levante* 4, 9–80.

Bietak, Manfred, Josef Dorner and Peter Jánosi. 2001. "Ausgrabungen in dem Palastbezirk von Avaris. Vorberichte Tell el-Dab'a/'Ezbet Helmi 1993–2000." *Ägypten und Levante* 11, 27–119.

Bietak, Manfred, Nanno Marinatos and Clairy Palyvou. 2000. "The Maze Tableau from Tell el Dab'a." In *WPT*, 77–90.

Bietak, Manfred, Nanno Marinatos and Clairy Palivou. 2007. *Taureador Scenes in Tell el-Dab'a (Avaris) and Knossos*. Vienna: Österreichischen Akademie der Wissenschaften.

Birtacha, Kiki and Manolis Zacharioudakis. 2000. "Stereotypes in Theran Wall Paintings: Modules and Patterns in the Procedure of Painting." In *WPT*, 159–172.

Blakolmer, Fritz. 1994. "Ikonographische Beobachtungen zu Textilkunst und Wandermalerei in der Bronzezeitlichen." *Jahreshefte des Österreichischen archäolischen Instituts in Wien, Beiblatt* 63, 1–28.

Blakolmer, Fritz. 1996. "Textilkunst und Wandmalerei in der frühen Ägäis. Ikonographische Evidenz aus dem 'Haus der Damen' in Akrotiri, Thera." In *Akten des 6. Osterreichischen Archäologentages. 3-5 Februar 1994, Universität*

Graz, eds. Thuri Lorenz, Gabriele Erath, Manfred Lehner and Gerda Schwarz. Graz: Instituts für klassische Archäologie der Karl-Franzens-Universität, 27–29.

Blakolmer, Fritz. 1997. "Minoan Wall Painting: The Transformation of a Craft into an Art Form." In *TEXNH*, 95–105.

Blakolmer, Fritz. 2012a. "Body Marks and Textile Ornaments in Aegean Iconography: Their Meaning and Symbolism." In *KOSMOS*, 325–333.

Blakolmer, Fritz. 2012b. "The Missing 'Barbarians': Some Thoughts on Ethnicity and Identity in Aegean Bronze Age Iconography." *Talanta* 44, 53–77.

Bloedow, Edward F. 1997. "Itinerant Craftsmen and Trade in the Aegean Bronze Age." In *TEXNH*, 339–447.

Borch, Christian. 2013. *The Politics of Crowds: An Alternative History of Sociology*. Cambridge: Cambridge University Press.

Borchardt, Ludwig. 1913. *Das Grabdenkmal des Königs Sahu-re, vol. II: Die Wandbilder. Ausgrabungen der deutschen Orient-Gesellschaft in Abusir 1902-1908*. Leipzig: J. C. Hinrichs.

Borchardt, Ludwig. 1929. "Die Entstehung der Teppichbemalung an altägyptischen Decken und Gewölben." *Zeitschrift für Bauwesen* 79, 111–115.

Borojevic, Ksenija and Rebecca Mountain. 2011. "The Ropes of Pharaohs: The Source of Cordage from "Rope Cave" at Mersa/Wadi Gawasis, Revisited," *Journal of Archaeological Research in Egypt* 47, 131–141.

Boulotis, Christos. 1979. "Zur Deutung des Freskofragmentes Nr. 103 aus der Tirynther Frauenprozession." *Archaeologisches Korrespondenzblatt* 9, 59–67.

Boulotis, Christos. 1987. "Nochmals zum Prozessionsfresko von Knossos: Palast und Darbringung von Prestige-Objekten." In *The Function of the Minoan Palaces. Proceedings of the Fourth International Symposium at the Swedish Institute in Athens, 10-16 June 1984*, eds. Robin Hägg and Nanno Marinatos. Stockholm: Paul Åströms Förlag, 145–156.

Boulotis, Christos. 2000. "Traveling Fresco Painters in the Aegean Late Bronze Age: The Diffusion Patterns of a Prestigious Art." In *WPT*, 844–858.

Brandt, Luise, Lena Tranekjer, Ulla Mannering, Maj Ringgaard, Karin Frei, Margarita Gleba and M. Thomas P. Gilbert. 2011. "Characterising the Potential of Sheep Wool for Ancient DNA Analyses," *Archaeological and Anthropological Sciences* 3, 209–221.

Brecoulaki, Hariclia, Sharon R. Stocker, Jack L. Davis and Emily Egan. 2015. "An Unprecedented Naval Scene from Pylos: First Considerations." *Mycenaean Paintings in Context: New Discoveries and Old Finds Reconsidered. Proceedings of an International Conference on Mycenaean Wall Paintings in Context, held 10-13 February 2011 at the National Hellenic Research Foundation* (Meletemata 72), eds. Hariclia Brecoulaki, Jack L. Davis, and Sharon R. Stocker. Athens: National Hellenic Research Foundation, 261–291.

Brecoulaki, Hariclia, Caroline Zaitoun, Sharon R. Stocker and Jack L. Davis. 2008. "An Archer from the Palace of Nestor: A New Wall-Painting Fragment in the Chora Museum." *Hesperia: Journal of the American School of Classical Studies at Athens* 77, 363–397.

Brogan, Thomas M., Philip P. Betancourt and Vili Apostolakou. 2012. "The Purple Dye Industry of Eastern Crete." In *KOSMOS*, 187–192.

Brown, Donald. 1991. *Human Universals*. New York: McGraw-Hill.

Brugsch, Emil. 1889. *Le Tente funéraire de la Princesse Isimkheb*. Cairo: Le Caire.

Brumfiel, Elizabeth M. 1991. "Weaving and Cooking: Women's Production in Aztec Mexico." In *Engendering Archaeology: Women and Prehistory*, eds. Joan M. Gero and Margaret W. Conkey. Oxford: Wiley-Blackwell, 224–251.

Brunton, Guy and Reginald Engelbach. 1927. *Gurob*. London: British School of Archaeology in Egypt.

Bruyère, Bernard. 1933. *Fouilles de Deir el Médineh 8.3 (1930)*. Cairo: Le Caire

Brysbaert, Ann. 2007. "Cross-Craft and Cross-Cultural Interactions during the Aegean and Eastern Mediterranean Late Bronze Age." In *Mediterranean Crossroads*, eds. Sophia Antoniadou and Anthony Pace. Athens: Pierides Foundation, 325–359.

Brysbaert, Ann. 2008a. "Painted Plaster from Bronze Age Thebes, Boeotia (Greece): A Technological Study." *Journal of Archaeological Science* 35, 2761–2769.

Brysbaert, Ann. 2008b. *The Power of Technology in the Bronze Age Eastern Mediterranean: The Case of the Painted Plaster*. London and Oakville, Conn.: Equinox Publishing.

Buchholz, Hans-Günter and Joseph Wiesner. 1977. *Kriegswesen: Schutzwaffen und Wehrbauten (Archaeologia Homericai 1.E)*. Göttingen: Vandenhoeck and Ruprecht.

Buchholz, Hans-Günter and Vassos Karageorghis. 1973. *Prehistoric Greece and Cyprus.* New York: Praeger.
Bulfinch, Thomas. [1855] 2006. *Bulfinch's Mythology*, repr. New York: Barnes and Noble Classics.
Burke, Brendan. 1997. "The Organization of Textile Production in Bronze Age Crete." In *TEXNH*, 413–424.
Burke, Brendan. 1999. "Purple and the Aegean Textile Trade of the Early Second Millennium B.C." In *Meletemata*, 75–82.
Burke, Brendan. 2005. "Materialization of Mycenaean Ideology and the Ayia Triada Sarcophagus." *American Journal of Archaeology* 109, 403–422.
Burke, Brendan. 2010. *From Minos to Midas: Ancient Cloth Production in the Aegean and in Anatolia* (Ancient Textiles Series 7). Oxford and Oakville, Conn.: Oxbow.
Burke, Brendan. 2012. "Looking for Sea-Silk in the Bronze Age Aegean." In *KOSMOS*, 171–177.
Button, Seth. 2007. "Mortuary Studies, Memory, and the Mycenaean Polity." In *Negotiating the Past in the Past: Identity, Memory, and Landscape in Archaeological Research*, ed. Norman Yoffee. Tucson: University of Arizona Press, 76–103.
Cameron, Mark A. S. 1967. "Notes on Some New Joins and Additions to Well-Known Frescoes from Knossos." In: *Europa: Studien zur Geschichte und Epigraphik frühen Aegaeis. Festschrift für Ernst Grumach*, ed. William C. Brice. Berlin: de Gruyter, 45–74.
Cameron, Mark A. S. 1968. "Unpublished Paintings from the 'House of the Frescoes' at Knossos." *Annual of the British School at Athens* 63, 1–31.
Cameron, Mark A. S. 1971. "The Lady in Red." *Archaeology Magazine* 24, 35–43.
Cameron, Mark A. S. 1975. "A General Study of Minoan Frescoes with Particular Reference to Unpublished Wall Paintings from Knossos," Ph.D. dissertation, University of Newcastle upon Tyne.
Cameron, Mark A. S. 1976. "On Theoretical Principles in Aegean Bronze Age Mural Restoration." *TUAS* 1, 20–41.
Cameron, Mark A. S. 1978. "Theoretical Interrelations among Theran, Cretan and Mainland Frescoes." In *TAW* I, 579–591.
Cameron, Mark A. S. 1984. "Section 3: The Frescoes." In *The Minoan Unexplored Mansion at Knossos*, by Mervyn R. Popham. Athens: The British School of Athens, 127–148.
Cameron, Mark A. S. 1987. "The 'Palatial' Thematic System in the Knossos Murals: Last Notes on Knossos Frescoes." In *The Function of the Minoan Palaces. Proceedings of the Fourth International Symposium at the Swedish Institute in Athens, 10–16 June 1984,* eds. Robin Hägg and Nanno Marinatos. Stockholm: Paul Åströms Förlag, 321–325.
Campbell, Thomas P. 1995-1996. "Introduction." *The Metropolitan Museum of Art Bulletin* (New Series) 53, 10–18.
Campbell, Thomas P. 2002. *Tapestry in the Renaissance: Art and Magnificence.* New York: Metropolitan Museum of Art; New Haven Conn.: Yale University Press.
Carington Smith, Jill. 1975. "Weaving, Spinning, and Textile Production in Greece: The Neolithic to Bronze Age." Ph.D. dissertation, University of Tasmania.
Carington Smith, Jill. 1992. "Textiles at Nichoria." In *Excavations at Nichoria in Southwestern Greece*, vol. 2, *The Bronze Age Occupation,* eds. William McDonald and Nancy Wilkie. Minneapolis: University of Minnesota Press, 674–711.
Carter, Howard and Percy E. Newberry. 1904. *Catalogue général des antiquités égyptiennes: The Tomb of Thoutmosis IV.* Westminster: A. Constable and Co.
Casson, Lionel. 1995. *Ships and Seamanship in the Ancient World.* Baltimore: Johns Hopkins University Press.
Catling, Hector W. and Alison Millett. 1965. "A Study in the Composition Patterns of Mycenaean Pictorial Pottery." *Annual of the British School at Athens* 60, 212–224.
Chadwick, John. 1988. "The Women of Pylos." In *Texts, Tablets, and Scribes: Studies in Mycenaean Epigraphy and Economy Offered to Emmett L. Bennett Jr.* (*Minos* Suppl. 10), eds. Jean-Pierre Olivier and Thomas Palaima. Salamanca: Universidad de Salamanca, 43–96.
Chantraine, Pierre. 1970. *Dictionnaire étymologique de la langue grecque histoire des mots* 2. Paris: Klincksieck.
Chapin, Anne. P. 1995. "Landscape and Space in the Aegean Bronze Age," Ph.D. dissertation, University of North Carolina at Chapel Hill.
Chapin, Anne P. 1997-2000 [2002]. "Maidenhood and Marriage: The Reproductive Lives of the Girls and Women from Xeste 3, Thera." *Aegean Archaeology* 4, 7–25.
Chapin, Anne P. 2004. "Power, Privilege, and Landscape in Minoan Art." In *XAPIΣ*, 47–64.
Chapin, Anne P. 2007. "Boys Will Be Boys: Youth and Gender Identity in the Theran Frescoes." In *Constructions of Childhood in Greek and Roman Antiquity* (*Hesperia: The Journal of the American School of Classical Studies at Athens* Suppl. 41), eds. Ada Cohen and Jeremy Rutter. Princeton: American School of Classical Studies at Athens, 229–255.

Chapin, Anne P. 2008. "The Lady of the Landscape: An Investigation of Aegean Costuming and the Xeste 3 Frescoes." In *Reading a Dynamic Canvas: Adornment in the Ancient Mediterranean World*, eds. Cynthia S. Colburn and Maura K. Heyn. Newcastle: Cambridge Scholars Publishing, 48–83.

Chapin, Anne P. 2009. "Constructions of Youth and Gender in Aegean Art: The Evidence from Crete and Thera." In *FYLO: Engendering Prehistoric 'Stratigraphies' in the Aegean and the Mediterranean. Proceedings of the International Conference, University of Crete, Rethymnon, Crete, 3-6 June 2005* (Aegaeum 30), ed. Katerina Kopaka. Liège: Université de Liège, 175–182.

Chapin, Anne P. 2010. "Frescoes." In *The Oxford Handbook of the Bronze Age Aegean (ca. 3000–1000 BC)*, ed. Eric Cline. Oxford: Oxford University Press, 223–236.

Chapin, Anne P. 2011. "Gender and Coalitional Power in the Miniature Frescoes of Crete and the Cycladic Islands." In *Proceedings of the 10th International Congress of Cretan Studies, Khania, Crete, 1-8 October 2006, Chania, Crete*, vol. A3, ed. Maria Andreadaki-Vlazaki. Chania: Philological Association "Chrysostomos," 507–522.

Chapin, Anne P. 2012. "Do Clothes Make the Man (or Woman?): Sex, Gender, Costume, and the Aegean Color Convention." In *KOSMOS*, 297–304.

Chapin, Anne P. 2014. "Aegean Painting in the Bronze Age." In *The Cambridge History of Painting in the Classical World*, ed. J. J. Pollitt. Cambridge: Cambridge University Press.

Chapin, Anne P. Forthcoming. "Mycenaean Mythologies in the Making: The Frescoes of Pylos Hall 64 and the Mycenae Megaron." In *METAPHYSIS: Ritual, Myth and Symbolism in the Aegean Bronze Age. 15th International Aegean Conference at the Institute for Oriental and European Archaeology, Department Aegean and Anatolia, Austrian Academy of Sciences and at the Institute of Classical Archaeology, University of Vienna on 22-25 April 2014* (Aegaeum)

Chapin, Anne P. and Maria C. Shaw. 2006. "The Frescoes from the House of the Frescoes at Knossos: A Reconsideration of Their Architectural Context and a New Reconstruction of the Crocus Panel." *Annual of the British School at Athens* 101, 57–88.

Çilingiroğlu, Çiler. 2005. "The Concept of 'Neolithic Package': Considering Its Meaning and Applicability." *Documenta Praehistorica* 32 (Neolithic Studies 12), 1–13.

Clarke, John R. 1975. "Kinesthetic Address and the Influence of Architecture on Mosaic Composition in Three Hadrianic Bath Complexes at Ostia," *Architectura: Zeitschrift für Geschichte der Baukunst* 5, 1–17.

Clarke, John R. 1979. *Roman Black-and-White Figural Mosaics*. New York: New York University Press for the College Art Association of America.

Cline, Eric H., ed. 2010. *The Oxford Handbook of the Bronze Age Aegean (ca. 3000–1000 BC)*. Oxford: Oxford University Press.

Coleman, Katherine. 1973. "Frescoes from Ayia Irini, Keos. Part I." *Hesperia: The Journal of the American School of Classical Studies at Athens* 42, 283–300.

Cosmopoulos, Michael. 2015. "A Group of New Mycenaean Frescoes from Iklaina, Pylos." *Mycenaean Paintings in Context: New Discoveries and Old Finds Reconsidered. Proceedings of an International Conference on Mycenaean Wall Paintings in Context, Held 10-13 February 2011 at the National Hellenic Research Foundation* (Meletemata 72), eds. Hariclia Brecoulaki, Jack L. Davis, and Sharon R. Stocker. Athens: National Hellenic Research Foundation, 249–259.

Costin, Cathy. 1991. "Craft Specialization: Issues in Defining, Documenting and Explaining the Organization of Production." In *Archaeological Method and Theory*, ed. Michael B. Schiffer. Tuscon: University of Arizona Press, 1–56.

Costin, Cathy. 1993. "Textiles, Women, and Political Economy in Late Prehispanic Peru." *Research in Economic Anthropology* 14, 3–28.

Crewe, Lindy. 1998. *Spindle Whorls: A Study of Form, Function and Decoration in Prehistoric Bronze Age Cyprus* (Studies in Mediterranean Archaeology Pocketbook 149). Jonsered: Paul Åströms Förlag.

Crowfoot, Grace. 1936. "Of the Warp-Weighted Loom." *Annual of the British School at Athens* 37, 36–47.

Crowley, Janice L. 1991. "Patterns in the Sea: Insight into the Artistic Vision of the Aegeans." In *Thalassa. L'Egée préhistorique et la mer. Actes de la troisième Rencontre égéenne internationale de l'Université de Liège, Station de recherches sous-marines et océanographiques (StaReSo), Calvi, Corse, 23-25 avril 1990* (Aegaeum 7), eds. Robert Laffineur and Lucien Basch. Liège: Université de Liège, Histoire de l'art et archèologie de la Grèce antique, 219–230.

Crowley, Janice L. 2012. "Prestige Clothing in the Bronze Age Aegean." In *KOSMOS*, 231–239.

Crowley, Janice L. 2013. *The Iconography of Aegean Seals* (Aegaeum 34). Leuven and Liège: Peeters.

Cunliffe, Richard John. 1924. *Lexicon of Homeric Dialect*. London: Blackie and Son.

Dabney, Mary. 1996. "Ceramic Loomweights and Spindle Whorls," in *Kommos I: The Kommos Region and Houses of the Minoan Town*, eds. Joseph W. Shaw and Maria C. Shaw. Princeton: Princeton University Press, 244-262.

Dalley, Stephanie. 1977. "Old Babylonian Trade in Textiles at Tell al Rimah." *Iraq* 39, 155-159.

Daressy, Georges. 1902. *Catalogue général des antiquités égyptiennes, 3: Fouilles de la Vallée des Rois.* Cairo: Le Caire.

Davey, Nicholas. 2002. "Hermeneutics and Art Theory." In *A Companion to Art Theory*, eds. Paul Smith and Carolyn Wilde. Oxford and Malden, Mass.: Blackwell.

Davies, Nina and Alan Gardiner. 1915. *The Tomb of Amenemhēt*. London: Egypt Exploration Society.

Davies, Norman de Garis. 1901. *Rock Tombs of Sheikh Saïd*. London: Egypt Exploration Society.

Davies, Norman de Garis. 1903. *Rock Tombs of El Amarna 1*. London: Egypt Exploration Society.

Davies, Norman de Garis. 1905. *Rock Tombs of El Amarna 2*. London: Egypt Exploration Society.

Davies, Norman de Garis. 1920. *The Tomb of Antefoker*. London: G. Allen & Unwin.

Davies, Norman de Garis. 1922. "The Egyptian Expedition 1921-1922: The Graphic Work of the Expedition." *Bulletin of Metropolitan Museum of Art* 17. 2, 50-56.

Davies, Norman de Garis. 1926. "The Egyptian Expedition 1924-1925: The Graphic Work of the Expedition." *Bulletin of Metropolitan Museum of Art* 21. 2, 41-45.

Davies, Norman de Garis. 1930. *The Tomb of Ken-amün at Thebes*. New York: Metropolitan Museum of Art.

Davies, Norman de Garis. 1933. *The Tomb of Nefer-hotep at Thebes*. New York: Metropolitan Museum of Art.

Davies, Norman de Garis. 1948. *Seven Private Tombs at Kurnah*. London: Egypt Exploration Society.

Davies, Norman de Garis and Nina Davies. 1933. *The Tomb of Menkheperrasonb, Amenmosĕ, and Another*. London: Egypt Exploration Society.

Davis, Ellen N. 1986. "Youth and Age in the Thera Frescoes." *American Journal of Archaeology* 90, 399-406.

Davis, Ellen N. 1987. "The Knossos Miniature Frescoes and the Function of the Central Courts." In *The Function of the Minoan Palaces. Proceedings of the Fourth International Symposium at the Swedish Institute in Athens, 10-16 June 1984*, eds. Robin Hägg and Nanno Marinatos. Stockholm: Paul Åströms Förlag, 157-161.

Davis, Ellen N. 1990. "The Cycladic Style of the Theran Frescoes." In *TAW III*, 214-222.

Davis, Ellen N. 2005. "The Organisation of the Theran Artists." In *Aegean Wall Painting*, 859-872.

Day, Peter M. and Maria Relaki. 2002. "Past Factions and Present Fictions: *Palaces* in the Study of Minoan Crete." In *Monuments of Minos*, 217-234.

Day, Peter M. and David E. Wilson. 2002. "Landscapes of Memory, Craft and Power in Prepalatial and Protopalatial Knossos." In *Labyrinth Revisited: Rethinking "Minoan" Archaeology*, ed. Yannis Hamilakis. Oxford: Oxbow Books, 143-166.

De Ridder, André. 1894. "Fouilles de Gha." *Bulletin de correspondance hellénique* 18, 271-310.

De Wilde, D. 2001. "Textile Remains on Vases from Tomb 1 and Tomb 2C." In *The Mycenaean Cemetery at Pylona on Rhodes* (British Archaeological Reports 988), ed. Efi Karantzali. Oxford: Archaeopress, 114-116.

Demakopoulou, Katie, ed. 1990. *Troy, Mycenae, Tiryns, Orchomenos. Heinrich Schliemann: The 100th Anniversary of His Death*. Athens: Ministry of Culture, Greek Committee ICOM; Berlin: Museum für Ur- und Frühgeschichte.

Dimopoulou-Rethemiotaki, Nota. 1993. "Πόρος–Κατσάμπας." Αρχαιολογικόν Δελτίον 48.2, 450-459.

Dimopoulou-Rethemiotaki, Nota. 2004. "Το επίνειο της Κνωσού στον Πόρο-Κατσαμπά." In *Knossos: Palace, City, State*, 363-380.

Donahue, Cristin J. 2006. "The Importance of Cloth: Aegean Textile Representation in Neopalatial Wall Painting." M.A. thesis, The Florida State University. http://etd.lib.fsu.edu/theses_1/available/etd-07062006-183736/unrestricted/donahue_thesis.pdf.

Dörpfeld, Wilhelm. 1885. "The Buildings of Tiryns." In *Tiryns: The Prehistoric Palace of the Kings of Tiryns*, by Heinrich Schliemann, 177-308. New York: C. Scribner's Sons.

Doumas, Christos. 1992. *The Wall-Paintings of Thera*. Athens: Thera Foundation.

Doumas, Christos. 2000. "Age and Gender in the Theran Wall Paintings." In *WPT*, 971-982.

Doumas, Christos. 2006. "Οι εργασίες του Ακρωτηρί κατά το 2006." *ALS* 4, 7-16.

Douskos, Iris. 1980. "The Crocuses of Santorini." In *TAW II*, 141-146.

Driessen, Jan. 2004. "The Central Court of the Palace at Knossos." In *Knossos: Palace, City, State*, 75-82.

Driessen, Jan and Colin F. Macdonald. 1997. *The Troubled Island: Minoan Crete before and after the Santorini Eruption* (*Aegaeum* 17). Liège: Université de Liège, Histoire de l'art et archéologie de la Grèce antique; Austin: University of Texas at Austin, Program in Aegean Scripts and Prehistory.

Duhoux, Yves. 1974. "Idéogrammes textiles du Linéaire B: *146, *160, *165 et *166." *Minos: Revista de filología egea* 15, 116–132.
Egan, Emily C. 2008. "The Relationship between Wall Painting and Vase Painting at Knossos During the Neo- and Final Palatial Periods," M.A. thesis, University of Cincinnati.
Egan, Emily C. 2012. "Cut from the Same Cloth: the Textile Connection between Palace Style Jars and Knossian Wall Painting." In *KOSMOS*, 317–324.
Egan, Emily C. 2015. "Working within the Lines: Artists' Grids and Painted Floors at the Palace of Nestor." In *Selected Papers in Ancient Art and Architecture* (Archaeological Institute of America Monographs 1), eds. Sarah Lepinski and Susana McFadden. Boston: Archaeological Institute of America, 188–204.
El Goresy, Ahmed. 2000. "Polychromatic Wall Painting Decorations in Monuments of Pharaonic Egypt: Compositions, Chronology and Painting Techniques." In *WPT*, 49–70.
Emery, Walter B. 1949. *Excavations at Saqqara: Great Tombs of the First Dynasty.* Cairo: Government Press.
Englebach, Reginald. 1924. *A Supplement to the Topographical Catalogue of the Private Tombs of Thebes (Nos. 253 to 334) with Some Notes on the Necropolis from 1913 to 1924.* Cairo: Print Office of the French Institute of Oriental Archaeology.
Evans, Arthur. 1909. *Scripta Minoa: The Written Documents of Minoan Crete with Special Reference to the Archives of Knossos,* vol. 1, *The Hieroglyphic and Primitive Linear Classes.* Oxford: Clarendon.
Evans, John. D. 1968. "Knossos Neolithic, Part II." *Annual of the British School at Athens* 63, 239–276.
Evely, R. D. G. 1999. *Fresco: A Passport into the Past. Minoan Crete through the Eyes of Mark Cameron.* Athens: British School at Athens; Athens: N. P. Goulandris Foundation–Museum of Cycladic Art.
Evely, R. D. G. 2000. *Minoan Crafts: Tools and Techniques. An Introduction 2* (Studies in Mediterranean Archaeology 92). Jonsered: Paul Åströms Förlag.
Frankfort, Henri. 1929. *The Mural Paintings of El-'Amarneh.* London: Egypt Exploration Society.
Frankfort, Henri and John D. S. Pendlebury. 1933. *The City of Akhenaten II.* London: Egypt Exploration Society.
Frei, Karin, Robert Frei, Ulla Mannering, Margarita Gleba, Marie-Louise Nosch and Henriette Lyngstrøm. 2009. "Provenance of Ancient Textiles: A Pilot Study Evaluating the Sr Isotope System." *Archaeometry* 51, 252–276.
Fu, Yong-Bi, Axel Diederichsen and Robin G. Allaby. 2012. "Locus-Specific View of Flax Domestication History." *Ecology and Evolution* 2, 139–152.
Furumark, Arne. [1941] 1972. *Mycenaean Pottery. I: Analysis and Classification; II: Chronology* (Skrifter utgivna av Svenska Institutet i Athen 20), repr. Stockholm: Svenska Institutet i Athen.
Fyfe, Theodore. 1902. "Painted Plaster Decoration at Knossos." *Journal of the Royal Institute of British Architects,* 3rd ser. X, 107–131.
Gadamer, Hans-Georg. 1989. *Truth and Method,* 2nd rev. ed., trans. Joel Weinsheimer and Donald G. Marshall. New York: Crossroad.
Gardiner, Alan and Arthur E. P. Weigall. 1913. *A Topographical Catalogue of the Private Tombs of Thebes.* London: Bernard Quaritch.
Gell, Alfred. 1989. *Art and Agency: An Anthropological Theory.* Oxford: Clarendon Press.
Georgiadis, Mercourios and Chrysanthi Gallou, eds. 2009. *The Past in the Past: The Significance of Memory and Tradition in the Transmission of Culture* (British Archaeological Reports 1925). Oxford: Archaeopress.
Glanville, S. R. K. 1972. *Wooden Model Boats.* London: British Museum.
Gleba, Margarita and Ulla Mannering, eds. 2012. *Textiles and Textile Production in Europe from Prehistory to AD 500.* Oxford: Oxbow Books.
Gombrich, Ernst H. 1960. *Art and Illusion: A Study in the Psychology of Pictorial Representation.* New York: Pantheon Books.
Gombrich, Ernst H. 1979. *A Sense of Order: A Study in the Psychology of Decorative Art.* Ithaca: Cornell University Press.
Good, Irene. 1995. "On the Question of Silk in Pre-Han Eurasia." *Antiquity* 69, 960–966.
Good, Irene. 2001. "Archaeological Textiles: A Review of Current Research." *Annual Review of Anthropology* 30, 209–226.
Good, Irene, J. Mark Kenoyer and Richard H. Meadow. 2009. "New Evidence for Early Silk in the Indus Civilization." *Archaeometry* 51, 457–466.
Graham, J. Walter. 1962. *The Palaces of Crete.* Princeton: Princeton University Press.

Graves, Robert. 1955. *The Greek Myths*, 2 vols. Baltimore: Penguin Books.
Gray, Dorothea. 1974. *Seewesen*, with a contribution, "Das Schiffsfresko von Akrotiri, Thera," by Spyridon Marinatos (*Archaeologia Homerica* 1.G). Göttingen: Vandenhoeck and Ruprecht.
Guralnick, Eleanor. 2000. "Proportions of Painted Figures from Thera." In *WPT*, 173–190.
Hackl, Rudolf. 1912. "Die Fussböden." In *Tiryns* II, 222–237.
Hadjidaki, Elpida and Philip P. Betancourt. 2005–2006. "A Minoan Shipwreck off Pseira Island, East Crete: Preliminary Report," *Eulimene* 6–7, 79–96.
Hägg, Robin. 1987. "On the Reconstruction of the West Façade of the Palace at Knossos." In *The Function of the Minoan Palaces, Proceedings of the Fourth International Symposium at the Swedish Institute in Athens, 10-16 June 1984*, eds. Robin Hägg and Nanno Marinatos. Stockholm: Paul Åströms Förlag, 129–134.
Hall, Henry R. 1909–1910. "An Addition to the Senmut-Fresco." *Annual of the British School at Athens* 16, 254–257.
Hallager, Eric. 2002. "One Linear A Tablet and 45 Noduli." *Creta Antica* 3, 105–109.
Halstead, Paul. 1981. "Counting Sheep in Neolithic and Bronze Age Greece." In *Pattern of the Past: Studies in Honour of David Clarke,* eds. Ian Hodder, Glynn Isaac and Norman Hammond. New York: Cambridge University Press, 307–339.
Halstead, Paul. 1991. "Lost Sheep? On the Linear B Evidence for Breeding Flocks at Knossos and Pylos." *Minos* 22, 343–365.
Halstead, Paul. 1993. "The Mycenaean Palatial Economy: Making the Most of the Gaps in the Evidence." *Cambridge Classical Journal* 38, 57–86.
Halstead, Paul. 1999a. "Missing Sheep: On the Meaning and Wider Significance of the Knossos Sheep Records." *Annual of the British School at Athens* 94, 145–166.
Halstead, Paul, ed. 1999b. *Neolithic Society in Greece.* Sheffield: Sheffield Academic Press.
Hamilakis, Yannis. 2002. "Too Many Chiefs?: Factional Competition in Neopalatial Crete." In *Monuments of Minos*, 179–199.
Hamilton, Edith. 1942. *Mythology: Timeless Tales of Gods and Heroes.* Boston: Little, Brown and Co.
Hankey, Vronwy and Peter Warren. 1974. "The Absolute Chronology of the Aegean Late Bronze Age." *Bulletin of the Institute of Classical Studies* 21, 142–152.
Hardy, Karen. 2008. "Prehistoric String Theory: How Twisted Fibres Helped to Shape the World." *Antiquity* 82, 271–280.
Hatt, Michael and Charlotte Klonk. 2006. *Art History: A Critical Introduction to Its Methods.* Manchester and New York: Manchester University Press.
Hayes, William C. 1937. *Glazed Tiles from a Palace of Ramesses II at Kantīr.* New York: Metropolitan Museum of Art.
Hayes, William C. 1959. *The Scepter of Egypt 2, The Hyksos Period and the New Kingdom (1675–1080 B.C.).* Cambridge, Mass.: Harvard University Press from the Metropolitan Museum of Art.
Hiendleder, Stefan, Bernhard Kaupe, Rudolf Wassmuth and Axel Janke. 2002. "Molecular Analysis of Wild and Domestic Sheep Questions Current Nomenclature and Provides Evidence for Domestication from Two Different Subspecies." *Proceedings of the Royal Society B: Biological Sciences* 269, 893–904.
Hillman, Gordon. 1975. "The Plant Remains from Abu Hureyra: A Preliminary Report." *Proceedings of the Prehistoric Society* 41, 70–73.
Hirsch, Ethel S. 1977a. *Painted Decoration on the Floors of Bronze Age Structures on Crete and the Greek Mainland* (*Studies in Mediterranean Archaeology* 53). Göteborg: Paul Åströms Förlag.
Hirsch, Ethel S. 1977b. "The Restoration of the Painted Courtyard Floor at Mycenae." *Antike Kunst* 20: 54–56.
Hirsch, Ethel S. 1980. "Another Look at Minoan and Mycenaean Interrelationships in Floor Decoration." *American Journal of Archaeology* 84, 453–462.
Hitchcock, Louise A. 2000. *Minoan Architecture: A Contextual Analysis* (*Studies in Mediterranean Archaeology* Pocketbook 155). Jonsered: Paul Åströms Förlag.
Hodder, Ian. 2012. *Entangled: An Archaeology of the Relationships between Humans and Things.* Malden, Mass.: Wiley-Blackwell.
Hodder, Ian and Scott Hutson. 2003. *Reading the Past: Current Approaches to Interpretation in Archaeology.* Cambridge: Cambridge University Press.
Hoffman, Marta. 1964. *The Warp-Weighted Loom: Studies in the History and Technology of an Ancient Implement* (*Studia Norvegica* 14). Oslo: Norsk folkemuseum.

Hood, Sinclair. 2005. "Dating the Knossos Frescoes." In *Aegean Wall Painting*, 45–81.
Hundt, Hans Jürgen. 1971. "On Prehistoric Textile Finds." *Jahrbuch Romisch-Germanisches Zentral-museum* 16, 59–71.
Hutchinson, Richard W. 1939–1940. "Unpublished Objects from Palaikastro and Praisos." *Annual of the British School at Athens* 40, 47–49.
Iakovides, Spyridon. 1979. "Thera and Mycenaean Greece." *American Journal of Archaeology* 83, 101–102.
Immerwahr, Sara A. 1973. *Early Burials from the Agora Cemeteries*. Princeton: American School of Classical Studies at Athens.
Immerwahr, Sara A. 1977. "Mycenaeans at Thera: Some Reflections on the Paintings from the West House." In *Greece and the Eastern Mediterranean in Ancient History and Prehistory. Studies Presented to Fritz Schachermeyr on the Occasion of His Eightieth Birthday*, ed. Konrad H. Kinzl. Berlin: Walter de Gruyter, 173–191.
Immerwahr, Sara A. 1990. *Aegean Painting in the Bronze Age*. University Park: Pennsylvania State University Press.
Jacobsen, Thorkild. 1970. "On the Textile Industry of Ur under Ibbi-Sin." In *Toward the Image of Tammuz and Other Essays on Mesopotamian History and Culture*, ed. William L. Moran. Cambridge, Mass.: Harvard University Press, 216–230.
Jéquier, Gustave. 1911. *Décoration égyptienne: Plafonds et frises végétales du nouvel empire thébain (1400 à 1000 avant J.-C.)*. Paris: Librairie centrale d'art et d'architecture.
Jones, Bernice R. 2001. "The Minoan 'Snake Goddess.' New Interpretations of Her Costume and Identity." In *POTNIA*, 259–268.
Jones, Bernice R. 2003. "Veils and Mantles: An Investigation of the Construction and Function of the Costumes of the Veiled Dancer from Thera and the Camp Stool Banqueter from Knossos." In *METRON*, 441–450.
Jones, Bernice R. 2007. "A Reconsideration of the Kneeling-Figure Fresco from Hagia Triada." In *Krinoi kai Limenes: Studies in Honor of Joesph and Maria Shaw*, eds. Philip P. Betancourt, Michael C. Nelson and Hector Williams. Philadelphia: Institute for Aegean Prehistory Academic Press, 151–158.
Jones, Bernice R. 2009. "New Reconstructions of the 'Mykenaia' and a Seated Woman from Mycenae." *American Journal of Archaeology* 113, 309–338.
Jones, Bernice R. 2012. "A New Discovery and Interpretation of the Fragmentary Figure Fresco from the House of the Ladies, Thera." Paper presented at the 113th Annual Meeting of the Archaeological Institute of America, Philadelphia, January 5–8.
Jones, Andrew. 2007. *Memory and Material Culture*. Cambridge and New York: Cambridge University Press.
Kaiser, Bernd. 1976. *Untersuchungen zum minoischen Relief*, Habelts Dissertationsdrücke: Reihe klassische Archäologie 7. Bonn: Habelt.
Kantor, Helene J. 1947. *The Aegean and the Orient in the Second Millennium B.C.* Bloomington, Ind.: Principia Press.
Karo, Georg. 1930–1933. *Die Schachtgräber von Mykenai*. Munich: Bruckmann.
Kemp, Barry J. and Robert S. Merrillees. 1980. *Minoan Pottery in Second Millennium Egypt*. Mainz: Philipp von Zabern.
Kemp, Barry J. and Gillian Vogelsang-Eastwood. 2001. *The Ancient Textile Industry at Amarna*. London: Egypt Exploration Society.
Killen, John. 1962. "The Wool Ideogram in Linear B Texts." *Hermathena* 96, 38–72.
Killen, John. 1964. "The Wool Industry of Crete in the Late Bronze Age." *Annual of the British School at Athens* 59, 1–15.
Killen, John. 1979. "The Knossos Ld (1) Tablets." In *Colloquium Mycenaeum: Actes du sixième Colloque international sur les textes mycéniens et égéens tenu à Chaumont sur Neuchâtel du 7 au 13 septembre 1975*, eds. Ernst Risch and Hugo Mühlestein. Geneva: Université de Neuchâtel, 151–181.
Killen, John. 1984. "The Textile Industry at Pylos and Knossos." In *Pylos Comes Alive: Industry and Administration in the Mycenaean Palace*, eds. Cynthia Shelmerdine and Thomas Palaima. Boston: Archaeological Institute of America, 49–64.
Killen, John. 1988. "Epigraphy and Interpretation of Knossos Woman and Cloth Records." In *Texts, Tablets, and Scribes: Studies in Mycenaean Epigraphy and Economy Offered to Emmett L. Bennett Jr.* (*Minos* Suppl. 10), eds. Jean-Pierre Olivier and Thomas Palaima. Salamanca: Universidad de Salamanca, 167–183.
Killen, John. 2007. "Cloth Production in Late Bronze Age Greece: The Documentary Evidence." In *Ancient Textiles: Production, Craft, Society*, eds. Carole Gillis and Marie-Louise Nosch. London: Oxbow, 50–58.
Kleiner, Fred S. 2012. *Gardner's Art Through the Ages: A Global History*, 14th ed. S. I.: Wadsworth.
Koehl, Robert B. 1986. "The Chieftain Cup and a Minoan Rite of Passage." *Journal of Hellenic Studies* 106, 99–110.
Koehl, Robert B. 2001. "The 'Sacred Marriage' in Minoan Religion and Ritual." In *POTNIA*, 237–243.

Konstantinidi, Eleni. 1995. "Ένθετα κοσμήματα ενδυμασίας της Εποχής του Χαλκού στην Ελλάδα." *Archaeologia* 54, 25–28.

Konstantinidi-Syvridi, Eleni. 2012. "A Fashion Model of Mycenaean Times: The Ivory Lady from Prosymna." In *KOSMOS*, 265–270.

Kraft, John C., Ilhan Kayan and Oğuz Erol. 1980. "Geomorphic Reconstructions in the Environs of Ancient Troy." *Science* 209, 776–782.

Kritseli-Providi, Ioanna. 1973. "Τοιχογραφία οκτωσχήμου ασπίδος εκ Μυκηνών." *Athens Annals of Archaeology* 6, 176–180.

Kritseli-Providi, Ioanna. 1982. Τοιχογραφίες του θρησκευτικού κέντρου των Μυκηνών. Athens: NP.

Kvavadze, Eliso, Ofer Bar-Yosef, Anna Belfer-Cohen, Elisabetta Boaretto, Nino Jakeli, Zinovi Matskevich and Tengiz Meshveliani. 2009. "30,000 Years Old Wild Flax Fibers – Testimony for Fabricating Prehistoric Linen." *Science* 325, 1359.

Kyriakidis, Evangelos. 1997. "Nudity in Late Minoan I Seal Iconography." *Kadmos* 36, 119–126.

Laffineur, Robert. 2000. "Dress, Hairstyle and Jewellery in the Thera Wall Paintings." In *WPT*, 890–906.

Laffineur, Robert. 2001. "Seeing Is Believing: Reflections of Divine Imagery in the Aegean Bronze Age." In *POTNIA*, 387–392.

Lamb, Winifred. 1919–1921. "Excavations at Mycenae: III. Frescoes from the Ramp House." *Annual of the British School at Athens* 24, 189–199.

Lamb, Winifred. 1921–1923a. "Excavations at Mycenae, VIII. The Palace. Frescoes from the Palace." *Annual of the British School at Athens* 25, 162–172.

Lamb, Winifred. 1921–1923b. "The Painted Stucco Floor." In "Excavations at Mycenae, VIII. The Palace." *Annual of the British School at Athens* 25, 193–195.

Landstrom, Björn. 1970. *Ships of the Pharaohs: 4000 Years of Egyptian Ship Building.* New York: Doubleday.

Lenuzza, Valeria. 2012. "Dressing Priestly Shoulders: Suggestions from the Campstool Fresco." *KOSMOS*, 255–264.

Letesson, Quentin and Klaas Vansteenhuyse. 2006. "Towards an Archaeology of Perception: 'Looking' at the Minoan Palaces." *Journal of Mediterranean Archaeology* 19, 91–119.

Levi, Doro. 1976. *Festòs e la civiltà minoica* I (Incunabula Graeca 60). Rome: Edizioni dell'Ateneo.

Lillethun, Abby. 2003. "The Recreation of Aegean Cloth and Clothing." In *METRON*, 463–472.

Lillethun, Abby. 2012. "Finding the Flounced Skirt (Back Apron)." In *KOSMOS*, 251–254.

Linnaeus, Carl. 1758. *Systema Naturae* I, 10th edn. Holmiae: Laurentii Salvii.

Lipke, Paul. 1984. *The Royal Ship of Cheops: A Retrospective Account of the Discovery, Restoration and Reconstruction* (British Archaeological Reports 225). Oxford: Archaeopress.

Long, Charlotte. 1974. *The Ayia Triadha Sarcophagus: A Study of Late Minoan and Mycenaean Funerary Practices and Beliefs.* Göteborg: Paul Åströms Förlag.

Lorimer, Hilda L. 1950. *Homer and the Monuments.* London: Macmillan.

Lubec, Gert, J. Holaubek, C. Feidle, B. Lubec and Eugen Strouhal. 1993. "Use of Silk in Ancient Egypt." *Nature* 362, 25.

Lupack, Susan. 2014. "Offerings for the *Wanax* in the Fr Tablets: Ancestor Worship and the Maintenance of Power in Mycenaean Greece." In *Ke-ra-me-ja: Festschrift in Honor of Dr. Cynthia Shelmerdine*, eds. Joann Gulizio and Dimitri Nakassis. Philadelphia: Institute for Aegean Prehistory Academic Press, 163–177.

MacDonald, Colin F. 2005. *Knossos.* Folio Society: London.

MacGillivray, J. Alexander. 2007. "Protopalatial (MM IB–MM IIIA): Early Chamber Beneath the West Court, Royal Pottery Stores, the Trial KV and the West and South Polychrome Deposits Groups." In *Knossos Pottery Handbook: Neolithic and Bronze Age (Minoan)* (British School at Athens Studies 14), ed. Nicoletta Momigliano. London: British School at Athens, 105–149.

Maeder, Felicitas. 2002. "The Project Sea-Silk – Rediscovering an Ancient Textile Material." *Archaeological Textiles Newsletter* 35, 9–11.

Maran, Joseph and Eftychia Stavrianopoulou. 2007. "Πότνιος Ἀνήρ: Reflections on the Ideology of Mycenaean Kingship." In *Keimelion: Elitenbildung und elitärer Konsum von der mykenischen Palastzeit bis zur Homerischen Epoche = The Formation of Elites and Elitist Lifestyles from Mycenaean Palatial Times to the Homeric Period. Akten des internationalen Kongresses vom 3. bis 5. Februar 2005 in Salzburg*, eds. Eva Alram-Stern and Georg Nightingale. Wien: Verlag der Österreichischen Akademie der Wissenschaften, 285–289.

Marcar, Ariane. 2004. "Aegean Costume and the Dating of the Knossian Frescoes." In *Knossos: Palace, City, State*, 225–238.
March, Jenny. 2009. *The Penguin Book of Classical Myths*. London and New York: Penguin.
Marinatos, Nanno. 1984. *Art and Religion in Thera: Reconstructing a Bronze Age Society*. Athens: D. & I. Mathioulakis.
Marinatos, Nanno. 1987a. "An Offering of Saffron to the Minoan Goddess of Nature: The Role of the Monkey and the Importance of Saffron." In *Gifts to the Gods. Proceedings of the Uppsala Symposium 1985*, eds. Tullia Linders and Gullög Nordquist. Uppsala: Uppsala Universitet, 123–132.
Marinatos, Nanno. 1987b. "Public Festivals in the West Courts of the Palaces." In *The Function of the Minoan Palaces. Proceedings of the Fourth International Symposium at the Swedish Institute in Athens, 10-16 June 1984*, eds. Robin Hägg and Nanno Marinatos. Stockholm: Paul Åströms Förlag, 134–143.
Marinatos, Nanno. 1988. "The Fresco from Room 31 at Mycenae: Problems of Method and Interpretation." In *Problems in Greek Prehistory*, eds. Elizabeth B. French and Ken A. Wardle. Bristol: Bristol Classical Press, 245–251.
Marinatos, Spyridon. 1933. "La Marine Créto-Mycénienne." *Bulletin de correspondance hellénique* 57, 170–235.
Marinatos, Spyridon. 1967. *Kleidung, Haar- und Barttracht* (Archaeologia Homerica 1, A'–B'). Göttingen: Vandenhoeck and Ruprecht.
Marinatos, Spyridon. 1973. "From the Miniature Fresco of Thera: A Detail of the Body-Shield." *Athens Annals of Archaeology* 6, 494–497.
Marinatos, Spyridon. 1974. *Excavations at Thera: Colour Plates and Plans*. Athens: Athens Archaeological Society.
Marinatos, Spyridon and Max Hirmer. 1960. *Crete and Mycenae*. London: Thames and Hudson.
Marketou, Toula. 2010. "Rhodes." In *The Oxford Handbook of the Bronze Age Aegean (ca. 3000-1000 BC)*, ed. Eric Cline. Oxford: Oxford University Press, 775–793.
Marketou, Toula. 2013. "The Art of Wall Painting at Ialysos on Rhodes from the Early 2nd Millennium BC to the Eruption of the Thera Volcano." In CHROSTERES/PAINTBRUSHES, Wall-Painting and Vase-Painting of the 2nd Millennium BC in Dialogue at Akrotiri, Thera, May 24–26, ed. Andreas Vlachopoulos. Athens: Society of the Promotion of Theran Studies, 104–109.
Mårtensson, Linda, Marie-Louise Nosch and Eva Andersson Strand. 2009. "Shape of Things: Understanding a Loom Weight." *Oxford Journal of Archaeology* 28, 373–398.
McCarter, Susan. 2007. *Neolithic*. New York: Routledge.
McCorriston, Joy. 1997. "The Fiber Revolution: Textile Extensification, Alienation, and Social Stratification in Ancient Mesopotamia." *Current Anthropology* 38, 517–549.
McDonald, William and George Rapp, Jr., eds. 1972. *Minnesota Messenia Expedition: Reconstructing a Bronze Age Regional Environment*. Minneapolis: University of Minnesota Press.
McEnroe, John C. 2010. *Architecture of Minoan Crete: Constructing Identity in the Aegean Bronze Age*. Austin: University of Texas Press.
McGovern, Patrick E. 1989. "Ceramics and Craft Interaction: A Theoretical Framework, with Prefatory Remarks." In *Cross-Craft and Cross-Cultural Interactions in Ceramics* (Ceramics and Civilization 4), eds. Patrick E. McGovern and M. D. Notis. Westerville, Ohio: American Ceramic Society, 1–11.
McKinley, Daniel. 1998. "Pinna and Her Silken Beard: A Foray into Historical Misappropriations." *Ars Textrina: A Journal of Textiles and Costumes* 29, 9–29.
Mee, Christopher. 1982. *Rhodes in the Bronze Age*. Warminster: Aris & Phillips.
Michailidou, Anna. 1992-1993. "'Ostrakon' with Linear A Script from Akrotiri (Thera): A Non-Bureaucratic Activity?" *Minos* 27–28, 7–24.
Militello, Pietro. 1998. *Haghia Triada I. Gli affreschi minoici di Haghia Triada*. Monografie della Scuola Archeologica Italiana di Atene e delle Missioni Italiene in Oriente 9. Padua: Bottega d'Erasmo.
Militello, Pietro. 2001. *Gli affreschi minoici di Festòs* (Studi di Archeologia Cretese 2). Padua: Bottega d'Erasmo.
Militello, Pietro. 2007. "Textile Industry and Minoan Palaces." In *Ancient Textiles*, 36–45.
Militello, Pietro. 2012a. "Emerging Authority: A Functional Analysis of the MM II Settlement of Phaistos." In *Back to the Beginning: Reassessing Social and Political Complexity on Crete during the Early and Middle Bronze Age*, eds. Ilse Schoep, Peter Tomkins and Jan Driessen. Oxford: Oxbow Books, 236–272.
Militello, Pietro. 2012b. "New Evidence for Textile Activity in Phaistos and Ayia Triada." In ΑΡΧΑΙΟΛΟΓΙΚΟ ΕΡΓΟ ΚΡΗΤΗΣ 2, Πρακτικά της Β Κριτική Συνάντησης, Ρέθυμνο 2010, eds. Michales Andrianachis, Petroula Varzalithou and Iris Tzachili. Rethymnon: Panepisthimio Kritis, 203–216.

Militello, Pietro. Forthcoming. *Festòs e Hahia Triada. La produzione tessile di età minoica.* Padua: Bottega d'Erasmo.
Miller, David L. 2013. *Introduction to Collective Behavior and Collective Action*, 3rd ed. S.l.: Long Grove, Ill.: Waveland Press.
Miller, Naomi F. 2006. "The Origins of Plant Cultivation in the Near East." In *The Origins of Agriculture: An International Perspective*, 2nd ed., eds. C. Wesley Cowan and Patty Jo Watson. Tuscaloosa, Ala.: University of Alabama Press, 39–58.
Molholt, Rebecca. 2011. "Roman Labyrinth Mosaics and the Experience of Motion." *Art Bulletin* 93, 287–303.
Möller-Wiering, Susan. 2006. "Tools and Textiles – Texts and Contexts. Bronze Age Textiles Found in Crete." The Danish National Research Foundation's Centre for Textile Research, University of Copenhagen. http://ctr.hum.ku.dk/tools/
Morford, Mark, Robert J. Lenardon and Michael Sham. 2013. *Classical Mythology*, 10th ed. Oxford: Oxford University Press.
Morgan, Lyvia. 1988. *The Miniature Wall Paintings of Thera: A Study in Aegean Culture and Iconography.* New York: Cambridge University Press.
Morgan, Lyvia. 1990. "Island Iconography: Thera, Kea, Milos." In *TAW* III, 252–266.
Morgan, Lyvia. 1998. "The Wall Paintings of the North-East Bastion at Ayia Irini, Kea." In *Kea-Kythnos: History and Archaeology. Proceedings of an International Symposium, Kea-Kythnos, 22-25 June 1994* (Meletimata 27), eds. Lina G. Mendoni and Alexander Mazarakis-Ainian. Paris: Diffusion de Boccard; Athens: Research Centre for Greek and Roman Antiquity, National Hellenic Research Foundation, 201–210.
Morgan, Lyvia. 2005a. "New Discoveries and New Ideas in Aegean Wall Painting." In *Aegean Wall Painting*, 21–44.
Morgan, Lyvia. 2005b. "The Cult Centre at Mycenae and the Duality of Life and Death." In *Aegean Wall Painting*, 159–171.
Morgan, Lyvia. 2007. "The Painted Plasters and Their Relation to the Wall Painting of the Pillar Crypt," with contributions by Mark Cameron. In *Excavations at Phylakopi in Melos 1974-1977* (Annual of the British School at Athens Suppl. 42), ed. Colin Renfrew. London: British School at Athens, 371–396.
Morgan Brown, Lyvia. 1978. "The Ship Procession in the 'Miniature Fresco'." in *TAW* I, 629–644.
Morgan, Mary and Dee Mosteller. 1977. *Trapunto and Other Forms of Raised Quilting.* New York: Scribner.
Morris, Christine. 2009. "Configuring the Individual: Bodies of Figurines in Minoan Crete." In *Archaeologies of Cult: Essays on Ritual and Cult in Crete in Honor of Geraldine C. Gesell* (Hesperia: The Journal of the American School of Classical Studies at Athens Suppl. 42), eds. Anna Lucia D'Agata and Aleydis Van de Moortel. Princeton: American School of Classical Studies at Athens, 179–187.
Morris, Sarah P. 1992. *Daidalos and the Origins of Greek Art.* Princeton: Princeton University Press.
Morrison, John Sinclair, John F. Coates and N. Boris Rankov. 2000. *The Athenian Trireme: The History and Reconstruction of an Ancient Greek Warship*, 2nd ed. Cambridge: Cambridge University Press.
Moulherat, Christophe and Youlie Spantidaki. 2007. "Preliminary Results from the Textiles Discovered in Santorini." In *NESAT IX, Archäologische Textilefunde, Braunswald, 18–21 May 2005*, eds. Antoinette Rast-Eicher and Renata Windler. Ennenda: ArcheoTex, 49–52.
Moulherat, Christophe and Youlie Spantidaki. 2008. "Première attestation de la laine sur le site protohistorique d'Akrotiri à Théra." In *Purpureae vestes II. Vestidos, textiles y tintes: Estudios sobre la producción de bienes de consumo en la Antigüedad. Actas del II symposium internacional sobre textiles y tintes del Mediterráneo en el mundo antiguo (Atenas, 24 al 26 de noviembre, 2005)*, eds. Carmen Alfaro and Lilian Karali. València Universitat de València. 37–42.
Moulherat, Christophe and Youlie Spantidaki. 2009a. "Cloth from Kastelli, Chania." *Arachne* 3, 8–15.
Moulherat, Christophe and Youlie Spantidaki. 2009b. "Archaeological Textiles from Salamis: A Preliminary Presentation." *Arachne* 3, 8–15.
Murra, John V. 1989. "Cloth and Its Function in the Inca State." In *Cloth and Human Experience*, eds. Annette B. Weiner and Jane Schneider. Washington, D.C.: Smithsonian, 275–302.
Murra, John V. and Craig Morris. 1976. "Dynastic Oral Tradition, Administrative Records and Archaeology in the Andes." *World Archaeology* 7, 270–279.
Murray, Suzanne Peterson. 2004. "Reconsidering the Room of the Ladies at Akrotiri." In *ΧΑΡΙΣ*, 101–130.
Müller, Kurt 1930. *Tiryns III: Die Architektur der Burg des Palastes.* Augsburg: Benno Filser.
Mylonas, George. 1966. *Mycenae and the Mycenaean Age.* Princeton: Princeton University Press.
Mylonas, George. 1973. Ο ταφικός κύκλος Β των Μυκηνών. Athens: Archaeological Society.
Naville, Edouard. 1906. *The Temple of Deir el Bahari* 5, London: Egypt Exploration Fund.

Nesbitt, Mark. 1995. "Plants and People in Ancient Anatolia." *Biblical Archaeologist* 58, 68–81.
Newberry, Percy E. n.d. *El Bersheh* I. London: Egypt Exploration Society.
Niemeier, Barbara and Wof-Dietrich Niemeier. 2002. "The Frescoes in the Middle Bronze Age Palace." In *Tel Kabri: The 1986-1993 Excavation Seasons* (Monograph Series 20), eds. Aharon Kempinski, Na'ama Scheftelowitz and Ronit Oren. Tel Aviv: Emery and Claire Yass Publications in Archaeology, Institute of Archaeology, Tel Aviv University, 254-298.
Niemeier, Wolf-Dietrich. 1985. *Die Palaststilkeramik von Knossos: Stil, Chronologie und historischer Kontext*. Berlin: Gebr. Mann Verlag.
Niemeier, Wolf-Dietrich. 1989. "Zur Ikonographie von Gottheiten und Adoranten in den Kultszenen auf minoischen und mykenischen Siegeln." In *Fragen und Probleme der bronzezeitlichen agäischen Glyptik*, CMS 3, 163–186.
Niemeier, Wolf-Dietrich. 1996. "On the Origin of Mycenaean Painted Plaster Floors." In *Atti e memorie del secondo congresso internazionale di micenologia Roma-Napoli, 14–20 ottobre 1991*, eds. Ernesto De Miro, Louis Godart and Anna Sacconi. Rome: Gruppo editoriale internazionale, 1249–1254.
Nosch, Marie-Louise. 2000. "The Organization of the Mycenaean Textile Industry." Ph.D. dissertation, University of Salzburg.
Nosch, Marie-Louise. 2001a. "The Geography of the *ta-ra-si-ja*." *Aegean Archaeology* 4, 27–44.
Nosch, Marie-Louise. 2001b. "The Textile Industry at Thebes in the Light of the Textile Industries at Pylos and Knossos." In *Festschrift in Honour of Antonin Bartoněk* (Studia Minora Facultatis Philosophica Universitatis Brunensis 6), eds. Irena Radová and Katarina Václavková-Petrovicová. Brno: Masarykova Univerzita, 177–189.
Nosch, Marie-Louise. 2003. "The Women at Work in the Linear B Tablets." In *Gender, Cult, and Culture in the Ancient World from Mycenae to Byzantium*, eds. Agnete Strömberg and Lena Larsson Lovén (Studies in Mediterranean Archaeology Pocketbook 166). Göteborg: Paul Åströms Förlag, 12–26.
Nosch, Marie-Louise. 2012. "From Texts to Textiles in the Aegean Bronze Age." In *KOSMOS*, 43–52.
Nosch, Marie-Louise and Massimo Perna. 2001. "Cloth in the Cult." In *POTNIA*, 471–477.
O'Brien, Joan B. 1993. *The Transformations of Hera: A Study of Ritual, Hero, and the Goddess in the Iliad*. Lanham: Rowman and Littlefield.
O'Connor, David. 2009. *Abydos: Egypt's First Pharaohs and the Cult of Osiris*. London: Thames and Hudson.
Oren, Eliezer, Jean-Pierre Olivier, Philip P. Betancourt, G. H. Meyer and J. Yellin. 1996. "A Minoan Graffito from Tel Haror (Negev, Israel)." *Cretan Studies* 5, 91–118.
Page, Denys L. 1963. *History and the Homeric Iliad*. Berkeley: University of California Press.
Palaima, Thomas G. 1991. "Maritime Matters in the Linear B Tablets." In *Thalassa: L'Egée préhistorique et la mer. Actes de la troisième rencontre égéenne internationale de l'Université de Liège, Station de recherches sous-marines et océanographiques (StaReSo), Calvi, Corse, 23-25 avril 1990* (Aegaeum 7), eds. Robert Laffineur and Lucien Basch. Liège: Université de Liège, Histoire de l'art et archéologie de la Grèce antique, 273–310.
Palaima, Thomas G. 1995. "The Nature of the Mycenaean *Wanax*: Non-Indo European Origins and Priestly Functions." In *The Role of the Ruler in the Prehistoric Aegean: Proceedings of a Panel Discussion Presented at the Annual Meeting of the Archaeological Institute of America, New Orleans, Louisiana, 28 December 1992, with additions* (Aegaeum 11), ed. Paul Rehak. Liège: Université de Liège, Histoire de l'art et archéologie de la Grèce antique; Austin: University of Texas at Austin, Program in Aegean Scripts and Prehistory, 119–139.
Palaima, Thomas G. 1997. "Potter and Fuller: The Royal Craftsmen." In *TEXNH*, 407–412.
Palyvou, Clairy. 2005. *Akrotiri Thera: An Architecture of Affluence 3,500 Years Old* (Prehistory Monograph 15). Philadelphia: Institute for Aegean Prehistory Academic Press.
Panagiotakopulu, Eva. 2000. "Butterflies, Flowers and Aegean Iconography: A Story about Silk and Cotton." In *WPT*, 585–592.
Panagiotakopulu, Eva, Philip Buckland, Peter Day, Christos Doumas, Anaya Sarpaki and P. Skidmore. 1997. "A Lepidopterous Cocoon from Thera and Evidence for Silk in the Aegean Bronze Age." *Antiquity* 71, 420–429.
Papadimitriou, Alkestis, Ulrich Thaler and Joseph Maran. 2015. "Bearing the Pomegranite Bearer: A New Wall Painting Scene from Tiryns," in *Mycenaean Paintings in Context: New Discoveries and Old Finds Reconsidered. Proceedings of an International Conference on Mycenaean Wall Paintings in Context, Held 10-13 February 2011 at the National Hellenic Research Foundation* (Meletemata 72), eds. Hariclia Brecoulaki, Jack L. Davis, and Sharon R. Stocker. Athens: National Hellenic Research Foundation, 173–211.

Papadopoulos, John K. 2007. "The Art of Antiquity: Piet de Jong and the Athenian Agora." In *The Art of Antiquity: Piet de Jong and the Athenian Agora*, ed. John K. Papadopoulos. Princeton: American School of Classical Studies at Athens, 1–32.

Papaodysseus, Constantin, Mihalis Panagopoulos, Panayiotis Rousopoulos, G. Galanopoulos and Christos Doumas. 2008. "Geometric Templates Used in the Akrotiri (Thera) Wall-Paintings." *Antiquity* 82, 401–408.

Papazoglou-Manioudaki, Lena. 1990. "Orchomenos." In *Troy, Mycenae, Tiryns, Orchomenos. Heinrich Schliemann: The 100th Anniversary of His Death*, ed. Katie Demakopoulou. Athens: Ministry of Culture, Greek Committee ICOM; Berlin: Museum für Ur- und Frühgeschichte, 130–136.

Parsons, Jeffrey R. and Mary Parsons. 1990. *Maguey Utilization in Highland Central Mexico* (Anthropological Papers 82). Ann Arbor: University of Michigan Museum.

Parsons, Mary. 1975. "The Distribution of Late Postclassic Spindle Whorls in the Valley of Mexico." *American Antiquity* 40, 209–215.

Payne, Sebastian. 1973. "Kill-Off Patterns in Sheep and Goats: The Mandibles from Aşvan Kale." *Anatolian Studies* 23, 281–303.

Pendlebury, John D. S., Thomas E. Peet, C. L. Woolley and Henri Frankfort. 1951. *The City of Akhenaten III, The Central City and the Official Quarters: The Excavations at Tell el-Amarna during the Seasons 1926-1927 and 1931-1936*. London: Egypt Exploration Society.

Pentelia, Maria C. 1993. "Spinning and Weaving: Ideas of Domestic Order in Homer." *American Journal of Philology* 114, 493–501.

Perlès, Catherine. 2001. *The Early Neolithic in Greece: The First Farming Communities in Europe*. Cambridge and New York: Cambridge University Press.

Persson, Axel W. 1931. *The Royal Tombs at Dendra near Midea*. Lund: C. W. K. Gleerup.

Peterson, Suzanne. 1981a. "Wall Paintings in the Aegean Bronze Age: The Procession Frescoes." Ph.D. dissertation, University of Minnesota.

Peterson, Suzanne. 1981b. "A Costuming Scene from the Room of the Ladies on Thera." *American Journal of Archaeology* 85, 211 (abstract).

Petrie, W. M. Flinders, Francis L. Griffith and Percy E. Newberry. 1890. *Kahun, Gurob, and Hawara*. London: K. Paul, Trench, Trübner.

Phelps, William, Yannos Lolos and Yannis Vichos, eds. 1999. *The Point Iria Wreck: Interconnections in the Mediterranean ca. 1200 B.C. Proceedings of the International Conference, Island of Spetses, 19 September 1998*. Athens: Hellenic Institute of Marine Archaeology.

Platon, Lefteris. 2002. "The Political and Cultural Influence of the Zakros Palace on Nearby Sites and in a Wider Context." In *Monuments of Minos*, 145–156.

Platon, Nicholas. 1964. "Ἀνασκαφαί Ζάκρου." Πρακτικά τῆς ἐν Ἀθήναις Ἀρχαιολογικῆς Ἑταιρείας, 142–168.

Platon, Nicholas. 1965. "Ἀνασκαφαί Ζάκρου." Πρακτικά τῆς ἐν Ἀθήναις Ἀρχαιολογικῆς Ἑταιρείας, 187–224.

Popham, Mervyn. 1967. "Late Minoan Pottery, a Summary." *Annual of the British School at Athens* 62, 337–351.

Popham, Mervyn. 1970. *The Destruction of the Palace at Knossos*. Göteborg: Paul Åströms Förlag.

Popham, Mervyn 1984. *The Minoan Unexplored Mansion at Knossos* (*Annual of the British School at Athens* Suppl. 17). London: Thames & Hudson.

Porter, Ray. 2000. "The Flora of the Theran Wall Paintings: Living Plants and Motifs – Sea Lily, Crocus, Iris and Ivy." In *WPT*, 603–630.

Porter, Bertha and Rosalind L. B. Moss. 1960–. *Topographical Bibliography of Ancient Egyptian Hieroglyphic Texts, Reliefs, and Paintings*, 2nd ed. Oxford: Clarendon.

Poursat, Jean-Claude. 1977. *Catalogue des ivoires mycéniens du Musée National d'Athènes*. Athens: École française d'Athènes.

Poursat, Jean-Claude, Chara Prokopiou and René Treuil. 2000. "Οι οικιακές δραστηριότητες στο Κυαρτίερ Μυ. Η άλεση και η υφαντική." In *Pepragmena H' Diethnous Kritologikou Synedriou, Irakleio, 9-14 Septemvriou 1996, vol. A3: Proïstoriki kai Archaia Elliniki Periodos*, eds. Alexandra Karetsou, Theocharis Detorakis and Alexis Kalokairinos. Herakleion: Etairia Kritikon Istorikon Meleton, 99–114.

Preziosi, Donald and Louise A. Hitchcock. 1999. *Aegean Art and Architecture*. Oxford: Oxford University Press.

Pulak, Cemal. 1998. "The Uluburun Shipwreck: An Overview." *International Journal of Nautical Archaeology* 27, 188–224.

Pullen, Daniel, ed. 2010. *Political Economies of the Aegean Bronze Age. Papers from the Langford Conference, Florida State University, Tallahassee, 22-24 February 2007.* Oxford: Oxbow Books.

Quibell, J. E. 1908. *Catalogue général des antiquités égyptiennes: Tomb of Yuaa and Thuiu.* Cairo: Le Caire.

Ramberg, Bjørn and Kristin Gjesdal. 2013. "Hermeneutics." In *The Stanford Encyclopedia of Philosophy* (Summer 2013), ed. Edward N. Zalta, http://plato.stanford.edu/archives/sum2013/entries/hermeneutics.

Reeves, Nicholas. 2007. *The Complete Tutankhamun. The King, The Tomb, The Royal Treasure.* London: Thames & Hudson.

Regensteiner, Else. 1970. *The Art of Weaving.* New York: Van Nostrand Reinhold.

Rehak, Paul. 1984. "New Observations on the Mycenaean 'Warrior Goddess.'" *Archäologischer Anzeiger*, 535-545.

Rehak, Paul. 1992. "Tradition and Innovation in the Fresco from Room 31 in the 'Cult Center' at Mycenae." In *EIKON: Aegean Bronze Age Iconography* (Aegaeum 8), eds. Robert Laffineur and Janice Crowley. Liège: Université de Liège, Histoire de l'art et archéologie de la Grèce antique, 39-62.

Rehak, Paul. 1996. "Aegean Breechcloths, Kilts and the Keftiu Paintings." *American Journal of Archaeology* 100, 35-51.

Rehak, Paul. 1997. "The Role of Religious Painting in the Function of the Minoan Villa: The Case of Ayia Triadha." In *The Function of the 'Minoan Villa'* (Skrifter utgivna av Svenska Institutet i Athen, 40), ed. Robin Hägg. Stockholm: Svenska Institutet i Athen; Paul Åströms Förlag, 163-174.

Rehak, Paul. 1998. "The Construction of Gender in Late Bronze Age Aegean Art: A Prolegomenon." In *Redefining Archaeology: Feminist Perspectives*, eds. Mary Casey, Denise Donlon, Jeanette Hope and Sharon Wellfare. Canberra: ANH Publications, 191-198.

Rehak, Paul. 1999. "The Mycenaean 'Warrior Goddess' Revisited." In *POLEMOS: Le contexte guerrier en Égée á l'âge du Bronze. Actes de la 7e Rencontre égéenne internationale Université de Liège, 14-17 avril 1998* (Aegaeum 19), ed. Robert Laffineur. Liège: *Université de Liège*, Histoire de l'art et archèologie de la Grèce antique; Austin: University of Texas at Austin, Program in Aegean Scripts and Prehistory, 227-239.

Rehak, Paul. 2002. "Imag(in)ing a Women's World in Bronze Age Greece: The Frescoes from Xeste 3 at Akrotiri, Thera." In *Among Women: From the Homosocial to the Homoerotic in the Ancient World*, eds. N. Rabinowitz and L. Auanger. Austin: University of Texas Press, 34-59.

Rehak, Paul. 2004. "Crocus Costumes in Aegean Art." In *ΧΑΡΙΣ*, 85-100.

Rehak, Paul. 2007. "Children's Work: Girls as Acolytes in Aegean Ritual and Cult" (ed. John G. Younger). In *Coming of Age: Constructions of Childhood in the Ancient World*, eds. Jeremy Rutter and Ada Cohen. Princeton: American School of Classical Studies at Athens, 205-225.

Rethemiotakis, Giorgos. 2001. *Minoan Clay Figures and Figurines from the Neopalatial to the Subminoan Period.* Athens: Archaeological Society at Athens.

Rethemiotakis, Giorgos. 2002. "Evidence on Social and Economic Changes at Galatas and Pediada in the New-Palace Period." In *Monuments of Minos*, 55-69.

Reusch, Helga. 1945. *Die kretisch-mykenische Textilornämentik,* Ph.D. dissertation, Friedrich-Wilhelms, University of Berlin.

Reusch, Helga. 1956. *Die zeichnerische Rekonstruktion des Frauenfrises im boötischen Theben.* Berlin: Akademie-Verlag.

Ribichini, Sergio and Paolo Xella. 1985. *La terminologia dei tessili di Ugarit* (Collezioni dei Studi Fenici 20), Rome: Consiglio nazionale delle ricerche.

Riefstahl, Elizabeth. 1944. *Patterned Textiles in Pharaonic Egypt.* New York: Brooklyn Institute of Arts and Sciences.

Robins, Gay. 1994. *Proportion and Style in Ancient Egyptian Art.* Austin: University of Texas Press.

Robkin, Ann Lou H. 1979. "The Agricultural Year, the Commodity SA and the Linen Industry of Mycenaean Pylos." *American Journal of Archaeology* 83, 469-474.

Rodenwaldt, Gerhart. 1911. "Fragmente mykenischer Wandgemälde," *Mitteilungen des Deutschen Archäologischen Instituts, Athenische Abteilung* 36, 221-250.

Rodenwaldt, Gerhart. 1919. "Mykenische Studien. I: Die Fussböden des Megarons von Mykenai." *Jahrbuch des Deutschen Archäologischen Instituts* 34, 87-106.

Rodenwaldt, Gerhart. 1921. *Der Fries des Megarons von Mykenai.* Halle: Niemeyer.

Rodenwaldt, Gerhart. 1926. "Mycenae: Report of the Excavations of the British School at Athens, 1921-1923." *Gnomon* 2, 241-247.

Rodenwaldt, Gerhart. 1941. "Mykenische Miscellen." In *Epitymbion Chrestou Tsountas* (Epimetron, Ethnological and Philological Society of Thrace). Athens: Athenai, s.n., 434-437.

Ruscillo, Deborah. 2006. "Faunal Remains and Murex Dye Production." In *Kommos: An Excavation on the South Coast of Crete by the University of Toronto under the Auspices of the American School of Classical Studies at Athens*, vol. V. *The Monumental Minoan Buildings at Kommos*, eds. Joseph W. Shaw and Maria C. Shaw. Princeton and Oxford: Princeton University Press, 776–844.

Russmann, Edna R. 2000. "The Egyptian Character of Certain Egyptian Painting Techniques." In *WPT*, 71–76.

Rutter, Jeremy. 2005. "Southern Triangles Revisited: Lakonia, Messenia and Crete in the 14th–12th Centuries BC." In *Ariadne's Threads: Connections between Crete and the Greek Mainland in Late Minoan III (LM IIIA2 to LM IIIC). Proceedings of the International Workshop Held at Athens, Scuola Archeologica Italiana, 5–6 April 2003*, eds. Anna Lucia D'Agata and Jennifer Moody. Athens: Scuola Archeologica Italiana di Atene, 16–64.

Ryder, Michael L. 1969. "Changes in the Fleece of Sheep Following Domestication." In *The Domestication and Exploitation of Plants and Animals*, eds. Peter J. Ucko and G. W. Dimbleby. Chicago: Aldine, 495–521.

Ryder, Michael L. 1983. *Sheep and Man*. London: Duckworth.

Ryder, Michael L. 1993. "Sheep and Goat Industry with Particular Reference to Textile Fibre and Milk Production." *Bulletin on Sumerian Agriculture* 7, 9–32.

Sakellarakis, Yannis and Efi Sapouna-Sakellaraki. 1997. *Archanes: Minoan Crete in a New Light*. Athens: Ammos Publications.

Sakellariou, Agnes. 1971. "Scène de bataille sur un vase mycénien en pierre?" *Revue archéologique* 1, 3–14.

Saleh, Mohammad. 1977. *Three Old-Kingdom Tombs at Thebes: The Tomb of Unas-Ankh No. 413, The Tomb of Khenty No. 405, The Tomb of Ihy No. 186* (Archäologische Veröffentlichungen 14). Mainz am Rhein: Philipp von Zabern.

Saliaka, Evi. 2008. "No. 159. Fragment of Wall Painting." In *From the Land of the Labyrinth: Minoan Crete, 3000-1100 B.C.*, eds. Maria Andreadaki-Vlazaki, Giorgos Rethemiotakis and Nota Dimopoulou-Rethemiotaki. New York: Alexander S. Onassis Public Benefit Foundation, 198–199.

Salmon-Minotte, Jack and Robert R. Franck. 2005. "Flax." In *Bast and Other Plant Fibres*, ed. Robert R. Franck. Cambridge: Woodhead, 94–175.

Sandars, Nancy K. 1985. *The Sea Peoples: Warriors of the Ancient Mediterranean*. London: Thames and Hudson.

Sapouna-Sakellaraki, Efi. 1971. Μινωικόν ζῶμα (*Bibliothēkē tēs en Athēnais Arhaiologikēs Etaireias* 71). Athens: Archaeological Society.

Sarpaki, Anaya. 2009. "Knossos, Crete: Invaders, 'Sea Goers,' or Previously 'Invisible,' the Neolithic Plant Economy Appears Fully-Fledged in 9,000 BP." In *From Foragers to Farmers: Papers in Honour of Gordon C. Hillman*, eds. Andrew S. Fairbairn and Ehud Weiss. Oxford: Oxbow Books, 220–234.

Säve-Söderbergh, Torgny. 1957. *Four Eighteenth Dynasty Tombs*. Oxford: Griffith Institute at the University Press.

Schallin, Ann-Louise and Petra Pakkanen, eds. 2009. *Encounters with Mycenaean Figures and Figurines: Papers Presented at a Seminar at the Swedish Institute at Athens, 27–29 April 2001* (Skrifter utgivna av Svenska Institutet i Athen 8°, 20). Stockholm: Svenska Institutet i Athen.

Schiaparelli, Erne+sto. 1927. *La Tomba intatta dell' Architetto Cha. Relazione sui Lavori della Missione archeological in Egitto* 2. Turin: Giovanni Chiantore.

Schick, Tamar. 1988. "Cordage, Basketry, and Fabrics," *Atiquot* 18, 31–43.

Schliemann, Heinrich. [1880] 1976. *Mycenae: A Narrative of Researches and Discoveries at Mycenae and Tiryns*, repr. New York: Arno.

Schliemann, Heinrich. 1881. *Orchomenos. Bericht über meine Ausgrabungen im böotischen Orchomenos*. Leipzig: F. A. Brockhaus.

Schoep, Ilse. 2002. *The Administration of Neopalatial Crete: A Critical Assessment of the Linear A Tablets and Their Role in the Administrative Process*. Salamanca: Ediciones Universidad Salamanca.

Schoep, Ilse. 2006. "Looking Beyond the First Palaces: Elites and the Agency of Power in EM III–MM II Crete." *American Journal of Archaeology* 110, 37–64.

Schofield, Louise. 2007. *The Mycenaeans*. Los Angeles: J. Paul Getty Museum.

Seager, Richard B. 1910. *Excavations on the Island of Pseira, Crete*. Philadelphia: University of Pennsylvania Museum.

Shank, Elizabeth. 2012. "The Jewelry Worn by the Procession of Mature Women from Xeste 3, Akrotiri." In *KOSMOS*, 559–565.

Shaw, Joseph W. 1984. "Excavations at Kommos (Crete) during 1982–1983." *Hesperia: The Journal of the American School of Classical Studies at Athens* 53, 251–287.

Shaw, Joseph, W. 2009. *Minoan Architecture: Materials and Techniques* (Studi di Archeologia Cretese 7). Padua: Centro di Archeologia Cretese and Bottega d'Erasmo.

Shaw, Joseph, W. 2015. *Elite Minoan Architecture: Its Development at Knossos, Phaistos, and Malia.* Philadelphia: Institute for Aegean Prehistory Academic Press.

Shaw, Joseph W. and Maria C. Shaw, eds. 2006. *Kommos: An Excavation on the South Coast of Crete by the University of Toronto under the Auspices of the American School of Classical Studies at Athens V. The Monumental Minoan Buildings at Kommos.* Princeton and Oxford, Princeton University Press.

Shaw, Maria C. 1970. "Ceiling Patterns from the Tomb of Hepzefa." *American Journal of Archaeology* 7, 25–30.

Shaw, Maria C. 1978. "A Minoan Fresco from Katsamba." *American Journal of Archaeology* 82, 27–34.

Shaw, Maria C. 1980. "Painted 'Ikria' at Mycenae?" *American Journal of Archaeology* 84, 167–179.

Shaw, Maria C. 1982. "Ship Cabins of the Bronze Age Aegean." *International Journal of Nautical Archaeology and Underwater Exploration* 11, 53–58.

Shaw, Maria C. 1993. "The Aegean Garden." *American Journal of Archaeology* 97, 661–685.

Shaw, Maria C. 1995. "Bull Leaping Frescoes at Knossos and their Influence on the Tell el-Dab'a Murals." In *Trade, Power and Cultural Exchange: Hyksos Egypt and the Eastern Mediterranean World 1800-1500 B.C. An International Symposium, November 3, 1993, Metropolitan Museum of Art* (Ägypten und Levante 5), ed. Manfred Bietak. Vienna: Verlag der Österreichischen Akademie der Wissenschaften, 91–120.

Shaw, Maria C. 1997. "Aegean Sponsors and Artists: Reflections of Their Roles in the Patterns of Distribution of Themes and Representational Conventions in the Murals." In *TEXNH*, 481–504.

Shaw, Maria C. 1998. "The Painted Plaster Reliefs from Pseira." In *Pseira II: Building AC (the "Shrine") and Other Buildings in Area A,* eds. Philip P. Betancourt and Costis Davaras. Philadelphia: University of Pennsylvania Museum, 55–76.

Shaw, Maria C. 2000a. "Anatomy and Execution of Complex Minoan Textile Patterns in the Procession Fresco from Knossos." In Κρήτη–Αίγυπτος: πολιτισμικοί δεσμοί τριών χιλιετιών. Μελέτες, ed. Alexandra Karetsou. Athens: Hypourgeio Politismou, 52–63.

Shaw, Maria C. 2000b. "Sea Voyages: The Fleet Fresco from Thera, and the Punt Reliefs from Egypt." In *WPT*, 267–282.

Shaw, Maria C. 2001. "Symbols of Naval Power at the Palace at Pylos: The Evidence from the Frescoes." In *ITHAKI: Festschrift für Jörg Schäfer zum 75. Geburtstag am 25. April 2001,* eds. Stephanie Böhm and Klaus-Valtin von Eickstedt. Würzburg: Ergon, 37–43.

Shaw, Maria C. 2003. "Grids and Other Drafting Devices in Minoan and Other Aegean Wall Painting: A Comparative Analysis Including Egypt." In *METRON*, 179–189.

Shaw, Maria C. 2005. "The Painted Pavilion of the 'Caravanserai' at Knossos." In *Aegean Wall Painting*, 229–240.

Shaw, Maria C. 2006. "Plasters from the Monumental Minoan Buildings: Evidence for Painted Decoration, Architectural Appearance, and Archaeological Event." In *Kommos: An Excavation on the South Coast of Crete by the University of Toronto under the Auspices of the American School of Classical Studies at Athens,* vol. V. *The Monumental Minoan Buildings at Kommos,* eds. Joseph W. Shaw and Maria C. Shaw. Princeton: Princeton University Press, 117–260.

Shaw, Maria C. 2010. "A Fresco of a Textile Pattern at Pylos: The Importation of a Minoan Artistic Technique." In *Cretan Offerings: Studies in Honour of Peter Warren* (British School at Athens Studies 18), ed. Olga Krzyszkowska. London: British School at Athens, 315–320.

Shaw, Maria C. 2012a. "New Light on the Labyrinth Fresco from the Palace at Knossos." *Annual of the British School at Athens* 107, 143–159.

Shaw, Maria C. 2012b. "Shields Made of Cloth? Interpreting a Wall Painting in the Mycenaean Palace at Pylos." In *KOSMOS*, 731–737.

Shaw, Maria C. and Kessa Laxton. 2002. "Minoan and Mycenaean Wall Hangings. New Light from a Wall Painting at Ayia Triada." *Creta Antica* 3, 93–104.

Shelmerdine, Cynthia W. 1985. *The Perfume Industry of Mycenaean Pylos.* Göteborg: Paul Åströms Förlag.

Shelmerdine, Cynthia W. 1987. "Industrial Activity at Pylos." In *Tractata Mycenaea. Proceedings of the 8th International Colloquium on Mycenaean Studies, Held in Ohrid, 15-20 September 1985,* eds. Petar Ilievski and Ljiljana Crepajac. Skopje: Macedonian Academy of Arts and Sciences, 333–342.

Shelmerdine, Cynthia W. ed. 2008. *The Cambridge Companion to the Aegean Bronze Age.* Cambridge: Cambridge University Press.

Shelmerdine, Cynthia W. 2013. "Crafts, Specialists, and Markets in Mycenaean Greece: Economic Interplay among Households and States," *American Journal of Archaeology* 117, 447–452.

Sherratt, Andrew. 1983. "The Secondary Products Revolution of Animals in the Old World." *World Archaeology* 15, 90–104.

Smith, Michael E. and Kenneth G. Hirth. 1988. "The Development of Prehispanic Cotton-Spinning Technology in Western Morelos, Mexico." *Journal of Field Archaeology* 15, 349–358.

Smith, William Stevenson. 1958. *The Art and Architecture of Ancient Egypt.* Harmondsworth: Penguin Books.

Smith, William Stevenson. 1965. *Interconnections in the Ancient Near East: A Study of the Relationships between the Arts of Egypt, the Aegean, and Western Asia.* New Haven: Yale University Press.

Snijder, G. A. S. 1936. *Kretische Kunst.* Berlin: Gebr. Mann.

Sollberger, Edmond. 1986. *Administrative Texts Chiefly Concerning Textiles* (L. 2752). Rome: Missione archeologica italiana in Siria.

Spantidaki, Stella. 2008. "Preliminary Results of the Reconstruction of Theran Textiles." In *Purpureae vestes II. Vestidos, textiles y tintes: Estudios sobre la producción de bienes de consumo en la Antigüedad. Actas del II symposium internacional sobre textiles y tintes del Mediterráneo en el mundo antiguo (Atenas, 24 al 26 de noviembre, 2005)*, eds. Carmen Alfaro and Lilian Karali. València: Universitat de València, 43–47.

Spantidaki, Youlie and Christophe Moulherat. 2012. "Greece." In *Textiles*, 185–200.

Spondylis, Elias. 2012. "A Minoan Shipwreck off Laconia." *Enalia* 11, 6–7.

Stein, Gil and M. James Blackman. 1993. "Specialized Craft Production in Mesopotamia." In *Economic Aspects of Water Management in the Prehispanic New World*, eds. Vernon L. Scarborough and Barry L. Isaac. Greenwich, Conn.: JAI Press, 29–59.

Stokstad, Marilyn and Michael Cothren. 2013. *Art History*, 5th ed. Boston: Pearson.

Strasser, Thomas F. 2010. "Location and Perspective in the Theran Flotilla Fresco." *Journal of Mediterranean Archaeology* 23, 3–26.

Strasser, Thomas F. and Anne P. Chapin. 2014. "Geological Formations in the Flotilla Fresco from Akrotiri." In *PHYSIS: Natural Environment and Human Interaction in the Prehistoric Aegean, 14th International Aegean Conference, Institut National d'Histoire de l'Art (INHA), University of Paris, 1 Panthéon-Sorbonne, 11-14 December 2012* (Aegaeum 37), eds. Gilles Touchais, Robert Laffineur, S. Andreou, Hara Procopiou, Francois Rougemont, E. Fouache. Liège: Université de Liège.

Stubbings, Frank. 1973. *Prehistoric Greece.* New York: John Day Co.

Stürmer, Veit. 2001. "'Naturkulträme' auf Kreta und Thera: Ausstattung, Definition und Funktion." In *POTNIA*, 69–75.

Swift, Ellen. 2009. *Style and Function in Roman Decoration: Living with Objects and Interiors.* Farnham, England; Burlington, Vt: Ashgate.

Szarzynska, Krystyna. 1988. "Records of Garments and Cloth in Archaic Uruk/Warka." *Altorientalische Forschungen* 15, 220–230.

Tartaron, Thomas. 2013. *Maritime Networks in the Mycenaean World.* Cambridge: Cambridge University Press.

Taylour, William. 1969. "Mycenae, 1968." *Antiquity: A Quarterly Review of Archaeology* 43, 91–97.

Televantou, Christina A. 1982. "Η γυναικεία ενδυμασία στην προϊστορική Θήρα." *Archaiologike Ephemeris* 113–135.

Televantou, Christina A. 1987. "Ἐγχάρακτες ἀπεικωνίσεις πλίων σε Θηραϊκή τοιχογραφία." *Athens Annals of Archaeology* 20, 115–122.

Televantou, Christina A. 1994. Ακρωτήρι Θήρας: Οι Τοιχογραφίες της Δυτικής Οικίας (Βιβλιοθήκη της εν Αθήναις Αρχαιολογικής Εταιρείας 143). Athens: Archaeological Society.

Televantou, Christina A. 1994–1995. "Ἐγχάρακτη παράσταση ἀκροβάτη σε Θηραϊκή τοιχογραφία," *Archaiologikon Deltion* 49–50, 13–22.

Televantou, Christina A. 2000. "Plates 1–3 (fold-out)." In *WPT*, vol. 3.

Terrace, Edward L. B. 1968. *Egyptian Paintings of the Middle Kingdom.* New York: George Braziller.

Thaler, Ulrich. 2012. "Going Round in Circles: Anmerkungen zur Bewegungsrichtung in mykenischen Palastmegara." In *Bild-Raum-Handlung: Perspektiven der Archäologie*, eds. Ortwin Dally, Susanne Moraw and Hauke Ziemssen. Berlin: De Gruyter, 189–214.

Todaro, Simona V. and Serena Di Tonto. 2008. "The Neolithic Settlement of Phaistos Revisited: Evidence for Ceremonial Activity on the Eve of the Bronze Age." In *Escaping the Labyrinth: The Cretan Neolithic in Context* (Sheffield Studies in Aegean Archaeology 8), eds. Valasia Isaakidou and Peter Tomkins. Oxford: Oxbow Books, 177–190.

Todaro, Simona V. and Serena Di Tonto. 2013. *The Phaistos Hills Before the Palace: A Contextual Reappraisal* (*Praehistorica Mediterranea* 5). Milan: Polimetrica.

Tomkins, Peter D. 2008. "Time, Space and the Reinvention of the Cretan Neolithic." In *Escaping the Labyrinth: The Cretan Neolithic in Context* (Sheffield Studies in Aegean Archaeology 8), eds. Valasia Isaakidou and Peter Tomkins. Oxford: Oxbow Books, 21–48.

Tomkins, Peter D. 2012. "Beyond the Horizon: Reconsidering the Genesis and Function of the 'First Palace' at Knossos (Final Neolithic IV–Middle Minoan IB)." In *Back to the Beginning: Reassessing Social and Political Complexity on Crete during the Early and Middle Bronze Age*, eds. Ilse Schoep, Peter Tomkins and Jan Driessen. Oxford: Oxbow Books, 32–80.

Trigger, Bruce G. 2006. *A History of Archaeological Thought*, 2nd ed. New York: Cambridge University Press.

Trnka, Edith. 2007. "Similarities and Distinctions of Minoan and Mycenaean Textiles." In *Ancient Textiles*, 127–129.

True, Marion. 2000. "The Role of Formal Decorative Patterns in the Wall Paintings of Thera." In *WPT*, 345–358.

Tsountas, Christos. 1886. "Ἀνασκαφαί Μυκηνῶν." *Praktika tes en Athenais Archaiologikes Etaireias*, 59–79.

Tsountas, Christos. 1887a. "Ἀρχαιότετες ἐκ Μυκηνῶν." *Archaiologike Ephemeris* 155–172.

Tsountas, Christos. 1887b. Μυκῆναι καὶ Μυκηναῖος Πολιτισμός. Athens.

Tzachili, Iris. 1990. "All Important yet Elusive: Looking for Evidence of Cloth-Making at Akrotiri." In *TAW II*, 380–389.

Tzachili, Iris. 1997. Υφαντική και Υφάντρες στο Προϊστορικό Αιγαίο 2000–1000 π. Χ./*Uphantikī kai uphantres sto proïstoriko Aigaio: 2000–1000 p. Ch.* Herakleion: Panepistīmiakes Ekdoseis Krītīs.

Tzachili, Iris. 1999. "Before Sailing: The Making of Sails in the Second Millennium B.C." In *Meletemata*, 857–862.

Tzachili, Iris. 2005. "*Anthodokoi Talaroi*: The Baskets of the Crocus Gatherers." In *Aegean Wall Painting*, 113–117.

Tzachili, Iris. 2007. "Weaving at Akrotiri, Thera: Defining Cloth-Making Activities as Social Process in a Late Bronze Age Aegean Town." In *Ancient Textiles*, 190–196.

Tzachili, Iris. 2012. "The Myth of Arachne and Weaving in Lydia." In *Textiles and Dress in Greece and the Roman East: A Technological and Social Approach. Proceedings of a Conference Held at the Department of History, Archaeology and Cultural Resources Management of the University of Peloponnese in Kalamata in Collaboration with the Department of History and Archaeology of the University of Crete on March 18-19, 2011*, eds. Iris Tzachili and Eleni Zimi. Athens: Ta Pragmata, 131–142.

Unruh, Julie. 2007. "Ancient Textile Evidence in Soil Structures at the Agora Excavations in Athens, Greece." In *Ancient Textiles*, 167–172.

Van Damme, Trevor. 2012. "Reviewing the Evidence for a Bronze Age Silk Industry." In *KOSMOS*, 163–169.

Van Dyke, Ruth M. and Susan E. Alcock, eds. 2003. *Archaeologies of Memory*. Oxford, Blackwell Publishing.

Van Zeist, Willem and J. A. H. Bakker-Heeres. 1975. "Evidence for Linseed Cultivation before 6000 BC." *Journal of Archaeological Science* 2, 215–220.

Vansteenhuyse, Klaas. 2002. "Minoan Courts and Ritual Competition." In *Monuments of Minos*, 235–248.

Vercoutter, Jean. 1956. *L'Égypte et le monde égéen préhellénique*. Cairo: Le Caire.

Vermeule, Emily, 1964. *Greece in the Bronze Age*. Chicago: University of Chicago Press.

Vlachopoulos, Andreas. 2003. "‹Βίρα–Μάινα›: Το χρονικό της συντήρησης μίας τοιχογραφίας από την Ξεστή 3 του Ακρωτηρίου." In *ΑΡΓΟΝΑΥΤΗΣ. Τιμητικός Τόμος για τον καθηγητή Χρίστο Ντούμα από τους μαθητές του στο Πανεπιστήμιο Αθηνών*, eds. Andreas Vlachopoulos and Kiki Birtacha. Athens: He Kathemerine, 505–526.

Vlachopoulos, Andreas. 2007. "*Mythos, Logos and Eikon*: Motifs of Early Greek Poetry in the Wall Paintings of Xeste 3 at Akrotiri, Thera." In *EPOS: Reconsidering Greek Epic and Aegean Bronze Age Archaeology. Proceedings of the 11th International Aegean Conference, Los Angeles, UCLA - The J. Paul Getty Villa, 20-23 April 2006* (*Aegaeum* 28), eds. Sarah P. Morris and Robert Laffineur. Liège: Université de Liège, Histoire de l'art et archéologie de la Grèce antique; Austin: University of Texas at Austin, Program in Aegean Scripts and Prehistory, 107–118.

Vlachopoulos, Andreas. 2008. "The Wall Paintings from the Xeste 3 Building at Akrotiri, Thera: Towards an Interpretation of Its Iconographic Programme." In *Horizon: A Colloquium on the Prehistory of the Cyclades*, eds. Neil Brodie, Jenny Doole, Giorgos Gavalas and Colin Renfrew. Cambridge: McDonald Institute for Archaeological Research, 491–505.

Vlachopoulos, Andreas. 2012. "Jewellery and Adornment at Akrotiri, Thera: The Evidence from the Wall Paintings and the Finds." In *KOSMOS*, 35–42.

Vlachopoulos, Andreas. 2015. "Detecting 'Mycenaean' Elements in the 'Minoan' Wall Paintings of a 'Cycladic' Settlement. The Wall Waintings at Akrotiri, Thera Within their Iconographic Koine." In *Mycenaean Paintings in Context: New Discoveries and Old Finds Reconsidered. Proceedings of an International Conference on Mycenaean Wall Paintings in Context, Held 10-13 February 2011 at the National Hellenic Research Foundation* (Meletemata 72), eds. Hariclia Brecoulaki, Jack L. Davis, and Sharon R. Stocker. Athens: National Hellenic Research Foundation, 37–65.

Vlachopoulos, Andreas, Lefteris Platon and Lisa Chrysikopoulou. 2011. "Μινωική βίλα Επάνω Ζάκρου–Ξεστή 3 Ακρωτηρίου Θήρας: εντοπίζοντας κοινά θεματολογικά, τεχνοτροπικά και τεχνολογικά στοιχεία σε δύο σύγχρονα μεταξύ τους εικονογραφικά προγράμματα." In *Proceedings of the 10th International Congress of Cretan Studies, Khania, Crete, 1-8 October 2006, Chania, Crete,* vol. A3, ed. Maria Andreadaki-Vlazaki. Chania: Philological Association "Chrysostomos," 437–459.

Völling, Elisabeth. 2008. *Textiltechnik im Alten Orient: Rohstoffe und Herstellung*, Würzburg: Ergon.

Wace, Alan J. B. 1919-1921. "Excavations at Mycenae: 4 – The Rhyton Well." *Annual of the British School at Athens* 24, 200–209.

Wace, Alan J. 1921-1923. "Mycenae: Report on the Excavations of the British School at Athens 1921–1923." *Annual of the British School at Athens* 25: 1–434.

Wace, Alan J. 1964. *Mycenae: An Archaeological History and Guide*. New York: Biblo and Tannen.

Wachsmann, Shelley. 2009. *Seagoing Ships and Seamanship in the Bronze Age Levant*, 2nd printing. College Station Tex.: Texas A&M University Press.

Wachsmann, Shelley. 2013. *The Gurob Ship-Cart Model and Its Mediterranean Context*. College Station Tex.: Texas A&M University Press.

Waetzoldt, Hartmuth. 1972. *Untersuchungen zur Neusumerischen Textilindustrie* (Studi Economici e Tecnologici 1). Rome: Centro per le antichità e la storia dell'arte del Vicino Oriente.

Waetzoldt, Hartmuth. 2007. "The Use of Wool for the Production of Strings, Ropes, Braided Mats and Similar Fabrics." In *Ancient Textiles*, 12–121.

Walberg, Gisela. 1976. *Kamares. A Study of the Character of Palatial Middle Minoan Pottery* (Acta Universitatis Upsaliensis, Boreas 8). Uppsala: Almqvist och Wiksell.

Walberg, Gisela. 1983. *Provincial Middle Minoan Pottery*. Mainz am Rhein: Philipp von Zabern.

Wallis, Henry. 1900. *Egyptian Ceramic Art*. London: Taylor and Francis.

Ward, Cheryl A. 2000. *Sacred and Secular: Ancient Egyptian Ships and Boats.* Boston: Archaeological Institute of America.

Ward, Cheryl A. 2012. "Building Pharaoh's Ships: Cedar, Incense and Sailing the Great Green." *British Museum Studies in Ancient Egypt and Sudan* 18, 217–232.

Ward, Cheryl A. and Chiaro Zazzaro. 2010. "Evidence for Pharaonic Seagoing Ships at Mersa/Gawasis, Egypt." *International Journal of Nautical Archaeology* 39, 1–17.

Warren, Peter. 1972. *Myrtos: An Early Bronze Age Settlement in Crete* (*Annual of the British School at Athens* Suppl. 7). Athens: British School at Athens.

Warren, Peter. 1975. *The Aegean Civilizations: The Making of the Past*. Oxford: Elsevier-Phaidon.

Warren, Peter. 1990. "Of Baetyls." *Opuscula Atheniensia* 18, 192–206.

Warren, Peter. 2005. "Flowers for the Goddess? New Fragments of Wall Painting from Knossos." In *Aegean Wall Painting*, 131–148.

Wedde, Michael. 2000. *Towards a Hermeneutics of Aegean Bronze Age Ship Imagery* (Peleus: Studien zur Archäologie und Geschichte Griechenlands und Zyperns 6). Mannheim und Möhnesee: Bibliopolis.

Weiner, Annette B. and Jane Schneider, eds. 1989. *Cloth and Human Experience* (Smithsonian Series in Ethnographic Inquiry). Washington, DC: Smithsonian Institution Press.

Weingarten, Judith. 1997. "Proportions and the Palaikastro Kouros: A Minoan Adaption of the First Egyptian Proportional Canon." In *Ancient Egypt, the Aegean, and the Near East: Studies in Honour of Martha Rhoads Bell*, eds. Martha Rhoades Bell, Jacke Phillips and Lanny Bell. San Antonio: Van Sicklen Books, 471–481.

Weingarten, Judith. 1999. "Male and Female S/He Created Them: Further Studies in Aegean Proportions." In *Meletemata*, 921–930.

Weingarten, Judith. 2000a. "Reading the Minoan Body: Proportions and the Palaikastro Kouros." In *The Palaikastro Kouros: A Minoan Chryselephantine Statuette and Its Aegean Bronze Age Context* (British School at Athens Studies 6), eds. Joseph A. MacGillivray, Jan M. Driessen and L. Hugh Sackett. London: The British School at Athens, 103–112.

Weingarten, Judith. 2000b. "Some Stamped Weights from Eastern Crete." In Πεπραγμενα Η' Διεθνους Κρητολογικου Συνεδριου 1996 Τομος Α.3. Herakleion, 485–495.
Westgate, Ruth. 2007. "Life's Rich Pattern: Decoration as Evidence for Room Function in Hellenistic Houses," In *Building Communities: House, Settlement and Society in the Aegean and Beyond. Proceedings of a Conference Held at Cardiff University, 17-21 April 2001* (British School at Athens Studies 15), eds. Ruth Westgate, Nick Fisher and James Whitley. London: The British School at Athens, 313–321.
Wetterstrom, Wilma. 1993. "Foraging and Farming in Egypt: The Transition from Hunting and Gathering to Horticulture in the Nile Valley." In *The Archaeology of Africa: Food, Metals and Towns*, eds. Thurston Shaw, Paul Sinclair, Bassey Andah and Alex Okpoko. London: Routledge, 165–226.
Whitelaw, Todd. 1983. "The Settlement at Fournou Korifi Myrtos and Aspects of Early Minoan Social Organization." In *Minoan Society. Proceedings of the Cambridge Colloquium 1981*, eds. Olga Krzyszkowska and Lucia Nixon. Bristol: Bristol Classical Press, 323–345.
Whittaker, Hélène. 2009. "Memory and Cultural Values in the Middle Helladic Period: Some Preliminary Thoughts." In *The Past in the Past: The Significance of Memory and Tradition in the Transmission of Culture* (British Archaeological Reports 1925), eds. Mercourios Georgiadis and Chrysanthi Gallou. Oxford: Archaeopress, 5-15.
Wilson, David E. 1984. "The Early Minoan II A West Court Houses at Knossos." Ph.D. dissertation, University of Cincinnati.
Wilson, Jennifer W. 2009. "Wall Paintings of Processions: The Implications for Gendered Activities in the Late Bronze Age." Ph.D. dissertation, University of Melbourne.
Wolff, Colette. 1996. *The Art of Manipulating Fabric.* Iola, Wisc.: Krause Publications.
Wright, Michael T. 2008. "Homer at Sea ('ιχθυόεντα κέλευθα)." In *Science and Technology in Homeric Epics* (History of Mechanism and Machine Science 6), ed. Stephanos A. Paipetis. Dordrecht: Springer Science and Business Media, 377–384.
Xenaki-Sakellariou, Agnes and Christos Chatziliou. 1989. "*Peinture en métal' à l'époque mycénienne.* Athens: Ekdotike Athenon.
Yoffee, Norman, ed. 2007. *Negotiating the Past in the Past: Identity, Memory, and Landscape in Archaeological Research.* Tucson: University of Arizona Press.
Younger, John G. 1980. *The Iconography of Late Bronze Age Seals: Studies in the Seals of the Aegean Bronze Age* I. Bristol: Bristol Classical Press.
Zohary, Daniel and Maria Hopf. 2000. *Domestication of Plants in the Old World: The Origins and Spread of Cultivated Plants in West Asia, Europe and the Nile Valley*, 3rd ed. Oxford: Oxford University Press.
Zohary, Daniel, Maria Hopf and Ehud Weiss. 2012. *Domestication of Plants in the Old World: The Origin and Spread of Domesticated Plants in Southwest Asia, Europe, and the Mediterranean Basin*, 4th ed. Oxford: Oxford University Press.